The Delivery of Regenerative Medicines and Their Impact on Healthcare

Edited by
Dr. Catherine D. Prescott
Professor Dame Julia Polak

CRC Press
Taylor & Francis Group
Boca Raton London New York

CRC Press is an imprint of the
Taylor & Francis Group, an **informa** business

CRC Press
Taylor & Francis Group
6000 Broken Sound Parkway NW, Suite 300
Boca Raton, FL 33487-2742

© 2011 by Taylor and Francis Group, LLC
CRC Press is an imprint of Taylor & Francis Group, an Informa business

No claim to original U.S. Government works

Printed in the United States of America on acid-free paper
10 9 8 7 6 5 4 3 2 1

International Standard Book Number: 978-1-4398-3606-4 (Hardback)

This book contains information obtained from authentic and highly regarded sources. Reasonable efforts have been made to publish reliable data and information, but the author and publisher cannot assume responsibility for the validity of all materials or the consequences of their use. The authors and publishers have attempted to trace the copyright holders of all material reproduced in this publication and apologize to copyright holders if permission to publish in this form has not been obtained. If any copyright material has not been acknowledged please write and let us know so we may rectify in any future reprint.

Except as permitted under U.S. Copyright Law, no part of this book may be reprinted, reproduced, transmitted, or utilized in any form by any electronic, mechanical, or other means, now known or hereafter invented, including photocopying, microfilming, and recording, or in any information storage or retrieval system, without written permission from the publishers.

For permission to photocopy or use material electronically from this work, please access www.copyright.com (http://www.copyright.com/) or contact the Copyright Clearance Center, Inc. (CCC), 222 Rosewood Drive, Danvers, MA 01923, 978-750-8400. CCC is a not-for-profit organization that provides licenses and registration for a variety of users. For organizations that have been granted a photocopy license by the CCC, a separate system of payment has been arranged.

Trademark Notice: Product or corporate names may be trademarks or registered trademarks, and are used only for identification and explanation without intent to infringe.

Library of Congress Cataloging-in-Publication Data

The delivery of regenerative medicines and their impact on healthcare / editors, Catherine Prescott and Dame Julia Polak.
    p. ; cm.
    Includes bibliographical references and index.
    Summary: "Based on input from an international panel of experts, this book provides first-hand experience of the challenges and opportunities facing the delivery of regenerative medicines to patients. It highlights key issues beyond science and clinical translation, such as finance and business models, intellectual property and regulatory landscapes as well as questions of how regenerative medicines will be evaluated for reimbursement. This book will become a pivotal reference to anyone within the healthcare sector interested in understanding and investing in the delivery of regenerative medicines to the benefit of patients"--Provided by publisher.
    ISBN 978-1-4398-3606-4 (hardcover : alk. paper)
    1. Tissue engineering--Economic aspects. 2. Regenerative medicine--Economic aspects. 3. Medical care--Economic aspects. I. Prescott, C. D. (Catherine D.) II. Polak, Julia M.
    [DNLM: 1. Regenerative Medicine. 2. Health Care Sector. WO 515]

R857.T55D45 2011
610.28--dc22
                                                                  2010026460

Visit the Taylor & Francis Web site at
http://www.taylorandfrancis.com

and the CRC Press Web site at
http://www.crcpress.com

Contents

Foreword vii
Preface ix
Acknowledgments xi
About the Editors xiii
Contributors xv

SECTION 1   Introduction

Chapter 1   What Is Regenerative Medicine? 3
                 Julia M. Polak

SECTION 2   Finance

Chapter 2   A New Political–Financial Paradigm for Medical Research:
                 The California Model? 11
                 Robert N. Klein and Alan Trounson

Chapter 3   Investment Models: Public Funding in Australia 35
                 Graham Macdonald

Chapter 4   Canada: Capitalizing on a 50-Year Legacy 43
                 Andrew Lyall

Chapter 5   Investing in Regenerative Medicine: What Drives Private Investors? ... 59
                 Catherine D. Prescott

Chapter 6   Public Investment Models: Coming out of the Closet and
                 Going Public! 67
                 Reni Benjamin

iii

SECTION 3 Business Models

Chapter 7   Cell-Based Products: Allogeneic 85
              Paul Kemp

Chapter 8   Autologous Cell-Based Products: Fulfilling the Promise of
              Cell Therapy 97
              Eduardo Bravo and Magdalena Blanco-Molina

Chapter 9   Business Models for Cord Blood 117
              Suzanne M. Watt

Chapter 10   Changing the Game of Drug Discovery 131
              John Walker

Chapter 11   Discovery of Small Molecule Regenerative Drugs 141
              Yen Choo

Chapter 12   Adoption of Therapeutic Stem Cell Technologies by Large
              Pharmaceutical Companies 153
              Alain A. Vertès

Chapter 13   Role of Tool and Technology Companies in Successful
              Commercialization of Regenerative Medicine 177
              Joydeep Goswami and Paul Pickering

Chapter 14   Key Considerations in Manufacturing of Cellular Therapies 189
              Robert A. Preti

Chapter 15   State of the Global Regenerative Medicine Industry 213
              R. Lee Buckler, Robert Margolin, and Sarah A. Haecker

SECTION 4 Intellectual Property

Chapter 16   Regenerating Intellectual Property: Europe after WARF 239
              Julian Hitchcock and Devanand Crease

Contents v

Chapter 17 Protecting Regenerative Medicine Intellectual Property in the
United States: Problems and Strategies 257

David Resnick, Ronald I. Eisenstein, and Joseph M. McWilliams

Chapter 18 Impacts of Indian Policies and Laws on Regenerative
Medicine Patent Applications 281

Prabuddha Ganguli

SECTION 5 Regulatory Landscape

Chapter 19 A CATalyst for Change: Regulating Regenerative Medicines in
Europe 295

Christopher A. Bravery

Chapter 20 United States Regulatory Reimbursement, Political
Environment, and Strategies for Reform 323

Michael J. Werner

SECTION 6 Reimbursement

Chapter 21 The Fourth Hurdle: Reimbursement Strategies for
Regenerative Medicine in Europe 337

François M. Meurgey and Micheline Wille

Chapter 22 Cellular Therapies and Regenerative Medicine: Preparing for
Reimbursement in the United States 351

Eric Faulkner

Chapter 23 Adoption and Evaluation of Regenerative Medicine and the
National Health Service 369

Margaret Parton

SECTION 7 Insurance and Risk Management

Chapter 24 Role of Insurance: If You Build It, Will They Insure It? 379

Matthew Clark

Index 391

Foreword

The regenerative medicine industry is set to revolutionize healthcare and has the potential to cure chronic diseases that are major economic burdens to healthcare systems worldwide. However, the delivery of regenerative medicines to benefit patients is a considerable challenge for an industry sector otherwise geared to the development and delivery of surgical procedures and traditional "blockbuster" medicines available as pills in bottles.

Cells are living products and so have relatively short shelf lives; they will need to be matched to suit individual patients and administered in clinical settings. Regenerative medicines will be expensive to develop and manufacture but are anticipated to have long-term benefits. The reimbursement sector will therefore be challenged with how to evaluate the cost-effectiveness of medicines whose benefits are measured over a long period of time. The rate at which these challenges can be surmounted will determine when regenerative medicines become routinely available to patients. The editors have successfully gathered input from a worldwide group of experts who share their first-hand experience of the challenges and opportunities facing the delivery of regenerative medicines to patients. This is the first time that such a publication brings together and highlights the key issues beyond science and clinical translation, relating to finance and business models, intellectual property and regulatory landscapes, as well as questioning how regenerative medicines will be evaluated for reimbursement. This book will become a pivotal reference to anyone in the healthcare sector interested in understanding and investing in the delivery of regenerative medicines to the benefit of patients.

Professor Lord Ara Darzi, KBE, HonFrEng FmedSci
Professor of Surgery, Oncology, Reproductive Biology, and Anesthetics
Imperial College, London

Preface

Regenerative medicines pose a whole new set of challenges to an industry sector otherwise geared to the development and delivery of traditional pharmaceuticals. However, significant strides have already been achieved regarding the many aspects of this nascent field.

This book is unique both in its focus and geographical perspective on these issues. The book covers a broad range of topics from how this new industry is being financed, the business models developed, the impact of a complex patent landscape, and an evolving regulatory environment, through to how these expensive products are viewed by the health insurance industry. Experts from all over the world, including leaders of public and private organizations, share their first-hand experience of the challenges and opportunities facing all aspects that underpin the delivery of regenerative medicines. This book aims to inform a wide audience, including members of the pharmaceutical and biotechnology industry, regional and central governments, investors, health insurers, and academics.

This book is very timely: in 2009, U.S. President Barack Obama relaxed federal funding restrictions for human embryonic stem cell research. The U.S. Food and Drug Administration approved the first clinical trial for an embryonic stem cell-derived therapy, and several large corporations have, for the first time, moved into the sector. The Regenerative Medicines Industry Group was launched in the U.K., and the Alliance for Regenerative Medicine was established in the United States.

To date, all other publications have focused on the science, technology, and ethics drawing predominantly on academic expertise. By contrast, this book addresses those issues that are important to the success of the business of regenerative medicines. In doing so, the editors have been rewarded with an enthusiastic response by top industry leaders from across the globe, affirming the importance and timeliness of this book. It is our understanding that this is the very first publication of its kind.

Acknowledgments

The editors are deeply indebted and grateful to everyone who helped bring this book to fruition, especially the contributors and publisher, Taylor & Francis Group. Special thanks go to Sandra Lock whose unstinting help, hard work, drive, and support ensured that we were able to complete the book. We also thank James Cameron, whose help and advice were absolutely invaluable.

About the Editors

Dr. Catherine D. Prescott has more than 25 years' experience in research, management, and business within the life-science and venture capital sectors. She is the founder–director of the consultancy firm Biolatris Ltd., co-founder and director of univerCELL-market.com (a global resource for the stem cell and regenerative medicine community), chair of the UK National Stem Cell Network Advisory Committee, a director of the East of England Stem Cell Network, and a member of the Life Sciences Advisory Board for the Worcester Polytechnic Institute (MA, USA).

Cathy is a "poacher turned game keeper"; prior to launching Biolatris, she worked on both sides of the biotechnology investment arena. After serving as Head of Drug Discovery for a start-up company (RiboTargets Ltd) she worked for several years as the science director for venture capital firm Avlar BioVentures. As a venture capitalist, Cathy gained considerable insight into the drivers for investment and how these impacted disruptive technologies such as regenerative medicines. As a consultant, Cathy continues to serve clients within the regenerative medicine community, as well as being actively engaged in developing innovative business and funding models. She was also formerly an assistant director at SmithKline Beecham and held post-doctoral fellowships at Max Delbrück Centre for Molecular Medicine (Berlin), Max Planck Institute for Molecular Genetics (Berlin), and Brown University (USA). Cathy holds a DPhil from Oxford University.

Professor Dame Julia Polak graduated from the University of Buenos Aires, Argentina, and obtained her postgraduate training in the UK. She is the founder and former director of the Tissue Engineering and Regenerative Medicine Centre, Imperial College and is now an emeritus professor from the Faculty of Medicine and resides in an office in the Department of Chemical Engineering. She is also a member of the Scientific Advisory Board of the Imperial College Institute of Biomedical Engineering and has recently been made a new member of the Stem Cell Advisory Board Panel of the joint MRC/UKSCF, Science Advisory Board, (October 2005), Panel of the new EPSRC Peer Review College (2006–2009), Panel of the MRC College of Experts (2006–2010) and Steering Group of the UK Stem Cell Immunology Programme (March 2006) and UK National Stem Cells Network Committee (October 2006). She is a council member of the Tissue Engineering Society International and the Academy of Medical Sciences (2002–2005)

and was also European Editor of Tissue Engineering (up until 2004). She is the author of 992 original papers, 118 review articles and editor/author of 27 books and is one of the most highly cited researchers in her field. She is a co-founder and director of an Imperial Spin Out Company called Novathera (now MedCell) dealing with Regenerative Medicine Products. She is also the recipient of a heart and lung transplant, in 1995, and into her 14th year post-transplant is one of the longest living survivors in the UK. She has received a number of honors and won a number of prizes.

Contributors

Dr. Reni Benjamin
Rodman & Renshaw
New York, New York, USA
rbenjamin@rodmanandrenshaw.com

Dr. Magdalena Blanco-Molina
Cellerix SA
Madrid, Spain
mblanco@cellerix.com

Dr. Christopher A. Bravery
Consulting on Advanced Biologicals Ltd.
London, United Kingdom
cbravery@advbiols.com

Eduardo Bravo
Cellerix SA
Madrid, Spain
ebravo@cellerix.com

R. Lee Buckler, B.Ed., LL.B.
Cell Therapy Group
Vancouver, British Columbia, Canada
lbuckler@celltherapygroup.com

Dr. Yen Choo
Plasticell Ltd.
Imperial BioIncubator
London, United Kingdom
yen@plasticell.co.uk

Dr. Matthew Clark
La Playa
Cambridge, United Kingdom
matthew.clark@laplaya.co.uk

Dr. Devanand Crease
Keltie
London, United Kingdom
dev.crease@keltie.com

Dr. Ronald I. Eisenstein
Nixon Peabody LLP
Boston, Massachusetts, USA
reisenstein@nixonpeabody.com

Eric Faulkner
RTI Health Solutions
Research Triangle Park, North Carolina, USA
efaulkner@rti.org

Dr. Prabuddha Ganguli
Vision IPR
Mumbai, India
pradbuddha.ganguli@gmail.com

Dr. Joydeep Goswami
Life Technologies
Carlsbad, California, USA
Joydeep.Goswami@lifetech.com

Dr. Sarah A. Haecker
Orasi Medical, Inc.
Minneapolis, Minnesota, USA
sarah.haecker@orasimedical.com

Dr. Julian Hitchcock
Field Fisher Waterhouse LLP
London, United Kingdom
julian.hitchcock@ffw.com

Dr. Paul Kemp
Intercytex
Manchester, United Kingdom
pkemp@intercytex.com and
   pauldavidkemp@yahoo.co.uk

Robert N. Klein, J.D.
The California Institute for
  Regenerative Medicine
San Francisco, California, USA
RKlein@cirm.ca.gov

Dr. Andrew Lyall
Stem Cell Network
451 Smyth Road, Room 3105
Ottawa, Ontario, Canada
dlyall@stemcellnetwork.ca

Professor Graham Macdonald
Biotechnology Consultant
Double Bay, New South Wales, Australia
grahjon@gmail.com

Dr. Robert Margolin
Genetics Policy Institute
Wellington, Florida, USA
rob@genpol.org

Dr. Joseph M. McWilliams
Partners Health Care Research Ventures
  & Licensing
Cambridge, Massachusetts, USA
joseph.m.mcwilliams@gmail.com

Dr. François M. Meurgey
TiGenix NV
Leuven, Belgium
francois.meurgey@Tigenix.com

Margaret Parton
NHS Technology Adoption Centre
Manchester Royal Infirmary
Manchester, United Kingdom
Margaret.Parton@cmft.nhs.uk

Dr. Paul Pickering
Life Technologies
Carlsbad, California, USA
paul.pickering@lifetech.com

Dr. Robert A. Preti
Progenitor Cell Therapy LLC
Allendale, New Jersey, USA
repreti@progenitorcelltherapy.com

Dr. David Resnick
Nixon Peabody LLP
Boston, Massachusetts, USA
dresnick@nixonpeabody.com

Dr. Alan Trounson
The California Institute for
  Regenerative Medicine
San Francisco, California, USA
ATrounson@cirm.ca.gov

Dr. Alain A. Vertès
London Business School, Sloan
  Fellowship
London, United Kingdom
avertes.sln2004@london.edu
and
F. Hoffmann-La Roche Ltd.
  Pharmaceuticals Division
Basel, Switzerland
alain.vertes@roche.com

John Walker
iPierian, Inc.
South San Francisco, California, USA
john.walker@ipierian.com

Dr. Suzanne M. Watt
Nuffield Department of Clinical
  Laboratory Sciences and NHS Blood
  and Transplant
John Radcliffe Hospital
University of Oxford
Oxford, United Kingdom
Suzanne.watt@nhsbt.nhs.uk

Michael J. Werner
Holland & Knight
Washington, DC, USA
michael.werner@hklaw.com

Dr. Micheline Wille
TiGenix NV
Leuven, Belgium
micheline.wille@Tigenix.com

Section 1

Introduction

1 What Is Regenerative Medicine?

Julia M. Polak

CONTENTS

1.1 Opportunities Offered by Regenerative Medicine 4
1.2 Challenges 5
    1.2.1 Cells 5
    1.2.2 Vascularization 5
    1.2.3 Immunology 5
    1.2.4 Imaging Methodologies 6
1.3 Regenerative Medicine: A New Business Model 6
1.4 Regulatory Hurdles 6
1.5 Clinical Applications 6
1.6 Cost and Funding 7
1.7 Conclusions 7
References 7

Regenerative medicine is a rapidly evolving multidisciplinary field that aims to replace, repair, or restore normal function to a given organ or tissue by delivering safe, effective, and consistent living cells either alone or in combination with especially designed materials. The field is a convergence of apparently separate therapeutic areas including cell therapy, tissue engineering (i.e., creation of in vitro tissues and/or organs for subsequent transplantations as fully functioning organs or as tissue patches), bioengineering, and gene therapy (Guillot et al. 2007). The multiple approaches to regenerative medicine include cell replacement (transplantation), repair (exogenous cell therapy), and regeneration (mobilization of endogenous pools of stem cells).

    The concept of tissue regeneration is by no means new—going back a long time as illustrated by the famous legend of Prometheus. Prometheus was a champion of human equality. He stole fire from Zeus and then gave it to the mortals. As a punishment for this crime, Zeus bound Prometheus to a rock and sent a giant eagle to eat his liver. However, his liver re-grew every night and the eagle had to return again and again.

    Tissue regeneration is also a primitive event, occurring in many organisms, such as newts, where it is well known that a sectioned limb will be completely regenerated within 6 to 8 weeks. In humans, solid organ transplantation and cell therapy have been practiced for many years; for example, kidney transplantation was first performed in 1954 and bone marrow transplantation has been performed since 1968 (Appelbaum 2008).

1.1 OPPORTUNITIES OFFERED BY REGENERATIVE MEDICINE

Regenerative medicine is likely to transform the way we practice medicine. With regenerative medicine, the repair of unhealthy tissue or restoration of bodily functions can be achieved by treating patients with cells. Cell therapy is likely to be a "once and for all" treatment, thereby differing entirely from current medical practice of using pharmacological or surgical procedures. With conventional pharmacological approaches, a patient is likely to require therapy for a considerable period, if not forever. Although cell therapy would appear to be expensive to produce/and or administer, the aim is to produce a permanent restoration of the lost function of an organ and/or tissue. Ultimately this is anticipated to be more economical and beneficial than current medical practice.

The opportunities for regenerative medicines are immense, especially in light of an ever-increasing aging population facing associated ailments. For example, cells can be used as vehicles for gene therapy (Kawamura et al. 2009) and cultured cells can be used to study in vitro a specific disease process or for drug development. The discovery of induced pluripotent stem cells (iPS) (see Chapter 10 by Walker) also offers the potential to produce disease models to support new drug discovery as well as patient-specific cells for therapy (Hollander and Wraith 2008). As regards biomaterials, again this field is intensely researched; the advent of nanotechnology has allowed the development of specially designed nanosurfaces that encourage cell attachment, cell growth, and differentiation (Hench and Polak 2002, Wise et al. 2009).

Worldwide research in the field is intense (Baker 2009) and several trials are currently progressing through the clinic (Green and Alton 2008, Newton and Yang 2009). For example, artificially constructed bladders have been successfully implanted into young children (Atala et al. 2006) and a trachea built from a patient's divided trachea and seeded with autologous mesenchymal cells was successfully transplanted back into the same patient (Macchiarini et al. 2008).

The mechanism of action of stem cell therapy is still being determined, but the general consensus suggests that the most likely mechanism may be through the release of cytokines and other growth-promoting molecules. Harnessing the potential of these biologic activities enables one to foresee a future where a "once and for all regenerative pill" may become available. If the field of regenerative medicine continues to progress at its current pace and becomes well established, it is likely to initiate a major revolution similar to that witnessed, for example, by the advent of monoclonal antibodies.

There are multiple coordinating efforts in this active multidisciplinary field such as the United Kingdom's National Stem Cell Network, the Alliance for Regenerative Medicine in the United States, and others. Furthermore, major pharmaceutical companies are actively investing in stem cell research (e.g., the Pfizer Regenerative Medicine Initiative in the United Kingdom and the United States, and the GSK alliance with the Harvard Stem Cell Institute, also in the United States. In the United

Kingdom, the first ever public–private partnership known as Stem Cells for Safer Medicine (SC4SM) has been set up to exploit human embryonic stem cells for drug safety testing.

1.2 CHALLENGES

Despite the promise of regenerative medicines, the challenges abound in this active but nascent field. Some examples are discussed below.

1.2.1 CELLS

There is no clear consensus as to which will ultimately be the most suitable cell type to be used and therefore research on all classes of stem cells, including embryonic and progenitor cells, remains intense (Guillot et al. 2007). It is apparent, however, that bone marrow stem cells are likely to reach the clinic sooner than other cell types. There remains the need to develop robust, effective, reproducible, and safe protocols, with well defined reagents for the differentiation of pure populations of cells (Wang et al. 2007). Furthermore, there is currently no clear consensus as to the number of cells needed, the mode/route of their delivery, and the appropriate time during the disease process for the cells to be administered or mobilized (Mason and Dunnill 2009).

In terms of the iPS cells, fundamental questions remain to be determined including whether iPS cells behave identically to ES cells. There is a need for comparative studies and development of more accurate cell markers and robust and automated cell expansion technologies.

1.2.2 VASCULARIZATION

Cell or construct implantation is currently limited by the inability to adequately vascularize the engrafted tissues. Issues of nutrient perfusion and mass transport limitations, especially oxygen diffusion, restrict the development of the construct and limit the ability for its *in vivo* integration. This field is intensely researched and includes the development of appropriate materials (many with nanosurfaces), microfabrication methodology, bioreactor development, endothelial cell seeding, and others (Lovett et al. 2009).

1.2.3 IMMUNOLOGY

Regenerative medicine uses a variety of autologous and allogeneic cell types (see Chapter 7 by Kemp, Chapter 8 by Bravo and Blanco-Molina, Chapter 9 by Watt, and Chapter 15 by Buckler et al.). Although recent examples suggest that bone marrow stem cells are likely to be the first to achieve reliable clinical applications, the use of allogeneic cells may ultimately be the answer. Embryonic stem cells are pluripotent, easily expandable, and may be differentiated into most or all cell types derived from the three germinal layers. Even so, the use of allogeneic cells is likely to encounter immunological hurdles. The subject is intensely researched and has recently been reviewed by Hollander and Wraith (2008).

1.2.4 IMAGING METHODOLOGIES

Regenerative medicine requires robust in vivo imaging techniques to track the administration, migration, integration, and fate of stem cells and monitor the effect this new form of treatment may exert on diseased tissue. Again, the field is thoroughly researched and has been recently reviewed by Newton and Yang (2009).

1.3 REGENERATIVE MEDICINE: A NEW BUSINESS MODEL

Surgery and drug therapy are currently accepted options for clinical practice. Large numbers of patients are treated with drugs that are typically self-administered. It is possible to foresee that with cell therapy selected patients will be treated by specialist involvement that will require the training of a new generation of medically qualified personnel and healthcare auxiliary staff.

1.4 REGULATORY HURDLES

Regenerative medicine is a new field and hence the regulatory landscape is still evolving. It is not yet clear whether regulatory agencies, including the U.S. Food & Drug Administration (FDA) and the European Medicines Agency (EMA), will consider stem cell therapy as a biological or a device (see Chapter 19 by Bravery and Chapter 20 by Werner). The FDA has set up the Office of Combination Products and the Office of Cellular, Tissue, and Cell Therapies. Furthermore, and unlike the landscape in a single country such as the U.S., the EMA in Europe may recommend guidelines, but whether member states will adhere to them remains an open question.

1.5 CLINICAL APPLICATIONS

It is clear that product consistency, uniformity, and stability are of paramount importance. Safety requisites should address toxicity, tumor formation (applicable only to embryonic stem (ES) cells), and immunogenicity. In instances in which transplanted cells become fully incorporated into tissue, unwanted and/or unexpected effects must be considered in advance. Cell therapy must offer a better clinical outcome than current therapies and must be cost effective in order to be accepted by healthcare sectors such as the National Health Service (see Chapter 21 by Meurgey and Wille, Chapter 22 by Faulkner, and Chapter 23 by Parton). Furthermore, it is important to develop suitable in vivo imaging methodologies to be able to track the migrations and final locations of transplanted cells.

Clinical candidates are currently undergoing trials; the most notable advances include those involved in cardiac repair and skin replacement (McNeil 2008). Furthermore, clinical trials must be carried out within acceptable clinical practice and due ethical considerations. Exalting the promise of regenerative medicine to vulnerable patients is unacceptable. The International Society of Stem Cell Research has recently issued useful guidelines in this regard (http://www.issrc.org).

1.6 COST AND FUNDING

Cell therapy is likely, at least initially, to be expensive. Both product development and clinical trials require considerable levels of funding. The cost of the product is considerable if one is to account for the cost of growth factors and small molecules needed for viable cell preparations, in addition to the cost of medical care, both direct (healthcare sector) and indirect (caregivers and others). Transplantation, storage, tracking, and administration all add to the costs. Reimbursement is a difficult issue that varies from country to country (see Chapter 21 by Meurgey and Wille and Chapter 22 by Faulkner).

Governments, charities, and private investors are currently providing some level of funding, but it is clear that more funding will be needed. In September 2009, the Technology Strategy Board launched an £18 million "RegenMed" programme of investment to support key areas of commercial research and development (R&D) and the development of R&D partnerships. The program is being developed in partnership with the Medical Research Council, the Engineering and Physical Sciences Research Council, and the Biotechnology and Biological Sciences Research Council, which will contribute an additional £3.5 million funding. The program will focus on regenerative medicine product development and validation; the development of tools and technologies required to underpin the regenerative medicine sector; and an understanding of the value systems and business models necessary for the delivery of regenerative medicines.

1.7 CONCLUSIONS

The field of regenerative medicine is here to stay, as exemplified by the nascent but exponential growth of examples of translation from bench to bed side (e.g., cardiac regeneration and bladder and tracheal implantation). The current hurdles are by no means insurmountable and therefore it is reasonable to assume that we can look forward to a more mature and highly rewarding field of endeavor.

REFERENCES

Appelbaum, F. R. 2008. Hematopoietic cell transplantation at 50. *New England Journal of Medicine* 357: 1472–1475.

Atala, A., S. B. Bauer, S. Soker, J. J. Yoo, and A.B. Retik. 2006. Tissue-engineered autologous bladders for patients needing cytoplasty. *Lancet* 367: 1241–1246.

Baker, M. 2009. How to fix a broken heart? Clues about how human hearts form hint at routes to cell-based therapies. *Nature* 460: 18–19.

Green, A. E. and E. Alton. 2008. Cardiac repair clinical trials. In *Advances in Tissue Engineering*, J. Polak et al., Eds., London: Imperial College Press, pp. 696–732.

Guillot, P. V., W. Cui, N. M. Fisk, and J. M. Polak. 2007 Stem cell differentiation and expansion for clinical applications of tissue engineering. *Journal of Cellular and Molecular Medicine* 11: 935–44.

Hench, L. L. and J. M. Polak. 2002. Third-generation biomedical materials. *Science* 295: 1016–1017.

Hollander, A. P. and D. C. Wraith. 2008. Stem cell immunology. In *Advances in Tissue Engineering*, J. Polak et al., Eds., London: Imperial College Press, pp. 199–213.

Kawamura, T. J., Y. V. Suzuki, S. Wang et al. 2009. Linking the p53 tumour suppressor pathway to somatic cell programming. *Nature* 460: 1140–1144.

Lovett, M., K. Lee, A. Edwards, and D. L. Kaplan. 2009. Vascularization strategies for tissue engineering. *Tissue Engineering Part B* 15: 353–370.

Macchiarini, P., P. Jungebluth, T. Go et al. 2008. Clinical transplantation of a tissue-engineered airway. *Lancet* 372: 2023–2030.

MacNeil, S. 2008. Tissue engineered skin comes of age. In *Advances in Tissue Engineering*, J. Polak et al., Eds., London: Imperial College Press, pp. 593–618.

Mason, C. and P. Dunnill. 2009. Quantities of cells used for regenerative medicine and some implications for clinicians and bioprocessors. *Regenerative Medicine* 4: 153–157.

Newton, R. and G.-Z. Yang. *In vivo* imaging for cell therapy. In *Cell Therapy For Lung Repair*, J. M. Polak, Ed., London: Imperial College Press (final proofs in preparation).

Wang, D., D. L. Haviland, A. R. Burns, E. Zsigmond, and R. A. Wetsel. 2007. A pure population of lung alveolar epithelial type II cells derived from human embryonic stem cells. *Proceedings of the National Academy Science of the USA* 104: 4449–4454.

Wise, J. K., A. L. Yarin, C. M. Megaridis, and M. Cho. 2009 Chondrogenic differentiation of human mesenchymal stem cells on oriented nanofibrous scaffolds: engineering the superficial zone of articular cartilage. *Tissue Engineering Part A* 15: 913–921.

Section 2

Finance

2 A New Political–Financial Paradigm for Medical Research
The California Model?

Robert N. Klein and Alan Trounson

CONTENTS

2.1 Introduction: Evaluation of Potential of California Model 12
2.2 Fundamental Concepts Driving Public Funding of Medical Research 12
2.3 U.S. History of Public Funding of Medical Research through Appropriations Process 13
2.4 Medical Research Produces the Intellectual Capital Infrastructure for Healthcare 14
2.5 Aligning Payments for Medical Research with Benefit Groups 14
2.6 Cost of Transformative Long-Term Research Should Be Spread over Benefitting Generations 15
2.7 Empowering a New Political and Funding Paradigm for Medical Research 15
2.8 Creating State Paradigm to Complement Federal Research Funding 17
    2.8.1 California Model 18
2.9 Basic Rationale of California Model 21
2.10 Optimizing Governmental Cash Flow of California Research Funding Model 21
2.11 Models Providing Enhanced Opportunities 23
2.12 Relationship of Research Complexity to Capital 24
2.13 Interface of Governmental Funding with Private Capital Markets 26
2.14 California Model for Funding Large-Scale Biotech Research 26
    2.14.1 Recourse Loans 27
    2.14.2 Non-Recourse Loans 27
2.15 Biotechnology's Full Engagement as Strategic Goal 27
2.16 Governmental Validation of Private Company Research 28
2.17 Global Funding Priorities for Medical Research 28
2.18 Financing to Reach Millennium Development Goals for Medical Objectives ... 29
2.19 Blending IFFI$_M$ and Proposition 71 Models 30
References 31
Glossary 32

2.1 INTRODUCTION: EVALUATION OF POTENTIAL OF CALIFORNIA MODEL

The California Model is an extraordinarily promising new paradigm for government funding of stem cell research and therapy development. It is structured to carry research project funding all the way to a Phase II human trial efficacy demonstration. While this model demonstrates numerous strategic advantages, its ultimate optimization in safely and expeditiously advancing stem cell therapies to patients is currently being tested in programs to integrate private capital and biotechnology enterprises with non-profit research institutions. All the performance milestones of the California agency and its scientific portfolio are extremely positive.

Over $1 billion (U.S.) in donor and institutional matching funds provides a strong external validation for the agency's programs and capital structure. Its seven international collaborative funding partners offer an independent international validation of its scientific quality and importance in contributing to the advancement of the translational frontier for stem cell research. Although the final verdict will take a number of years, there is strategic value in examining the strength of the California Model's capital structure and organizational independence—all subject to executive branch and legislative oversight and audits.

At its conclusion, a recent study funded by the National Science Foundation (NSF) stated, "California has established itself as a major center for stem cell research. Recruitment of world-class stem cell scientists from across the globe has been a direct result of CIRM* funding." (Adelson and Weinberg 2010). The study summarizes Proposition 71's impact† by stating: "In its short history, the CIRM has taken on a vigorous life of its own. It is apparent that the shift of a major focus for stem cell research to California will have a significant effect into the future on the geographic distribution of biological science and biotechnology infrastructure in the United States; on the location of university, biotechnology, and pharmaceutical research and start-up firms; and on the investment of venture capital. Evidence for this is the $300 million the CIRM has invested in stem cell facilities, already leveraged to more than $1 billion in linked donations."

2.2 FUNDAMENTAL CONCEPTS DRIVING PUBLIC FUNDING OF MEDICAL RESEARCH

The scientific mission and its discoveries targeted to reduce human suffering from disease and injury, produce the *Intellectual Capital* of a society needed to enable and

---

* The California Institute for Regenerative Medicine, San Francisco, California. See homepage on the Internet.
† Eighty patient advocacy groups united behind Proposition 71. Selective examples include the American Diabetes Association, National Coalition for Cancer Research, Parkinson's Action Network, Alzheimer's Association, California Council, American Nurses Association of California, California Medical Association (representing 35,000 doctors), Cancer Research and Prevention Foundation, Christopher Reeve Paralysis Foundation, Cystic Fibrosis Research, Inc., Elizabeth Glaser Pediatric AIDS Foundation, Juvenile Diabetes Research Foundation, Michael J. Fox Foundation for Parkinson's Research, Prostate Cancer Foundation, and Sickle Cell Disease Foundation of California.

protect the right of the individual to live a healthy life. With a highly mobile world population, a society must organize to protect human health aggressively or face:

1. A rapid and continuous series of pandemics and health disasters
2. Rising levels of chronic disease
3. Widespread impacts of environmentally induced disease from industrial pollution

The current system for funding society's Intellectual Social Capital for healthcare is based upon an industrial capital system that is inefficient, frequently counterproductive, and inappropriate to deliver on the fundamental Intellectual Capital requirements and opportunities of 21st century medicine. Industrial Capital values direct financial returns; this system is not designed to capture the societal benefits of longer productive lives or reduced governmental healthcare costs. Nor is it organized to capture the benefits to individuals of reduced pain, a broader spectrum of physical activity, or a healthier more vibrant life, unless the individual has an unlimited ability to pay. Even then, with an unlimited financial capacity, the capital system for medical research is not producing the breadth of medical options that would be available under alternative financial structures that support research and therapy development. *The intent of the public financial funding model described in this chapter is not to replace the existing system, but rather to supplement it with a series of financial structures that align the interests of society and the individual with the financial systems driving the direction and breadth of medical research.*

2.3 U.S. HISTORY OF PUBLIC FUNDING OF MEDICAL RESEARCH THROUGH APPROPRIATIONS PROCESS

While primary U.S. medical research public funding has come through the federal government's annual or biannual appropriations process, states have also followed this model. A reliance on the appropriations process for funding has historically led to major swings in research funding. Negative economic cycles, wars, and other financial stresses that force an intense competition for annual appropriations generate an extremely high level of uncertainty in the funding patterns for U.S. medical research.

Predictably, massive federal deficits, trade imbalances, and constraints on global financing of governmental needs will soon re-establish severe restrictions on U.S. federal funding of medical research. For current appropriations, the "pay–go" system (Wikipedia 2009) that requires revenue increases or spending cuts to authorize any supplemental expenditures by the U.S. Congress will necessarily severely constrain any future increases in U.S. medical research funding and/or any renewal of the 2009 stimulus-driven increases to the budget of the National Institutes of Health (NIH; Adelson and Weinberg 2010).

The fundamental question is whether current government appropriations are the best approach to future medical research funding—*in any country*. Should and can the burden of medical research funding be carried by current taxpayers? Should medical research compete for funding against critical current needs for operating costs of public clinics and public hospitals and/or medical reimbursements under Medicare or other national healthcare systems? Is medical research an operating cost of the country or society?

2.4 MEDICAL RESEARCH PRODUCES THE INTELLECTUAL CAPITAL INFRASTRUCTURE FOR HEALTHCARE

The public funding premise of this chapter is founded on the concept that medical research produces a vital *intellectual capital infrastructure* that determines the advances on the frontiers of healthcare for any nation and/or the world. Indeed, biotech and pharma industries have their core financial values organized around a system of patents and licenses of intellectual capital. In the 20th century, states and nations that invested heavily and early in their *Physical Infrastructures* propelled their societies to great prosperity. These infrastructure investments—roads, railways, bridges, harbors—were major determinants of the speed of economic development and the sustained competitive capacity of these states and nations. It is the thesis of this chapter that the Intellectual Capital Infrastructure in each of the core areas of society's development sectors—specifically including healthcare—will be the primary determinants of economic and social prosperity in the 21st century.

Intellectual Capital is not an annual disposable good or expense like operating costs normally funded through annual appropriations. When capital expenditures compete directly against critical operating costs within the healthcare system, the capital options can generally be expected to fare poorly because of the urgent and non-negotiable nature of current care demands of patients with life-threatening conditions. Medical research should not compete against healthcare operating costs for scarce, current operating appropriations of the government. Intellectual Capital investments in medical research represent a long-term capital asset of society that should be funded under a separate system from critical, current healthcare.

2.5 ALIGNING PAYMENTS FOR MEDICAL RESEARCH WITH BENEFIT GROUPS

Any process of appropriations or funding that draws down current funding resources to pay for *Intellectual Medical Research Capital* creates a misalignment between the intended medical benefit group and the group paying for the investment. Consider the Salk vaccine as an example: it created massive improvements in health and cost savings through the avoidance of broad scale polio over the last 50 years (Thompson and Duintjer Tebbens 2006). For the U.S. alone, in the late 1950s, it was estimated that by 2005 it would cost $100 billion per year just to maintain polio victims in iron lungs housed in hotels specifically developed to meet the scale of victims anticipated (Thompson and Duintjer Tebbens 2006). Clearly, American society has benefited over a number of generations from the successful research investment in Intellectual Capital made in the 1950s; yet the cost of developing the vaccine was borne solely by the generation of that time.

2.6 COST OF TRANSFORMATIVE LONG-TERM RESEARCH SHOULD BE SPREAD OVER BENEFITTING GENERATIONS

To accomplish this, the research investment should be funded through long-term capital financing structures such as state, national, or international bonds that amortize the cost over the benefitting generations. By utilizing bonds that spread the cost over 30 to 50 years, the critical mass of financial assets that can be marshaled in the near-term increases enormously. As discussed below, California's Proposition 71, a $6 billion initiative approved by the voters in 2004, demonstrates the power of this concept, even at a state level, to lift an entirely new field of Medical Intellectual Capital—Stem Cell Research—from an exploratory phase into an intense medical revolution. Proposition 71 also demonstrates the positive ripple effect that can occur when one jurisdiction undertakes to align the research cost structure with the benefitting group. Once a major state or nation demonstrates a commitment to raise vast sums of capital through long-term bonds, other states and nations will be encouraged, if not compelled, to raise their investments in Intellectual Capital to remain competitive in the future research advances and commercialization of this broad-based *Intellectual Capital Asset*: the development of stem cell therapies for chronic diseases and injuries.

2.7 EMPOWERING A NEW POLITICAL AND FUNDING PARADIGM FOR MEDICAL RESEARCH

By changing the political and economic structures for medical research funding to align the medical benefit group with the payer group, through the utilization of long-term capital funding bonds, the politics of medical research funding profoundly changes. Healthcare constituencies have historically been deeply fractured by the competitive conflict between funding of current medical care and long-term medical research. In the competition for funding of current medical care, hospital suppliers and the medical and nursing professions, along with advocates for low-income, underserved groups, are aligned together. In competing for the same funds, scientific and medical researchers along with a portion of the patient advocacy organizations will vie politically for specific research agendas and targets. Patient advocacy organizations are further fractured into specific advocacy initiatives focused around their own specific disease interests.

When the funding structure changes to long-term bonds authorized through the state initiative process or other state bond approval political processes presented to voters, the healthcare constituencies are united in support and the historical fractures are healed for these specific efforts. When the cost of the medical research is to be funded by long-term bonds, the hospitals and medical professionals no longer have their direct operating cost budgets threatened competitively in the appropriation process. It is in their collective interest that the voters approve the bonds, by a direct ballot process, so that this capital resource demand is separately satisfied. The healthcare constituencies know that if the bonds fail, the capital demands for research will fall back upon the appropriations process.

When the funding mechanism for medical research requires a public vote for a bond authorization and an objective, balanced peer review process to award and fund the best medical science is assured across the entire spectrum of disease, patient advocacy groups can be united behind a singular unified effort (Health.org) rather than dissipating their individual strength in fighting for their specific medical appropriations programs that address their unique diseases. Even when the appropriation process, as with the federal National Institutes of Health (NIH) funding for research, claims to fairly cover the entire spectrum of medical research, embedded institutional resource allocation prejudices reflected in the historical allocation of funds may play a distorting role.

Unless there are informal agreements to reallocate resources among the individual institutes of the NIH, for example, the congressional appropriation process carries grossly different benefits for competing disease advocacy organizations. This results in supplementary appropriation "set-aside" or "earmarking" competitions between intensely competitive disease advocacy organizations. These politically costly struggles consume substantial political capital that otherwise could be used to increase the overall scientific medical funding for research, therapy development, and clinical trials to implement new discoveries. Until the appropriation funding process for medical research is substantially supplemented by a long-term bond-type funding program through an independent agency, preferably with a separate governing board, the intense battles for earmarked appropriations will not be significantly mitigated.

There are endless examples of these battles for special medical research appropriations for cancer, heart disease, Alzheimer's disease, and every other major and/or orphan disease. The examination of even a single example demonstrates clearly how harnessing this intense effort by patient advocacy organizations into a unified effort can empower a new scientific medical funding paradigm for stem cell research.

One such example occurred in 2002. President Bush had instructed the Republican leadership in the House of Representatives and the Senate to shut down all of the appropriation committees of both houses of Congress as to any appropriation increases or renewals. No new appropriations were to be approved by committees outside of the core budget to run the U.S. government and huge special appropriations to fund the new Homeland Security Agency, and the prospective war in Iraq. By blocking the committee approval of several bills that would have renewed the Supplemental Mandatory NIH appropriation for Type I Juvenile Diabetes research, the NIH Type I research appropriations would have been reduced for this disease by over 30%. These deep cuts would have shut down vital research to mitigate complications and/or funding to advance pending clinical trials. Concurrently, the expiring Type II Diabetes appropriation funding of diabetes clinics for Native Americans, where over 50% of the resident population of many reservations were experiencing Type II Diabetes, would have led to tragic complications and unnecessary deaths among those disease victims. Without this funding, these Native American clinics on reservations would have been closed.

To remedy this crisis, a combined, stand-alone Supplemental Mandatory Appropriations Bill for $1.5 billion was created at the 11th hour to renew these special targeted medical appropriations. To pass such an appropriations bill that does not go through any congressional committee, a unanimous vote of the House of Representatives and the Senate is required. No current congressional members or staff could ever recall this occurring; however, this bill passed both houses unanimously after extraordinary legislative advocacy of the National Juvenile Diabetes Research Foundation in key congressional districts across the nation.

Through the personal contacts of individual advocate families, the last U.S. Senate holdout, the incoming Republican U.S. Senate Budget Chairman, Senator Nichols of Oklahoma, experienced a flood of calls from corporate leaders (from his home state) that rose to such an extreme level that the switch boards in his state U.S. Senate Office and in his Washington U.S. Capitol Office were at times shut down due to an overload for two days before the final vote. When combined with the bipartisan U.S. Senate leadership that supported the bill—Democratic Senators Harry Reid and Max Baucus, and Republican Senators Orrin Hatch and Arlen Specter (then Republican)—Congress demonstrated a rare bipartisan unity behind medical research funding by unanimously passing this stand-alone legislation, even in the face of a major new war. Patient advocacy had again demonstrated its tremendous strength.

This example illustrates the political strength that is available when the nation's patient advocacy groups unite behind a single bond funding program that must be approved by the voters within a state or nation; the unifying power of their advocacy, combined with reuniting the entire healthcare constituency, presents a powerful and effective voting and advocacy force to empower a new funding paradigm.

2.8 CREATING STATE PARADIGM TO COMPLEMENT FEDERAL RESEARCH FUNDING

California's Proposition 71 was designed to create a paradigm change in governance and funding structures, to launch a new field of medical research—stem cell therapies—and to provide the funding platform to carry that research safely at an unprecedented speed through the 5- to 15-year development process to initial human efficacy trials. The voters of California approved $6 billion ($3 billion in the principal amount of bonds and $3 billion to pay the interest) over approximately 35 years. *This funding model was not designed as an interim replacement for the NIH. In fact, it contemplates the NIH as a long-term funding partner.* Although Proposition 71 filled a critical gap and continues to fund embryonic stem cell research outside the funding authority of the NIH, one of its core purposes is to establish a funding system for medical research that is within the governmental powers of some states and/or foreign states, provinces, and/or nations via collaborative funding agreements. The U.S. Congress and Executive Branch cannot readily duplicate the California Model under the federal governmental system.

The primary and complementary role of the California funding agency is to drive discoveries from stem cell research to the clinic (Trounson, Klein, and Murphy 2008). Funding from the NIH generally is not targeted or designed to carry discoveries through the entire development pipeline to the clinic. At the end of 2009, CIRM, the California agency, had allocated approximately $1 billion to research and facilities. The distribution of these funds was as follows:

- $320 million for facilities and equipment ($50 million for shared laboratory grants and $270 million for major facilities grants)

- $388 million for basic research, training grants, research development and tools projects, and research faculty funding
- $310 million for translational medicine to take discoveries to the clinic

The California agency was able to financially leverage the building of the 12 new stem cell research facilities in California with US$540 million from private donors, and a further sum of about US$340 million in institutional support in commitments for facilities construction, initial faculty hiring, and equipment funding for the institutes. Combined with the state agency funding, the 12 California facilities have therefore been supported with approximately $1.2 billion for facilities, faculty, and equipment alone. Table 2.1 summarizes the major facilities grants.

2.8.1 CALIFORNIA MODEL

The California Model is intended to change the nature, the structure, and the speed at which scientific discoveries can be made and delivered to patients. The six key components of the model are described below.

1. **Creating an independent agency**—The initiative, through a state constitutional and statutory amendment, created within the state government an independent agency governed by a 29-member board (Cal. Health & Saf. Code §125290.20(a)) composed of medical school deans (6) (principally appointed by their University of California chancellors); executive officers of scientific research institutions, research hospitals, and universities (7); patient advocates (10); and biotech industry representatives (4). All board members (other than the five appointed by the UC Chancellors) must be appointed by California's State Constitutional Executive Officers and/or legislative leaders, according to detailed specifications covering expertise and scientific and/or medical experience and leadership. These members serve for 6- to 8-year terms (Cal. Health & Saf. Code §125290.20(c)) and they are not subject to removal, except for statutory violations. The Governing Board elects its Chairman and two Vice Chairmen from additional patient advocates nominated by the governor, lieutenant governor, treasurer, and controller (Cal. Health & Saf. Code §125290.20[a]). The second Vice Chairman is selected by the board from among its membership at large.
2. **Funding derived from bonds**—The initiative's funding for research and facilities is derived from general obligation bonds of the state of California, not from appropriations of the "State's General Fund." Constitutionally, bonds of the state have their debt service paid from General Fund revenues immediately after the state's commitments to education are met from the top 40% of state revenues (Cal. Const. Art. XVI, §8(a); §1). This constitutional priority provides extraordinary stability to the state's bond debt service payments, enabling the state to issue bonds even during difficult economic cycles. The initiative directs the state to "capitalize" the first five years of interest payments in the initial bond issues, thereby relieving the State General Fund of debt service payments for five years (Cal. Health & Saf. Code §125291.45(c)).

TABLE 2.1
CIRM Major Facilities Grant Program: Final Summary

Institutions (by CIRM category), Institutes, Centers of Excellence, and Special Programs Institutes	Total Project Cost	CIRM Award	Donor and Institutional Project Funds	Other Donor and Institutional Funds for Recruitment and Other Capital Costs	Total Project and Other Funding	Size of Facility (gross square feet)	Size of Research Team at Capacity	Total PIs and Research Staff in Stem Cell Program May 2008
Stanford[a]	$200,000,000	$43,578,000	$156,422,000	$25,450,000	$225,450,000	200,000	612	196
San Diego Consortium[c]	115,202,026	43,000,000	72,202,026	40,000,000	163,202,026	101,667	247	109
UC San Francisco[a]	94,514,740	34,862,400	59,652,340	40,900,000	135,414,740	74,832	245	124[b1]
USC[a]	82,610,000	26,972,500	55,637,500	60,000,000	142,610,000	87,537	234	66
UC Davis[a]	61,770,588	20,082,400	41,688,188	37,100,000	98,870,588	54,227	132	84
UC Irvine[a]	60,457,400	27,156,000	33,301,400	21,500,000	81,957,400	100,635	165	36
UC Los Angeles[a]	41,834,478	19,854,900	22,979,578	40,000,000	81,834,478	34,587	68	114[b2]
Center of Excellence								
UC Berkeley[a]	78,610,000	20,183,500	58,426,500	14,000,000	92,610,000	59,600	224	28
Buck Institute[d]	70,080,747	20,500,000	49,580,747	21,600,000	91,680,747	65,708	128	18
Special Programs								
UC Santa Clara[d]	12,896,500	7,191,950	5,704,550	13,400,000	26,296,500	19,829	68	18
UC Merced[a]	7,458,000	4,359,480	3,098,520	800,000	8,258,000	8,140	36	8
UC Santa Barbara[a]	6,352,400	3,205,800	3,146,600	7,750,000	14,102,400	16,581	50	25
TOTALS	$832,786,879	$270,946,930	$561,839,949	322,500,000	1,155,286,879	823,343	2,209	826

PIs = principal investigators. UC = University of California. USC = University of Southern California.

[a] Research investigations to be completed in 2010.
[b1] Space will be retained for stem cell research at the research staffing capacity previously utilized.
[b2] Space will remain fully utilized in their corporation, new faculty, at UCLA.
[c] San Diego Consortium includes UC San Diego, Salk Institute, Sanford-Burnham Medical Research Institute, and Scripps Research Institute; building to be completed in 2012. The consortium building does not include the space at each individual institution campus that continues to be dedicated to stem cell research.
[d] Research investigations to be completed in 2012.

3. **Large-scale, long-term portfolios**—The $3 billion in bond principal authorized by the public in the 2004 election created a minimum critical portfolio funding scale intended to generate a national-scale research program for stem cell scientists and clinicians within California. Historically, large-scale, long-term portfolios of medical research have high statistical opportunities for success because of broad risk diversification—a critical strategic requirement for innovative new fields of medical research. Additionally, with $3 billion, even if spread over 10 to 12 years, the annual funding portfolio could realistically engage scientists across the entire state; and, with other states and countries engaged through collaborative funding agreements, the agency could provide a broad platform for synergy and real-time, iterative scientific advances, each reinforcing the field's momentum.
4. **Unlimited term**—The term of the California initiative is unlimited (Cal. Const. Art. XXXV). The initiative is established within the California Constitution as a state agency with no time limitation. Before considering loan repayments, including principal, interest, and stock warrant revenue, the original general obligation bond funding for the agency would be exhausted around 2017 unless the California public viewed the performance of the agency's funded research to merit approval for an additional bond authority.
5. **Horizontally integrated pipeline from basic science through Phase II trials**—The agency has an authorized staff of 52, including the chairman and the statutory Board Vice Chairman. The president of the agency creates a strategic plan, subject to the Governing Board's approval, which evolves with the progress of scientific and clinical discovery. The intent is to create a horizontally integrated pipeline from basic science through FDA-approved Phase IIA or IIB clinical trials to verify efficacy. All grants and loans under this strategic plan must obtain recommendations from a confidential peer review of the Grants Working Group (GWG) populated by panels composed of 15 U.S. scientists and clinicians from other states and nations and 7 patient advocates from the Governing Board. Recommendations then must be submitted to the governing board for discussion of confidential or proprietary information in executive session followed by a final debate and approval in public session.
6. **Collaborative funding agreements to enable globalization of effort**—In order to facilitate the globalization of the Californian research endeavors in stem cell research, CIRM has linked together with many of the world leading researchers in collaborative research with California colleagues. Agreements with public funding agencies in Great Britain, Spain, Japan, Canada, Germany, China, and the state of Victoria Australia enable scientists from these countries to submit joint applications for funding with those selected and then supported by CIRM and the country involved. These joint project grants effectively break down scientific barriers between countries and enable the world's premier scientists and clinicians to work together for the common good. CIRM has a similar arrangement with the state of Maryland and the International Juvenile Diabetes Research Foundation. These arrangements further leverage the Californian public investment in achieving goals for new clinical treatments and cures.

2.9 BASIC RATIONALE OF CALIFORNIA MODEL

The California Model assumes that with outstanding scientific talent and facilities, the character of the capital funding source becomes a primary determinant in the potential for medical discovery and advances in implementing those scientific discoveries. In designing a capital funding structure to fund medical research, the Initiative's six central key structural features were organized to meet the following five strategic objectives:

1. **Structure must protect funding**—The organizational structure must protect the source of the funding from real and perceived potential pressures and distortions to the scientific discovery process.
2. **Critical long-term funding**—A long-term commitment of the funding source is critical to provide adequate assurances to attract the best scientific talent and to permit complex long-term scientific challenges to be undertaken.
3. **Stability of funding critical**—The stability of the funding—its insulation from interruption—is critical to provide the security to embark on challenging, innovative research with a long development path and attract major philanthropic, biotechnology, and institutional matching fund commitments.
4. **Financial scale**—The capital must reach a financial scale sufficient to drive a critical mass of core research in the field into a portfolio of translational therapies that result in a number of novel and efficacious treatments.
5. **Objective resource allocation**—The resource allocations system for the capital must be based on objective scientific and medical criteria that permit research to be funded for a horizontally integrated pipeline through Phase II human proof of concept trials, rather than an allocation system that funds only discrete increments of discovery, preclinical development, and human trial processes.

After these criteria are met, the California Model proposes that scientific and medical advances can be driven from basic concept discovery grants through (1) preclinical proof of concept; (2) evidence of safety; and (3) early indications of benefit and efficacy (Phase I/IIA or B human clinical trials). A high level of predictability of a continuing chain of funding is essential, as is a development program that requires the research to meet robust peer review milestones and standards. This generates a *continuous* funding stream up to proof of human efficacy, the threshold criteria for consideration by most venture capital and/or commercial support sources. This capacity to fund proof of human efficacy represents a critical strategic advantage rarely available through public funding models for scientific research.

2.10 OPTIMIZING GOVERNMENTAL CASH FLOW OF CALIFORNIA RESEARCH FUNDING MODEL

To strengthen governmental support for the California funding model through bonds, the cash flow costs and benefits should be organized in the original financial structure to minimize or offset general fund payments of bond debt service in the years before net state medical costs savings become available to offset significant general obligation bond debt service payments. Generally, in the first five to seven years of a major medical research program in a broad-based field of high potential, the only state governmental revenue flows from state income and sales taxes generated by the research facilities construction, research expenditures, and the normal economic multipliers on those expenditures. In the United States, because of the strength of private philanthropy, these revenue benefits are multiplied by matching funds donated by individuals and institutions.

In California, for example, $100 million in new state tax revenue is projected to be received by the end of the fifth year of the agency's *full* strength funding operations that started in 2006[*] due to funding delays arising from ideologically driven constitutional litigation (*California Family Bioethics Council v. California Institute for Regenerative Medicine*). These revenues represent economic activity driven only by $320 million in Proposition 71 funding advanced under the first $1 billion in agency funding commitments. The revenues are, however, enhanced by private donor and institutional matching funds of $844 million for facilities construction, equipment, and new faculty hiring that will be expended during this period under matching fund commitments contractually pledged in exchange for funding from the California Institute for Regenerative Medicine (CIRM) (*2008 Annual Report*).

The cash flow impact on California's General Fund is also mitigated by the Initiative's requirement that all interest payments on the bonds during the first 5 years will be capitalized in the bonds (paid by bond proceeds). The new state tax revenues are therefore available to pay debt service on the bonds arising in years 6 and later (Cal. Health & Saf. Code §125291.45(c)). Current projections through year 10 suggest that bond payments by the General Fund to the middle of year 9 will be almost completely offset by the initial $100 million in tax revenue generated by the end of year 5 plus supplemental tax revenue in years 6 through 8. If matching funds continue to be committed, at even 25% of the rate to date, General Fund expenditures for debt service could actually be offset for several additional years, before considering actual medical services cost savings for California.

The design of the Proposition 71 initial cash flow plan did not project any intellectual property revenue share collections from royalties or licensing fee participations *until the end of year 14*. However, some initial medical savings from research advances and therapy developments were anticipated by year 10 at the minimal level necessary to offset bond debt service payments at that point. In fact, an FDA-approved Phase I human trial of a therapy developed in part with CIRM funding has recently been concluded successfully and demonstrated strong initial efficacy, even as a Phase I trial. If efficacy continues to be demonstrated for treating polycythemia vera and primary myelofibrosis, the economic savings are expected to reach $100 million (2010 Report for CIRM by LECG, LLC).

An analysis is currently in progress to project the potential savings and the portion of that savings that will reduce California's government healthcare costs. In addition, because the therapy allows patients to return to work full time, additional state tax revenues will be generated by the therapeutic results. These savings, if realized,

---

[*] In 2006, $50,000,000 of initial reported funding was raised from private placement of bonds during litigation.

would already substantially exceed the original projections for this very early stage of Proposition 71 funding, even though these conditions affect only approximately 12,000 Californians. Intellectual property state revenue participations would be in addition to the numbers cited above. Furthermore, the second clinical trial, arising from CIRM-funded research started in 2010 and it is expected that a third human trial may receive FDA approval in 2011.

Apart from these initial indications of potential revenue and/or medical savings (from avoided costs) for California, more than 400 scientific papers were published during the first 36 months of research funding (CIRM Announcement 2009). The discoveries and knowledge represented in those papers creates a portfolio of work that provides substantial promise of improvements in the current treatment of chronic disease along with new therapies. While the actual cash flow savings and/or inflows generated by therapy development and new discoveries for California will not be definitive—even preliminarily—for 4 to 5 years (at the earliest), the current research portfolio includes 14 disease teams that have provided to the independent peer review and the Governing Board "compelling and reproducible evidence" that "demonstrates that the proposed therapeutic has disease- (or injury-) modifying activity" and that "there is reasonable expectation that an IND filing" for a Phase I human trial "can be achieved within 4 years [48 months] of the project start date." (CIRM Press Release, October 28, 2009; CIRM Request for Application 09-01, Disease Research Team Award). In short, the research portfolio of CIRM is on track or ahead of schedule in demonstrating a credible case that new tax revenues and initial governmental medical savings can reach the minimum levels during the first 10 years of a bond-funded program, to offset a substantial portion, if not all, of the early debt service payments. This approach, again, relies upon the initial five years being structured on an interest-only basis, with this debt service capitalized within the original bond issues.

2.11 MODELS PROVIDING ENHANCED OPPORTUNITIES

By supporting the biotechnology industry with grants, and loans (when a company budget request is in excess of $3 million), CIRM is further leveraging public funds to enhance the ability of the for-profit sector to develop new therapies, new instrumentation, methods, and reagents and to more effectively chaperone translational and clinical programs through regulatory agencies such as the FDA for clinical trials. CIRM looks forward to developing constructive partnerships with other major stakeholders in the pharmaceutical and finance industries.

The California CIRM model has not been functional long enough to determine the success of the integrated academic and biotechnology team approach to translational research. However, it is clear that scientists who have engaged with CIRM and are building impressive inter-institutional and international teams that include one or several biotechnology partners and companies are also seeking academic and medical partnership expertise to enhance their intellectual competiveness. This is well demonstrated in the successful CIRM Disease Team Program of preclinical research awarded in October 2009 (Press Release, April 8, 2008). The spillover benefits include support for growth of the biotech industry, jobs associated with the new research facilities, and increased competitiveness of CIRM-supported scientists for national grants.

It is not uncommon that major grants are awarded to institutions by pharmaceutical companies for first right of access to research developments and discoveries, particularly those with intellectual property rights attached. These awards are useful in underwriting work that otherwise cannot be adequately funded by public agency grants. These may be seen at times to be very successful but more frequently do not deliver a constant source of new discoveries that are useful to the companies.

Organizations that fund a wide variety of research projects, particularly those that fund the translation, preclinical, and early clinical phases of research, are attractive to major pharmaceutical companies because they source a larger population of scientists and hence ideas; the research is further down the pipeline of application and hence closer to a potential product for application. Also the work has been comprehensively reviewed and managed for success and hence more likely to lead to a successful product. As a result, many of these companies are looking at some kind of partnership arrangements with publically funded organizations such as CIRM. The object is for the companies to access high-value clinical opportunities, and the interest of the funding body is to connect end-users to the teams that have made progress toward the clinic but still require substantial financing to undertake the expensive phase IIB/IV trials needed to finally enable the community to access these new developments.

The possible development of reinsurance funds under which health plans contribute from healthcare savings as a result of progress to cures of disease brought about by stem cell research warrants further examination. Such funds should attract government contributions and could be used to offset some of the development costs of clinical trials or to contribute to cost claims of new stem cell therapies. It seems unlikely that all the potential clinical developments will be able to attract the large quantum of finance necessary for completion of late stage clinical trials. At risk are orphan diseases, conditions that have low cost recovery because they are rare, or a simple cell therapeutic cure that can be delivered as an outpatient's procedure. While the costs of clinical trials remain extremely high there will be many examples of effective therapies with an insufficient return to attract private investment. Solutions for these problems are needed in the near future.

2.12 RELATIONSHIP OF RESEARCH COMPLEXITY TO CAPITAL

The California Model was designed to empower greater levels of research complexity than would normally be feasible through traditional models, governmental or private industry funding. As a starting point for analysis by private capital, *there is an inverse relationship between the complexity of scientific research and the tolerance of private capital for risk*. Particularly in a new medical research field like stem cell medicine, government capital must normally fund research until early Phase II human trial efficacy is demonstrated. That governmental funding role is especially critical during a downturn in the global financial cycle. Despite a few notable exceptions to this position, the private biotech companies funding major preclinical research and

Phase I clinical trials for cellular therapies (especially those derived from human embryonic stem cells) obtained their primary capital bases prior to 2005.

In the current economic climate and for the foreseeable future, the complex development paths for cellular therapies will rely upon governmental sources to carry them through preclinical and early stage clinical trials. To optimize the research potential through this difficult developmental period, governmental funding sources can provide large-scale grants or loans that permit and/or encourage multi-institutional teams that will often include private companies. By building multi-institutional teams that target Phase I and/or Phase II clinical trials, from the starting point of an identified Phase I IND (Investigational New Drug) clinical target, the scope of the skill set and experience level of the entire team can increase significantly, but the complexity of the management challenge and the scale of the financial investment are substantially increased.

Under the California Model, the grant or loan portfolio size is significant enough to tolerate risk increments in the range of $20 million to $40 million because that range well represent less than 10% of the respective grant or loan portfolios before counting matching funds or loan repayments. This permits optimization of the team composition and tolerates a risk scale that the private sector would infrequently embrace at the IND definition point, even with preliminary preclinical evidence that an IND approval by the FDA could be achieved within 48 months. The California agency created a specific funding model to match this risk spectrum, with the justification that the higher level of integrated expertise early in the preclinical process will expedite therapy development and reduce risks in the Phase I and II human clinical trials. Few private companies have been established in this early stage preclinical and clinical profiled space over the past 2 years, and this is not expected to change until significant commercial product successes occur.

As discussed, generally, here in Section 7.1.1(6), international scientific collaboration is an important goal of the California Model. The creation of Disease Team program grants in the $20 million range (the California team portion) for preclinical and therapy development research in pursuit of a Phase I IND approval builds an attractive scale for international scientific collaboration. As a validation of this concept, CIRM has signed bilateral agreements with seven nations to advance international scientific collaboration and accelerate potential stem cell therapy development. Active programs have been launched or are in the process of initial funding rounds with five of the seven governments. Agreements are in place with scientific funding organizations in the United Kingdom, Spain, Japan, Canada, Germany, China, and the state of Victoria, Australia. Scientists in these world-leading stem cell research nations can file team applications with their California counterparts; research grant awards approved for a jurisdiction are funded by that jurisdiction. The scale of the portfolio that permits large-scale grants and the broad-based developments of scientific capacity in California, with the assurance of long-term stable funding, incentivizes and enables a level of international collaboration on translational medicine that has rarely been achieved. After the threshold transactional costs of building a funding relationship have been invested, additional collaborative relationships to perform complementary research in immunology and/or basic science, for example, can also be advanced with smaller scale grants.

When nations can verify a stable, long-term funding source on a major scale, there is a strategic value in building a scientific collaboration, especially where the funding jurisdiction represents a global center of outstanding scientific capacity. Proposition 71 and the California Model permitted the California agency to meet these strategic utility criteria. In the first year of this program of international collaboration, over $58 million in international funding and leverage has been obtained. Dissolving the artificial national geographic funding boundaries (that have historically prevented the world's best scientists and clinicians from building international teams to advance critical therapy development for chronic disease) represents an additional strategic advantage of the financial funding structure under the California Model.

2.13 INTERFACE OF GOVERNMENTAL FUNDING WITH PRIVATE CAPITAL MARKETS

If governmental funding is to maximally leverage its impact on stem cell research, it must create a capital framework that recruits private capital into shared risk relationships at the earliest possible stage of research. While private capital will not generally undertake early stage development projects, on cellular therapies in particular, prior to a positive Phase IIA or Phase IIB human efficacy trial, private capital can be induced to participate in early stage stem cell therapy preclinical risks, if there is a credible funding access to government capital that can leverage their private capital assets. To the extent that private capital can predictably evaluate the opportunity to diversify its portfolio risks with substantial government leverage, private capital can justify spreading significant funding into a number of early stage stem cell investments, with a reasonable expectation that some small percentage of a large portfolio will be successful.

Government funding leverage for private capital also provides a major benefit in averaging down the capital carrying costs on complex, long-term therapy development projects. If the entire cost had to be carried at venture capital internal rates of return, a complex project with a long development horizon would, as a general rule, immediately be eliminated from the eligible investment list (see Chapter 5 by Prescott). Given the high-risk premiums assigned to even real property mortgage securities, starting with the 2008 economic cycle, novel stem cell therapies will predictably need to be funded by social capital (public financing) from governmental units that can internalize and capture medical savings across a broad cross-section of their populations.

2.14 CALIFORNIA MODEL FOR FUNDING LARGE-SCALE BIOTECH RESEARCH

For major funding opportunities with biotech companies, the California Model of Proposition 71 employs a loan structure rather than a grant approach. The intent of the loan model is to recycle state research funding to drive a broader and longer-term portfolio. Two types of loans are provided: (1) recourse (company-backed) loans, and (2) non-recourse (product-backed) loans with payback requirements conditioned on producing a commercial product.

2.14.1 RECOURSE LOANS

Under a recourse loan of up to 10 years, principal and interest accrue for 5 years, unless an acceleration liquidity event (e.g., cash sale of the company) triggers an accelerated payment. Extensions beyond 5 years require partial prepayments of accrued interest, annually. The recourse loan carries a repayment obligation regardless of whether the research project financed is successful. This type of loan allows recourse to the company as a general obligation and it carries a 10 to 75% stock warrant obligation adjusted for the financial strength and track record of the company.

2.14.2 NON-RECOURSE LOANS

A non-recourse loan must be repaid only if the project financed is successfully commercialized by the company and/or sold and commercialized by a successor in interest. The non-recourse loan attaches only to revenues of the company's research product funded by the loan and derivative products from that research. This loan carries a stock warrant obligation from 50 to 100%, adjusted based on the company's co-investment in the research. Again, if the product is not successful, neither principal nor interest of the non-recourse loan needs to be repaid, but the agency retains the contract right to the stock warrants. All interest and principal payments accrue for 5 years, unless a repayment major liquidity event triggers acceleration of repayment. The loan, with interim payments, can be extended up to a 10-year total term.

While the CIRM loan program is in its start-up phase, the long-term benefits of recycling any substantial portion of state government funding would provide a major strategic value in funding a broader disease portfolio and permitting larger scale funding for any specific project. The commitment to any individual project can reach sizable proportions when a Phase I preclinical therapeutic research project leading to a Phase I human trial approval is followed by Phase I and Phase IIA or IIB clinical trial funding.

A loan task force of the Governing Board, with substantial lender and venture capital public testimony along with a PricewaterhouseCoopers independent study, found that even with a very high percentage of non-performance on the loan portfolio, the interest and stock warrant revenue on the minority performing share of the portfolio could result in doubling of the portfolio from payback revenues every ten years (PricewaterhouseCoopers 2008). Even if the program were half as successful as projected, the recycling benefits would be significant.

2.15 BIOTECHNOLOGY'S FULL ENGAGEMENT AS STRATEGIC GOAL

Ultimately, to engage the best scientific minds in California with the greatest therapy development experience, private sector biotech companies must be fully engaged as central participants in the California Model. While private sector capital risk sharing is important strategically, the experiences of private sector personnel in managing therapeutic products through the FDA process to the patient and commercialization is a critical human resource asset necessary to successfully develop a portfolio of stem cell therapies for chronic disease and injury. Beyond participating with CIRM as principal investigators (PIs) through the loan model, for larger scale CIRM requests for applications (RFAs), private companies can also participate on teams with non-profit research institutions as co-PIs or as contractual collaborators. Private companies can also apply directly as PIs for smaller scale grants.

2.16 GOVERNMENTAL VALIDATION OF PRIVATE COMPANY RESEARCH

As CIRM seeks to recruit greater private company participation, it becomes clear that as private companies receive public grant approvals or loan approvals from CIRM, the "validation value" of CIRM's peer review and board approval may be substantial. After a public approval, companies often receive significant new expressions of private capital interests and/or their stock valuations or stock values are expected to increase. At this point, information to prove this theory is merely anecdotal, because neither a large enough pool of companies nor a long enough validation period for verification yet exists. The anecdotal evidence is, however, promising.

2.17 GLOBAL FUNDING PRIORITIES FOR MEDICAL RESEARCH

Chronic disease is a global burden. In 2004, the Priority Medicines Project of the World Health Organization (WHO) outlined priorities for future public funding for research and development of new drugs and vaccines. Using burden-of-disease rankings, the project identified 20 major diseases that account for 60% of the total disease burden worldwide, measured in disability-adjusted life years (DALYs). After adjusting with information on the most vulnerable groups—women, children, and the elderly—and neglected (mostly tropical) diseases, a list of the 10 highest priorities was developed (WHO 2004):

- Infections caused by antibacterial-resistant pathogens
- Pandemic influenza
- Cardiovascular disease
- Diabetes types 1 and 2
- Cancer
- Acute stroke
- HIV/AIDS
- Tuberculosis
- Neglected diseases (including but not limited to sleeping sickness (trypanosomiasis), Buruli ulcer, leishmaniasis, and Chagas disease
- Malaria

It is important to note that five of these were included in the first 14 CIRM Disease Team stem cell grants and loans. Disease Team grants or loans approved by the Governing Board are represented by the priority research areas listed including:

- Glioblastoma, brain tumor, cancer (two grants)
- Type I diabetes
- Leukemia and cancer (two grants)
- HIV/AIDS (two grants)
- Acute stroke
- Cancer stem cells
- Cardiovascular disease

Additionally, in the most advanced economies, up to 75% of healthcare costs are consumed by chronic diseases, dominantly represented above. Certainly, there is a global consensus on the severity of the human and financial burdens imposed by these chronic diseases, but funding for research to cure or substantially mitigate these diseases remains largely segregated along national and/or regional jurisdictional lines. This territorial, fractured approach to medical research funding is dysfunctional if our goal is to build the finest global teams to advance medical research in these critical areas of patient suffering and massive governmental cost burdens.

2.18 FINANCING TO REACH MILLENNIUM DEVELOPMENT GOALS FOR MEDICAL OBJECTIVES

One of the most promising new sources of funding for addressing the Millennium Development Goals to eliminate chronic disease has followed the bond financing model. To front-end load the financial resources available for immunization efforts against infectious disease in the developing world, bond financing against a chain of future government financial pledges has emerged as one of the most effective new financial tools.

While remarkable, innovative examples of donations and creative approaches have been devised by individual countries, achieving an effective global funding scale quickly may best be served by studying the International Finance Facility for Immunization (IFFIm). The creation of this financing authority was announced in 2005 by Gordon Brown, then British Chancellor of the Exchequer, and Bill Gates, then Chairman of Microsoft. As of 2008, IFFIm benefitted from more than $5 billion in pledges from at least eight nations.* This model relies on international bonds backed by the pledges of the participating nations; bond payments are spread over a period of 20 years, matching the principal amortization payment schedule on the bonds.

The bond funding structure for the IFFIm is worthy of immediate focus as a model for it could certainly be brought to a much higher scale quickly. Although the funds are utilized for immunizations, the goal is to eliminate the target diseases, just as smallpox was eradicated globally in 1979. These expenditures for immunization are therefore more of a capital investment in international health, with a goal of permanently securing global health by providing long-term protection against the risks and costs of infectious disease. In that context, the cost of the program could have properly be amortized by bonds spreading the cost of the program for the groups

---

* The donors are the United Kingdom, France, Italy, Spain, Sweden, Norway, South Africa, and the Netherlands. http://www.iff-immunisation.org/donors.html

that benefit globally. The current funding structure does not align the contributing nations and the direct beneficiary nations, but the funding structure arguably leverages the foreign aid structures of the major nations, capturing a human health capital asset—the permanent freedom of the world's peoples from these deadly diseases.

2.19 BLENDING IFFIm AND PROPOSITION 71 MODELS

The current global financial crisis and the resulting national and international debt burdens arising from recovery stimulus programs and financial bailouts will constrain many national and regional government medical research funding options over the next several decades. The United States and European governments in particular will face ever increasing and tighter financial discipline in funding medical research. The U.S. Congress should expect a "pay–go" system under which no appropriation can be increased or renewed without cutting another competing government program an equal amount or increasing taxes in an offsetting amount. Many European Union countries may arrive at similar difficult budgetary tradeoffs.

Given the crushing weight of rising national medical costs, the global challenge will be how to fund a quantum increase in medical research as the best hope to reduce the future health burden while meeting the extraordinary current demands of rising healthcare costs. This conflict over resource choices should be expected to be especially severe in the United States.

If the leading nations that contribute to the World Bank were to recognize the value of the California Model and agree to finance substantial increases in global medical research through bonds, a major supplementary funding source for stem cell research—indeed all medical research—could be mobilized rapidly. The World Bank currently acts as the financial advisor and the treasury manager to IFFIm. Rather than having the bonds backed by a pool of nations' credits or the individual credit of a pledging nation, a World Bank guarantee would clearly enhance the efficiency of the borrowing structure. An international peer review panel could allocate the research funding derived from the bonds, with a recusal of the scientists from judging any applicant in which they had a professional, financial, personal, or institutional relationship within the past 3 to 5 years.

For California, these rules, while stricter than NIH guidelines for conflict, have worked well to protect the quality and preserve the integrity of the peer review. An additional Board requirement excludes any scientist from California from participating in peer review. A high sensitivity to conflicts of interest is a recommended feature of any peer review system; and, it should enhance efforts to recruit a large number of nations as financial contributors to a research funding mechanism of this type.

For countries in the European Union, this program should be highly attractive, since Eurostat ruled in the fall of 2005 that each country would bear only a budgetary charge for the current year's pledge to IFFIm instead of the following 15 to 19 years of their commitments encumbered by the financing. It is doubtful that budgetary funding in the U.S. would follow this model, but deferred start dates and long-term funding commitments spread over 20 to 40 years should be easier to obtain than major upfront appropriations spread over 5 years. For example, setting the starting contribution at year 7 with a stream of continuing pledges running through year 30 could substantially enhance the potential for a country to commit to the program. It will be critical that these bonds be understood to fund critical Intellectual Capital, not operating deficits.

Like California's plan, the first 5 to 7 years might feature a capitalized interest structure and deferred principal payments to better align the start of the benefit period of medical savings and new tax revenue with the beginning of interest and principal payments. A stable 15- to 20-year funding stream for the international funding agency would have to be established and highly defined governing board selection criteria would need to separate expertise and mission commitment from political office seekers.

A prototype program of $5 billion to $10 billion might test this translation of the California and/or IFFIm Models on an international application for stem cell research. If successful, the stem cell research prototype could reasonably be transformed into a general medical research funding model with a global commitment at the $50 billion to $100 billion level. If a country's scientists could participate only when the nation made a financial commitment to the common effort based on a proportion of its gross domestic product (GDP), the participation level might include a broad array of nations. The best scientists of the world funded adequately on effective global teams, could conceivably shorten or mitigate the suffering and cost of the WHO's list of the planet's most deadly diseases. A historic reduction in the future of human suffering is possible, perhaps even predictable, if novel financial structures permit concentrated major medical research funding up front. On November 7, 2006, when the first $1 billion in IFFIm bonds were sold, Gordon Brown and Bill Gates said, "We need more minds devoted to finding creative solutions. By matching the power of medical advance with innovative finance we can fill the gap between what we are capable of and what we are willing to do- and unleash the power of human ingenuity and goodness to save millions of lives" (*Independent* 2006). They also quoted Mahatma Gandhi, "The difference between what we do and what we are capable of doing would suffice to solve most of the world's problems."

REFERENCES

Adelson and Weinberg, 2010.
Cal. Health & Saf. Code §125290.20(a).
Cal. Health & Saf. Code §125290.20(c).
Cal. Health & Saf. Code §125291.45(c).
Cal. Const. Art. XVI, §8(a) and §1.
Cal. Const. Art. XXXV.
California Family Bioethics Council v. California Institute for Regenerative Medicine, 147 Cal.App. 4th 1319, 2007.
California Institute for Regenerative Medicine, San Francisco, CA. Homepage on the Internet.
California Institute for Regenerative Medicine, Request for Application 09-01, Disease Research Team Award.
California Institute for Regenerative Medicine, Government Announcement, December 21, 2009. Available from http://www.cirm.ca.gov/Collaborative_funding.
California Institute for Regenerative Medicine, Government Announcement, December 21, 2009. Available from http://www.cirm.ca.gov/meetings/pdf/2008/050608_item_3c.pdf.

California Institute for Regenerative Medicine, Government Announcement, December 9, 2009. Available from http://www.cirm.ca.gov/announcement_120909_400th.
California Institute for Regenerative Medicine, Press Release, October 28, 2009. Available from http://www.cirm.ca.gov/PressRelease_102809.
California Institute for Regenerative Medicine, *CIRM Interim Economic Impact Review: Addendum on CIRM Facilities.* September 19, 2008, p. 5. http://www.cirm.ca.gov/pub/pdf/EcoEval_091008_Addendum.pdf.
California Institute for Regenerative Medicine, *2008 Annual Report*, p 21.
California Institute for Regenerative Medicine, Press Release, April 8, 2008. Available from http://www.cirm.ca.gov/PressRelease_040808.
HealthVote.org 2004 Propositions. http://www.healthvote.org/index.php/for_against/C28.
Independent, U.K. http://www.independent.co.uk/opinion/commentators/gordon-brown-and-bill-gates-how-to-help-the-worlds-poorest-children-423250.html.
International Finance Facility for Immunization (IFFIm), 2004. www.iffimmunisation.org/pdfs/Iffim_booklet_EN.pdf.
PricewaterhouseCoopers, 2008.
Thompson K. and Duintjer Tebbens R. Retrospective cost effectiveness analyses for polio vaccination in the United States, *Risk Analysis* 26, 1423–1440, 2006. Available from http://www3.interscience.wiley.com/journal/118562964.
Trounson, A., Klein, R.N., and Murphy, 2008.
WHO, 2004, Priority Medicines for Europe and the World. WHO/EDM/PAR/2004.7.
Wikipedia, 2009. www.rules.house.gov/111/.../111_hres_ruleschnge_smmry.pdf; also available from http://en.wikipedia.org/wiki/PAYGO.

GLOSSARY

CIRM: the California Institute for Regenerative Medicine.
CIRM Center of Excellence: a CIRM Major Facility in which researchers conduct two of the three types of researchperformed in a CIRM Institute.
CIRM Institute: a CIRM Major Facility in which researchers conduct basic research, translational research, and clinical research at the same institution.
CIRM Major Facilities Grant Programs: a program established by CIRM to fund, with public and private dollars, the construction of major research facilities in the State of California in order to conduct stem cell and related research.
CIRM Special Program: a CIRM Major Facility in which researchers conduct one of the three types of research performed in a CIRM Institute.
Disease Team Program: a program established by the California Institute for Regenerative Medicine to fund teams of researchers who are focused on a particular disease and who have demonstrated to the Grants Working Group, an independent scientific peer review group, a reasonable expectation that an Investigational New Drug Application [Phase 1 FDA-approved human trial] can be filed within four years of the project start date.
Governing Board: the Governing Board of the California Institute for Regenerative Medicine, also know as the Independent Citizens' Oversight Committee, which is charged with the approval of all grants, loans, standards, policies, and regulations for CIRM and with overseeing the operation of CIRM and the distribution of its grant and loan funds.

Industrial Capital: the physical infrastructure necessary for industrial/commercial development and commerce.

Initiative Proposition 71: the California Stem Cell Research and Cures Act, the ballot measure which established the California Institute for Regenerative Medicine and authorized the issuance of $3 billion in bonds to fund stem cell research and related research and facilities.

Intellectual Capital: the intellectual infrastructure necessary for scientific and technological advancement, including new discoveries that add to Intellectual Capital Infrastructure and that can be patented, traded, sold, and amortized as a long-term capital asset.

Intellectual Capital Asset: the scientific and technical knowledge and discoveries upon which scientific and technological advances are based, including intellectual property that can be patented, traded, sold and amortized as a long-term capital asset.

Intellectual Capital Infrastructure: the overarching base of scientific and technical knowledge and discoveries upon which scientific and technological advances are based.

Intellectual Medical Research Capital: the biomedical, scientific and clinical knowledge and discoveries upon which the development of new drugs, therapies and medical treatments are based.

Investigational New Drug: an investigational new drug application for Phase 1 human safety trial that is made to the Food and Drug Administration under section 505(i) of the Federal Food, Drug, and Cosmetic Act (21 U.S.C. 505(i)).

Millennium Development Goals: the medical objectives established by the World Health Organization to eliminate chronic disease.

Physical Infrastructure: the physical assets of a society, such as roads, bridges, water delivery systems, sewers, etc.

Supplemental Mandatory NIH Appropriation: the mandatory and directed appropriation for a specific and limited range of research sponsored by the National Institutes of Health, as an authorized and appropriate supplement to the NIH budget.

State General Fund: the fund into which unrestricted general tax and general revenues received by the State of California are deposited. Bonds issued with a debt service pledge from the State General Fund carry the full faith and credit guarantee of the State of California.

3 Investment Models
Public Funding in Australia

Graham Macdonald

CONTENTS

3.1 Historical Background 35
3.2 Government Incentives for Innovation and Entrepreneurialism 36
3.3 Commercial Stem Cell Research 36
3.4 Australian Stem Cell Centre 37
    3.4.1 Governance 37
    3.4.2 Objective 37
    3.4.3 Reporting Responsibilities 38
    3.4.4 Organization of Research 39
    3.4.5 Progress 39
    3.4.6 Recent Developments 40
    3.4.7 Current Status 41
References 41

3.1 HISTORICAL BACKGROUND

Australia's investment in biomedical research, public and commercial, has historically been low in relationship to other economically comparable countries. Over the past 30 years, measures to bridge the gap in government funding have placed the country at about the middle level among advanced economies whether calculated as percentage of GDP or per capita expenditure. In contrast, business investment in life sciences continues to be low when placed in the context of Australia's success in this area in terms of publications in the international literature. Since 1990, federal and state governments have taken up the theme of the "knowledge economy" with enthusiasm, and biotechnology in all its forms is seen as a leading example on which to build a national innovation capability.

Australia was not alone in this view and federal governments, in particular, saw the "new economy" as something whose development was urgent if Australia was to remain competitive in a changing international market. In reality, conservative and radical governments alike have assigned a higher economic priority to exploiting the country's rich mineral resources and have been less financially or legislatively supportive of industries with arguably more significant long-term value-added elements. "Innovation" is also a catchword for governments and several reports have been commissioned to study how Australia can be transformed into a country whose economic future is tied more intimately to the nation's creative abilities (held to be considerable) and willingness to undertake medium-term financial risk investment—two hitherto unimpressive efforts.

Two reports commissioned by the federal government addressed the relationship between health and biomedical research and investment on the one hand (Australian Government Department of Health and Aging 1998), and the development of a culture based on innovation (Australian Government Department of Innovation, Industry, Science and Research 2009) on the other. The reports are consistent in their emphasis on education, strong government funding for research, risk taking, and realistic financial and career rewards for companies and individuals exhibiting more adventurous behavior.

3.2 GOVERNMENT INCENTIVES FOR INNOVATION AND ENTREPRENEURIALISM

Since 1983, the commonwealth government has developed several programs to fill the gap between the fiscal needs of companies and institutes commercializing intellectual property and their ability to raise finance through the private market or the academic research funding systems. (Similarly to the old United Kingdom system, the Australian government has funded basic and clinical biomedical science research through the National Health and Medical Research Council [NH&MRC] and non-biomedical research through the Australian Research Council [ARC]).

A series of commonwealth programs have developed schemes to provide pre-seed and seed funding for biotechnology companies and institutes that support translational research. In addition, in 1991, the government established a system of cooperative research centers (CRCs), collaborative partnerships between publicly funded researchers and end users that attract matching contributions from the federal government. This is a well-funded system that joins academic researchers from several institutions with one or more industrial partners to form continua from basic research to commercialization. The centers encompass a wide range of disciplines, including biotechnology and translational research.

No CRC was established on the basis of pluripotent cell research although close links exist between stem cell research groups and CRC partners in areas such as smart membranes and polymers and corneal lens development for delivering stem cell treatments for corneal scarring.

The Howard Liberal and National Party coalition government that introduced these programs lost power in November 2007. The succeeding Labor government terminated many of the programs (but not the CRC scheme) and in its May 2009 national budget, substituted an overarching plan called Superscience that approaches the development of a knowledge-based economy from a wide perspective and includes biotechnology among its priorities.

3.3 COMMERCIAL STEM CELL RESEARCH

Four Australian companies were established to commercialize stem cell research: Bresagen, Stem Cell Sciences (SCS), Embryonic Stem Cells International (ESI), and Mesoblast. ESI was spun off from Monash University and transferred its operations

Investment Models

to Singapore prior to the commencement of most of the government programs. The company was supported at its inception by the Victorian government via a small start-up grant.

At least one company in the stem cell field (Mesoblast) received a Commercial Ready grant while the scheme was in operation, which formed the major income for the 2005–2006 year. SCS received no government support. Bresagen benefited from the R&D start program, which contributed substantially to its budget in 2002–2003. It is clear that all four companies relied almost entirely on private investment to start and maintain their businesses, subsequently taking advantage of R&D tax concessions. Like most Australian biotechnology companies, their positions were precarious and the only one of the four companies in the hands of its originators is Mesoblast.

3.4 AUSTRALIAN STEM CELL CENTRE

In January 2002, the Commonwealth Department of Industry Science and Technology (DIST) called for applications under the Backing Australia's Ability scheme for funding under a Biotechnology Centre of Excellence (BCE) program. The winning bid was for the National Stem Cell Centre (NSCC). Initial funding was to be $43.55 million from the commonwealth, subsequently supplemented by a further $55 million under the BCE agreement, $11.375 million under the Victorian Science and Technology Infrastructure Scheme, and a combined Victorian and federal grant of $5.5 million because NSCC was a major national research facility. The commonwealth's contribution was shared between the DIST and the ARC. The vision of the BCE program was "to create critical mass in research and to establish Australia as a regional and world centre for biotechnology innovation and application" (Australian Department of Industry, Tourism and Resources).

The NSCC was renamed the Australian Stem Cell Centre (ASCC) and effectively began operations in 2003. It remains the only organization funded as a BCE. Similar centres were established in other technological areas such as the National Information and Communications Technology Australia (NICTA).

3.4.1 GOVERNANCE

The ASCC was established under a deed of agreement between the DIST (at the time of execution renamed the Department of Industry, Tourism and Resources or DITR), ARC, and ASCC, which was established as a company limited by guarantee with governance entrusted to an independent board of directors with an elected chairperson. The government reserved the right to appoint a nominee to the board; the members and associated institutions were not represented. In 2007, a second node opened within the University of Queensland but governance remained within a single board of directors that did not include representatives of member institutions.

3.4.2 OBJECTIVE

The ASCC's objective was "to undertake research of the highest quality in the stem cell field in order to discover and ultimately commercialise new therapies for human disease" to be achieved by:

- Creating a world-class Australian research engine focused on identifying, undertaking, and supporting excellence in stem cell research and related technologies
- Identifying and rapidly responding to global changes in stem cell research and industry ensuring the centre's competitive position
- Successfully commercializing outcomes of stem cell research in Australia
- Being globally recognized as an ethical best-in-class Biotechnology Centre of Excellence
- Attracting and retaining leading scientists, students, and management expertise to build capability and sustainability

The deed of agreement stipulated that the ASCC "…will function as a distinct and independent centre with its own identity and with the scale, resources and reputation to attract high-calibre researchers from both within Australia and overseas."

3.4.3 Reporting Responsibilities

Notwithstanding this, the commonwealth required the ASCC to operate within constraints applicable to corporate bodies under Australia's corporation legislation (mainly the Corporations Act 2001 as well as the reporting and accountability demands applicable to recipients of government funds. The ASCC was incorporated as a company limited by guarantee with seven members, and with members' liability limited to an amount of $100 each. Each member was an Australian research university or institute prominent in the biomedical research sector. A further four research universities and institutes agreed formally to participate in the centre's activities. These participants, together with the members, became the "stakeholders" of the Centre.

As is usual for limited companies, the members could not influence the board's functions but had the right, acting in concert, to dismiss the board. The deed of agreement did not specify the relationship of the board, the centre's members, and the stakeholders but the relationship came to be interpreted as a "hands off" arrangement whereby members were only able to question the board about its performance and that of the centre in a discussion session that followed the centre's annual general meeting.

The Centre was required to report four times a year to the government and to publish a public report of its activities annually. In addition it was required to submit to a science panel review of its activities at intervals of three years to be incorporated in the body of a wider ranging review of governance, accountability, administration, and finances.

Internally, principal investigators were funded by the centre on the basis of competitive assessment of statements of work, written in an industrial format including key performance indicators (KPIs), timelines, and indices of commercial potential. They submitted quarterly reports to the centre executive along with reports at the expiries of project funding. At least one of the internal reviews drew attention to this reporting load as being excessive.

3.4.4 Organization of Research

Initially, all ASCC research was conducted by extramural teams within universities and institutes. By 2006, several principal investigators originally on the staff of Monash University were appointed to intramural positions as ASCC employees. At the same time, the number of projects was reduced from 32 to 19, of which 5 were intramural and the rest at other institutions including Monash University. The ASCC occupied offices in a purpose-built technology sector of the university's main campus. Its laboratories were located contiguous to those of the Monash Immunology and Stem Cell Laboratory where the largest number of residual extramural workers were housed.

The changes in the research portfolio were in accord with an executive decision to direct research toward a commercial end with a program in hematology, specifically the growth and supply of pluripotent cell-derived blood cellular replacement products. Most of the other research supported the commercial project with work involving the development of "smart" biomembranes, stem cell bone marrow niches, mesenchymal stem cell differentiation, and small molecular differentiation factors along with work on basic pluripotent cell biology. Other fields supported where a particularly strong external team was identified included breast and lung stem cells and kidney regeneration based on adult stem cells.

In addition to the research teams, the ASCC established two core laboratory facilities, one at the University of Queensland and one at Monash University, to provide basic laboratory support for the research teams. The core facilities provided cell lines and media and worked on project development with specific investigators. Products from these laboratories were the only commercial outputs from the centre and were marketed by a scientific supply company.

3.4.5 Progress

The scientific output of the Centre in terms of publications, presentations at meetings, inclusion of principal investigators in international conference committees, and invited presentations met or exceeded the KPIs after the first two years of the centre's operations. Indecision about the cell type to be commercialized and technical challenges delayed the achievement of commercial objectives.

It would appear that the Centre's original objective of achieving financial independence within 9 years on the basis of commercializing its intellectual property was unrealistic. The usual development time for a synthetic pharmaceutical molecule, while shorter than previously, is still in the range of 10 to 12 years. For cellular therapies, there are few of the guiding regulatory frameworks, clinical development pathways, and established manufacturing standards that are integral parts of the pharmaceutical development model (see Section 5 of this book). It was to be expected that getting a stem cell-derived product to market would demand a significantly longer timeline.

Although the report of the second review in 2008 remains confidential, its recommendations have been released and can be interpreted as indicative of areas of inadequacy of the centre at the time of review. The main issues appeared to be the closed culture of the centre, poor communication levels between different projects (even those where synergies could be identified that would have strengthened both research efforts), and poor relationships with would-be collaborators. The main commercialization project involving transfusible leucocytes needed more rigorous due diligence as it progressed. The centre was urged to widen its views of what elements of its work had potential commercial value. Its policy on intellectual property ownership needed liberalization so that returns could be shared in proportion to each party's contribution.

None of these flaws could be attributed back to the deed of agreement and the terms of the Centre's responsibilities to the commonwealth. The heavy internal reporting schedules in retrospect, however, are likely to have been counterproductive. They occupied time better devoted to advancing the science and encouraged successive reports to be "cut and pasted" from predecessors. Responsibility for compiling the reports may well have been delegated down to junior team members who lacked insight into the overall research strategy and direction.

3.4.6 Recent Developments

The review may or may not have been responsible for events around the time of its delivery to the government. In short, the CEO left the ASCC in August 2008 and the board resigned en masse soon after that. The Centre's members, with the agreement of the commonwealth, formed an interim board composed of the deputy vice chancellors for research from the two major member universities (Queensland and Monash), the business manager of a Melbourne research institute, and an independent chairman. This group began the process of addressing the problems revealed in the report and selected a new substantive board that took over responsibility for the Centre in May 2009.

The research portfolio was reorganized into four major research themes encompassing 32 project modules. Research groups without previous connections with the Centre incorporated and some that were previously dropped from the portfolio were involved again. It was decided not to pursue an aggressive commercial policy. Rather than target early stage biology for commercialization, the decision was to concentrate on staff education and awareness of intellectual property and its management and to provide high-level guidance to ASCC researchers across the board.

The centre embarked on this new research policy at the beginning of July 2009 so its outcomes cannot yet be assessed. Research worker acceptance has been enthusiastic and morale has become higher. Major elements of the transfusible blood cell product project continue under a tighter regime of milestones and project due diligence. Several modules address the development of commercially valuable non-therapeutic products such as reporter cell lines for drug development, the development of small molecules to modulate expansion and differentiation of pluripotent cells, and manipulation of induced pluripotent cells with approaches to measuring their suitability as therapeutic agents.

3.4.7 Current Status

The deed of agreement that funds the ASCC and its activities will expire at the end of June 2011. It is probable that stem cell research funding will then become part of the normal mechanism of the national competitive funding system and university research. It remains to be seen to what extent stem cell therapies and applications will be supported through these avenues.

REFERENCES

Australian Government Department of Health and Aging (1998), The Virtuous Circle: Working Together for Health and Medical Research (The Wills Report). http://www.health.gov.au/internet/main/publishing.nsf/Content/hmrsr.htm

Australian Department of Industry, Tourism and Resources, *Annual Report 2002–2003*, Secretary's Report, p.2. http://www.innovation.gov.au/General/Corporate/Pages/AnnualReport20022003.aspx

Australian Government Department of Innovation, Industry, Science and Research (2009) Powering Ideas: An Innovation Agenda for the 21st Century (The Cutler Report). http://www.innovation.gov.au/innovationreview/Documents/PoweringIdeas_fullreport.pdf

4 Canada
Capitalizing on a 50-Year Legacy

Andrew Lyall

CONTENTS

4.1 The Stem Cell Network 45
    4.1.1 Translational Research 46
    4.1.2 Development of a Highly Qualified Workforce 48
    4.1.3 Networking and Partnering 48
4.2 Aggregate Therapeutics Experiments 49
4.3 The Next Decade 54
References 57

Canada has an exceptional history of advancing the field of stem cell research. In fact, a strong case can be made that it all started here in 1961 when Drs. James Till and Ernest McCulloch, a physicist and a hematologist at the Princess Margaret Hospital in Toronto, were the first in the world to identify and characterize stem cells. Over the next few years, they published a series of seminal papers (Becker et al. 1963) that defined the lexicon of stem cell research still in use 50 years later. It is an achievement for which they were recognized with the Lasker Prize in 2005.

    The laboratory of Till and McCulloch spawned a whole generation of Canadian stem cell research leaders. According to an April 2002 *Nature Immunology* review examining hematopoietic stem cells, Canadians were the first authors of 19 of the 32 most influential articles published on the subject between 1960 and 2000 (Betting on HSC 2002). In the past 20 years, Canadians have made other significant advances including the identification of cancer stem cells by Dr. John Dick in 1994 (Lapidot et al. 1994) and their correlates in solid brain tumors by Dr. Peter Dirks in 2004 (Singh et al. 2004), neural stem cells by Dr. Sam Weiss in 1992 (Reynolds and Weiss 1992), retinal stem cells by Dr. Derek van der Kooy in 2000 (Tropepe et al. 2000), and skin stem cells by Dr. Freda Miller in 2001 (Toma et al. 2001). Drs. Janet Rossant and Andras Nagy made their groundbreaking contribution by fully reconstituting a mouse from a single embryonic stem cell, thus proving the cells' inherent and defining characteristic of pluripotency (Nagy et al. 1993). Today these stem cell pioneers, together with internationally respected researchers such as Drs. Connie Eaves, Michael Rudnicki, Guy Sauvageau, and Mick Bhatia, form an outstanding core of basic scientific expertise that few other countries can match.

TABLE 4.1
Disease Incidence and Healthcare Cost in Canada

Disease Indication	Incidence	Annual Direct Healthcare ($)	Annual Indirect Economic Burden ($)
Acute myocardial infarction	60,000 new cases per annum	$1 billion	$3 billion
Diabetes	2 million Canadians	$3.5 billion	$1.3 billion
Stroke	55,000 new cases per annum	$700 million	$1.5 billion
Macular degeneration	200,000 new cases pper annum	No data available	$2.6 billion
Multiple Sclerosis	35,000 to 55,000 Canadians	$200 million	$600 million
Hemophilia A	2,500 Canadians	$200 million	No data available
Parkinson's disease	100,000 Canadians	$100 million	$500 million

Sources: Health Canada (2002); Dawson et al. (2002); Brown et al. (2002); Canadian Burden of Illness Study Group (1998); Heemstra et al. (2005); Health Canada and Parkinson Society Canada (2003).

Canada's impact has not been limited to basic science. In 2000, clinicians at the University of Alberta published "The Edmonton Protocol" in the *New England Journal of Medicine,* providing the first proof of principle that Type 1 diabetes could be treated effectively by using replacement beta cells from donated pancreases (Shapiro et al. 2000). This innovation has spurred global investment in many companies to explore the potential of deriving functional beta cells from stem cells. Canadian clinicians are now developing new protocols to treat Parkinson's disease, multiple sclerosis, graft-versus-host disease, stroke, pulmonary hypertension, neuroblastoma, and more.

Allied to this historical leadership, Canada's publicly funded health care system has enabled the establishment of tissue and tumor banks and access to large population data sets that provide ongoing and invaluable research tools. It has driven the establishment of a national clinical trials consortium that has enabled rapid testing and enrollment for potential new treatments. Furthermore, decisions to invest in new therapeutic approaches are not solely driven by the lowest cost solution and are also made in the context of the economic impacts of healthy and productive populations. Some sense of the potential impact on the healthcare system and total disease incidence in Canada can be gleaned from Table 4.1. While it is unlikely that stem cell research will provide cures for all forms of each disease, nevertheless progress on even some of the indications will have a profound impact.

The global excitement engendered by the first derivation of human embryonic stem cells in the late 1990s became the impetus for Canada's leading researchers to create a national science initiative and capitalize on the Till and McCulloch legacy. Led by Dr. Ron Worton, who trained as a post-doctoral researcher with the two pioneers 40 years earlier, they saw an opportunity to position Canada at the forefront of the translation of scientific discoveries into clinical medicine. The outcome was the establishment of the Stem Cell Network (SCN) in 2001.

Spurred on by an international board drawn from industry, academia, and patient groups, the SCN quickly identified a number of parallel strategies to achieve this overarching objective. One was to catalyze translational research and fund initial clinical developments. A second was to sponsor specific research initiatives that would bring stem cell biology and pharmacological approaches closer together (with the long-term aim of engaging the pharmaceutical industry in the field). Section 4.1 elucidates in much more detail on how the network went about undertaking this, and places those activities in the context of other related investments of public funds in Canada. The third strategy was to directly sponsor commercial activity, the result of which was the creation of Aggregate Therapeutics Inc. (ATI), a company that pooled the stem cell intellectual properties (IPs) coming from 37 research labs at 16 leading Canadian universities and research hospitals. The lessons learned from the creation of ATI are examined in Section 4.2. Last, but by no means least, was the strategy to invest significantly in research into the ethical, legal, and social issues arising from stem cell research, with the objective of understanding the issues early in the process. While not explored in this chapter, this research has made an important contribution to the shaping of public policy and the regulatory environment in Canada and abroad, while allowing legal scholars such as Prof. Tim Caulfield and Dr. Bartha Knoppers to become acknowledged international authorities in the field.

4.1 THE STEM CELL NETWORK

The Stem Cell Network's (SCN's) goal is to accelerate the development of stem cell-based clinical applications and commercial products in Canada while providing a focal point for academic input into the development of related public policy. The network's primary source of funding is the Canadian government's Networks of Centres of Excellence program that has committed a total of CAD $82 million to SCN between 2001 and 2015. The current $6.4 million in annual funding represents 5% to 10% of the total investment in regenerative medicine in Canada in any given year. More than half of the research funding for stem cell research in Canada ($30 million to $60 million per year) is won competitively through open research grant competitions run by Canada's federal granting councils [Canadian Institutes of Health Research (CIHR), Natural Sciences and Engineering Research Council (NSERC), Social Sciences and Humanities Research Council (SSHRC), and Genome Canada]. With the exception of CIHR's competition for regenerative medicine (worth about $10 million to $12 million per year), none of the funding is directed specifically to stem cell research.

Since 2000, the Canada Foundation for Innovation, a national infrastructure agency, has directed approximately $120 million (http://www.innovation.ca/en/projects-funded) into regenerative medicine facilities across the country, ranging from new stem cell research laboratories to cGMP cell therapy manufacturing. The foundation funds have been matched by another $180 million in contributions from provincial governments and university and hospital foundations. While other countries are beginning to make such investments, in most cases the Canadian investments in infrastructure are already online or nearing completion. The balance of funding comes from provincial funding agencies such as the Ontario Institute of

Cancer Research, charitable foundations such as the Juvenile Diabetes Foundation and Canadian Cancer Society, personal philanthropy, and corporations within and outside of Canada.

SCN brings together over 170 organizations with an interest in advancing stem cell research in Canada, including more than 100 research groups in over 60 universities and research hospitals (half outside of Canada), 40 corporations, 11 government departments, and 55 other non-governmental organizations. SCN uses its funding to marshal and align the objectives, priorities, and investments of these organizations in three areas fundamental to realizing the full potential of this extraordinary technology: (1) support for translational research, (2) the development of highly qualified people, and (3) networking and establishing relevant partnerships of all the key players.

4.1.1 Translational Research

SCN spends approximately two-thirds of its funding in supporting collaborative research projects of academic scientists and their commercial or not-for-profit partners. The majority of the investment is in its core research portfolio, which comprises large multi-year, multi-disciplinary initiatives that pursue integrated and goal-directed programs to address major clinical and/or commercial outcomes. Currently SCN funds 12 such projects at a level of about $1 million per project over 3 years. Clearly, these funding levels alone are insufficient to move therapies into Phase I trials. They have, however, been very important in the Canadian context by providing the incentive, the opportunity, and the glue to unite the independently funded research efforts of multiple investigators and their partners with a common purpose and focus.

This impact has been strengthened by the decision of SCN to focus its funding on research projects that have reached sufficient maturity and risk falling into what has been called the "translational gap." In Canada, as in many other jurisdictions, a significant funding deficit occurs in the stage between early discovery and the onset of clinical application. Into that gap fall an array of product development and preclinical questions that are generally too late to be the topics of academic grants and too early for full industry sponsorship. In an emerging area of science such as stem cell research, this translational gap is particularly perilous. By structuring its research funding to fill this gap, SCN acts as catalyst and enabler, supporting academic researchers that engage industry partners, physicians, and hospitals to advance potential therapeutic applications into early phase clinical trials (see Figure 4.1).

In the early days of the SCN, partners typically became engaged by lending their expertise to address specific development issues or by guiding the research to more commercially or clinically relevant questions. As the SCN has developed and the value of its national, collaborative and cross-sectoral approach became evident, partner cash and in-kind investments to address translational gap questions also have grown. Relationships are building and partners and academics are collaborating earlier in the development cycle. As a result, when the products currently crossing the gap are ready to move into full-scale commercial or clinical development, partners with the knowledge, the means, and the will to make the more sub-

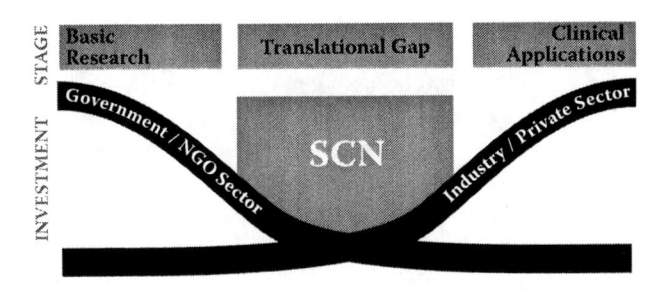

FIGURE 4.1 Translational funding gap between therapeutic discovery and clinical use.

stantial investment required are already at the table. The following case studies illustrate this well.

Case Study 1—In 2005, the leadership of SCN saw an opportunity to combine a number of technologies and start using stem cells as assays in high-throughput screens to identify potential new clinical indications for existing drugs. Using SCN funding for this purpose, Dr. David Kaplan, a researcher at the Hospital for Sick Children in Toronto, found that a drug currently approved for use as an immunosuppressant also appeared to have a marked and targeted effect on the specific cancer stem cells believed to cause neuroblastoma, which accounts for nearly 10% of all childhood cancers and is the most common tumor in babies under 1 year of age. With the support of SCN and several research foundations, Dr. Kaplan was able to rapidly validate the findings in appropriate animal models and move into a compassionate care trial at the children's hospital within two years. Two years after that, following further validation studies and regulatory approvals, Solving Kids' Cancer, a major North American charity, provided substantial funding to allow this protocol to move into a full Phase I trial.

Case Study 2—In 2003, The SCN funded a multi-disciplinary team to examine several different approaches to the treatment of stroke using stem cells. The network assembled a range of non-profit partners to provide additional financial support for the project team. Drawing on the team's expertise and resources, Dr. Samuel Weiss, a Gairdner Award winner at the University of Calgary, was able to rapidly test a new strategy to stimulate endogenous stem cells to promote brain repair following stroke. By 2005, the approach had been validated in the laboratory and licensed to Stem Cell Therapeutics, a new Canadian company that has since raised over $20 million and, as of late 2009, carried the product through to a Phase IIb trial. Meanwhile, the SCN has supported further research to examine the extension of the protocol to treat multiple sclerosis and is hopeful that a Phase I trial will be initiated shortly.

In both studies, the SCN played a crucial role in aligning the research priorities of the project participants. It provided funding to kick-start the work and undertake the less glamorous experiments that were crucial for clinical advancement but not necessarily novel or exciting enough to secure basic science funding. As a result, within 4 years of SCN funding, two stem cell-based therapies have been made available to Canadians. Should the trials prove successful, partners are in place to carry these treatments into the market.

4.1.2 Development of a Highly Qualified Workforce

There was also an early acknowledgment by SCN's leadership that training a new generation of graduate students, post-doctoral researchers, and technicians in the art of translating science into clinical applications and commercial products would be essential to establishing a regenerative medicine industry in Canada and continuing the country's international competitiveness in the stem cell field. As a result, SCN has invested significantly in training programs and requires meaningful participation of young scientists in the design and management of its research program and in all project meetings. It has organized thematic workshops and courses and exposed trainees to the various ethical, legal, and social issues associated with stem cell research and regenerative medicine. It has placed trainees at the centers of its annual scientific meetings and has partnered with other organizations to offer industry internships along with management and leadership training in the near future. Over the network's first nine years, almost 1,000 personnel have participated in SCN-supported training activities and events.

The network also recognized the need to expose graduates from all disciplines to a common body of knowledge in regenerative medicine. To that end, SCN has partnered with researchers at the University of Toronto to develop a multi-disciplinary graduate level course in regenerative medicine. The course is comprised of 30 lectures delivered in person or by interactive webcast and is eligible for a full credit toward a graduate degree at over 20 universities and hospitals across Canada. The lecture topics cover all aspects of regenerative medicine, ranging from causes of organ failure to different therapeutic approaches and innovative technologies and the contextual ethical, legal, social, and business issues. Enrollment increases every year and the course is now close to capacity, despite its heavy workload and the unfamiliarity of the topics to many participants.

4.1.3 Networking and Partnering

The premise of SCN is that all organizations with vested interests in the advancement of stem cell research can accelerate their progress by working together. Furthermore, SCN views networking and partnering as constituting the primary mechanism to enable it to achieve its strategic objectives and ensure a permanent legacy to Canadians from their public investment in the field. The network invests heavily in interdisciplinary workshops, inter-laboratory exchanges, and a host of other multi-sector networking activities including an annual scientific meeting for approximately 400 participants. The network partners across all sectors and areas of its mandate and the nature of its interactions are effectively summarized in Table 4.2.

In short, SCN represents one national approach to investing in stem cell research. Unlike many other jurisdictions, Canada's strategy is largely driven from the bottom up. It plays a central role, balancing the need to provide vision and leadership with a desire to advance the field based on consensus and partnership. It focuses on enabling therapies and related technologies to bridge the translational gap. Its success to date is reflected in the United Kingdom's Pattison report which commented that, while

TABLE 4.2
Stem Cell Network Partnerships and Interactions

Sector	Major Areas of Common Purpose or Interest	Nature of Interactions and Support
Industry	Commercially focused strategic research programs Early protection of IP Facilitation of research contracts Licensing IP Enabling regulatory environment Researcher, TTO, investor education on commercialization	Co-funding of research Participation in design of network projects Co-funding of IP protection Use of IP toolkits Support of and participation in SCN training program Sponsorship and participation in SCN AGM Participation in public policy and egulatory workshops
Not-for-profit	Disease focused research projects Graduate training Enabling regulatory environment Outreach Public support for hESC research Setting policy frameworks	Co-funding of common research projects Co-funding of training initiatives International patient FAQ Workshops Support for stem cell charter Collaborators on SCN project teams focused on ethical, legal, and social issues
Government	Input into regulatory frameworks Commercialization of IP Economic development in Canada Maximum use of government research facilities	Co-funding of commercialization pilot Distribution of IP toolkit Use of government laboratories
Academia	Research excellence Commercialization of IP Funding graduate students Additional stem cell research infrastructure Maximizing use of facilities	TTO time and resources to support commercialization pilot Co-funding of trainees through focused foundation fund raising Drawing on national research base to maximize use of specific facilities (e.g., cGMP cell handling facility in Montréal)
International networks	Development of graduate students Outreach Managing networks Research problems best approached on international scale Enabling regulatory frameworks	Addressing stem cell "tourism" International graduate exchanges Common workshops Sharing lessons learned ICSCN International stem cell forum and initiative

Canada's spending on stem cell research just made the top 10 globally, it was used more effectively than spending elsewhere because of its highly coordinated approach.

4.2 AGGREGATE THERAPEUTICS EXPERIMENTS

Despite this substantial foundation, the corporate stem cell and regenerative medicine sector has remained largely undeveloped in Canada until the past 5 years.

Many of the reasons are the same as those identified elsewhere in this book and apply to this sector across all jurisdictions (Rowley and Martin 2009): much of the science remains early and unproven; no proven and repeatable business model yet exists; the IP landscape is fragmented; and a paucity of investment in biotech has continued for much of the past decade. In addition, the clinical regulatory approval pathway remains relatively untested and is only beginning to emerge as a separate consideration from the public debate on the derivation and use of human embryonic stem cells. Other factors are more typically Canadian: the small and risk-averse nature of a venture capital community bred on the rapid time-to-market model of the IT sector; a pharma R&D presence too small to anchor a strong biotechnology industry; and a scarcity of management experienced in growing successful biotechnology enterprises.

After the SCN launched its first round of research projects, it turned its attention to how best to facilitate economic development in the regenerative medicine space. By late 2003, SCN concluded that traditional models were unlikely to work and that a radical new approach was needed for Canada to be competitive. The result was the creation of Aggregate Therapeutics Inc. (ATI), a virtual accelerator company with a national scope that could pool and develop all the IP, knowledge, and know-how coming out of Canada's world-leading laboratories. At the time, the rationale for this approach seemed sound.

First, it was well understood that many stem cell-based products for transplantation would require a combination of cells, biomaterials (likely a matrix or scaffold), and growth factors. Devices would also need to be developed in parallel. Pooling the IP from the multiple research teams already collaborating on potential therapies through SCN seemed a straightforward mechanism to maintain freedom to operate, avoid difficult issues around stacking royalties, and build a strong product-based patent estate. Furthermore, it was considered that this broad view of the potential Canadian patent estate would allow ATI to be more strategic in its patent filing, with tighter and more effective claims. It was also thought that the critical mass of the company would give it a greater ability to defend its own patent estate and in-license additional pieces on favorable terms.

While Canada has one of the world's largest biotechnology sectors, many of its companies are small and underfunded. As a result, the CEO is often either a founding scientist without substantial business experience or a more seasoned professional working part-time across multiple companies. Neither is the ideal scenario for building world-class companies. It was expected that the critical mass afforded by the creation of ATI would allow the recruitment of a full-time CEO, well-versed in the stem cell space. Not only would this bring valuable product development expertise to all aspects of Canada's public investment in stem cells, it would also bring more commercial rigor and focus to the basic research undertaken in the labs of participating researchers.

The pooling concept was also thought of as a novel way to defray investor risk. Given Canada's track record of discovery in stem cells, the creation of ATI was intended to allow investors to bet on the field rather than having to pick a couple of research laboratories. Allied to this was the notion that the critical mass of ATI would allow it to develop a product pipeline from an early stage, again reducing investment risk should the lead product fail to meet its milestones.

Finally, it was hypothesized that it would prove easier to raise investment capital in sufficient amounts to make Canada a very meaningful player in the stem cell industry by developing a strategy to build a company with a strong patent portfolio and capacity for product development, world-class management, and the opportunity to defray risk.

Over 2004 and 2005, SCN worked hard to build support for the concept and in early 2006 ATI was established by 16 founding universities and research hospitals and 37 founding scientists. Essentially, ATI was structured so that all the founders became equal shareholders in exchange for providing rights of first refusal on their unencumbered technologies. As technology was in-licensed by the founders to ATI, they would participate through additional equity and/or traditional licensing considerations such as royalty streams or milestone payments.

The initiative became a proving ground for an unparalleled level of national collaboration among universities, hospitals, investigators, research councils, and economic development agencies as they piloted this new model for commercializing publicly funded research in Canada. For example, SCN conducted Canada's first national technology transfer consultation process. Early on, it met with all member institutions and investigators to develop the concept. Later, in the detailed design stage, a working group of eight major technology transfer offices (TTOs) was tasked to provide input into the commercialization model and negotiate collaborative IP agreements. These agreements standardized legal templates to accelerate licensing or negotiations related to sponsored research. Captured within an IP tool kit, they are national in scope and participation, spanning hospital, university, and multiple institutional IP policies. Separate disclosure and stand-still agreements also became the bases for a unique partnership of TTOs and stem cell researchers to review early technology disclosures and provide financial and intellectual support for IP filing. These templates remain in use today and their influence now extends far beyond ATI and the stem cell field. They can all be downloaded from the SCN's website.

The SCN also undertook the first comprehensive national IP review focused on a specific field of science and/or technology. Initially, the ATI team compiled an inventory of some 150 IP families held by Canadian researchers or institutions associated with SCN. These IP families were reviewed over a 3-year period and 45 regenerative medicine technologies were identified for further evaluation based on four criteria (IP position, market size, therapeutic area, and stage of development) to assess their commercial potential. Eighteen technologies met ATI's criteria for commercial potential. By the end of 2006, nine technologies and related patent estates had been in-licensed to the company.

Most importantly, the ATI management team created and implemented a translational development pipeline or funnel model that applies industrial discipline to early-stage technology growth. A major benefit of conducting a national review of stem cell related IP was the input it provided to designing the commercialization model for ATI. The funnel model, depicted in Figure 4.2, essentially applies clinical development discipline to the black box of preclinical therapeutic development. It establishes four development stages or tiers prior to clinical development and

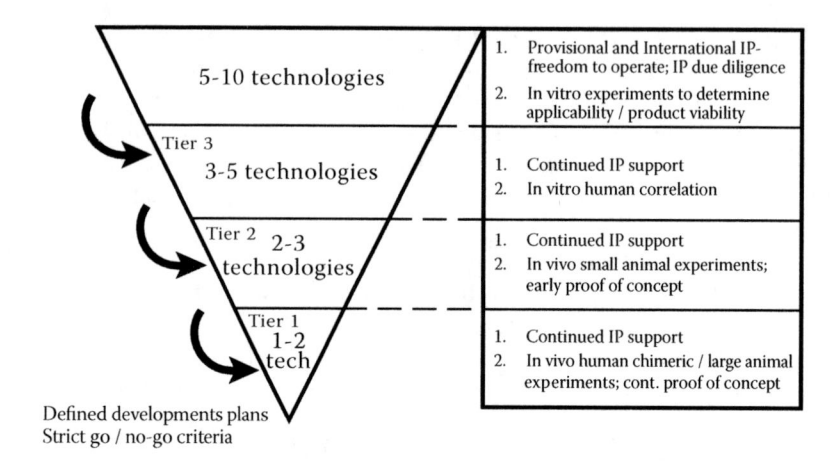

5-10 technologies	1. Provisional and International IP-freedom to operate; IP due diligence
2. In vitro experiments to determine applicability / product viability
Tier 3
3-5 technologies	1. Continued IP support
2. In vitro human correlation
Tier 2
2-3 technologies	1. Continued IP support
2. In vivo small animal experiments; early proof of concept
Tier 1
1-2 tech	1. Continued IP support
2. In vivo human chimeric / large animal experiments; cont. proof of concept

Defined developments plans
Strict go / no-go criteria

FIGURE 4.2 ATI funnel model.

commercialization against which to assess early-stage technologies. As technologies were optioned or in-licensed, they entered the funnel and product development plans were defined with clear go or no-go criteria. The concept was that as the technologies matured, the company would fund further sponsored research in collaboration with its academic partners or bring them in-house for further due diligence and development studies. The benefit of this model was that it enabled company management and its academic partners to take an objective and realistic view of the status of technology development, form consensus on strategies to address potential technology pitfalls, and agree on the gating criteria for further investment at every stage.

The process quickly proved beneficial to participating researchers and institutions, with the early feedback from ATI management causing many to refine their research plans to address specific commercially relevant questions. Equally important, it allowed TTOs and researchers to abandon lines of enquiry that were unlikely to prove clinically or commercially rewarding and focus their efforts on opportunities with a greater probability of success.

The initiative was financed through 2006 and 2007 by SCN and grants from the CIHR, NSERC, MaRS Discovery District (a commercialization initiative based in Toronto), and the regional economic development agencies for Western Canada and Quebec. (Indeed, this initiative was notable for being the first time two regional economic development agencies collaborated on a national initiative to generate long-term regional benefits). However, it had been made clear by all parties at the outset that, as a private corporation, ATI needed private investment as a long-term source of financing; however, the investment community, while long on praise for what had been accomplished, was reluctant to invest. Beyond the generally challenging financing environment for biotechnology, a number of reasons specific to ATI also contributed to its failure to secure private investment, all of which ultimately flowed from the fundamental underlying pooling concept.

First, a number of smaller start-ups had been funded prior to the establishment of ATI. While the strength of ATI's position was its right of first refusal on future unencumbered IP coming out of its researcher laboratories over 5 years, there was a perception that Canada's best IP may already have been licensed to these start-up companies and that what was left over was low-quality IP that no-one else was ready to invest in. As it turned out, this was a somewhat unfair criticism. Some interesting IP was found in the labs of researchers who previously had not been inclined to think about the clinical application of their work. Other commercially relevant IP had previously been optioned or licensed but had reverted to the institutions when the companies that had licensed them failed for unrelated reasons. Nevertheless, the perception stuck.

Second, the IP review found that most of the IP available to ATI was still in the earliest stages of development and, in most cases, was five to 10 years from market. Pooling these technologies in themselves did not reduce that time to market. In fact, it simply established a larger and more expensive patent estate and, consequently, a higher burn rate than a traditional smaller start-up would have had. While investors appreciated that the quality of management attracted by ATI was capable of taking the best IP and significantly reducing the product development timeframe, their preference was for effort to focus on a single family of technologies in a traditional start-up model, with lower investment than that required by the ATI model. In short, while the hypothesis was that risk would be defrayed by betting on the field, the practice was that investors preferred to defray risk by focusing the efforts of the management team on the technology closest to market.

The third challenge was that the ATI model did not fit with the fundraising models of venture capital (VC) investors. Typically VC investors operate 10-year closed funds. This means that they need to liquidate the investments they make within a 3- to 7-year timeframe, so that the capital gains can be distributed to their own investors at the end of each 10-year cycle. As a result, venture capitalists were more interested in funding potential companies that could be spun out of ATI's development funnel and fit their normal mode of operations rather making a long-term commitment to ATI.

The final challenge proved to be the complexity of the corporate structure. At incorporation, ATI had 53 common shareholders, in addition to which SCN held a special voting share that allowed it to appoint two-thirds of the board and vote two-thirds of the common shares. This was to allow SCN to control ATI while its primary source of financing was public funds, redeemable at ATI's option on securing $5 million of private investment. While the strong bonds already forged through its research and training programs made it possible for SCN to effectively manage this stakeholder community, investors were concerned about the cost benefit of maintaining all these relationships and the challenges of acting quickly with so many founders to manage. As a result, they expressed a preference for a more traditional corporate structure.

Although several different financing strategies were pursued, it became evident as markets crashed around the world in the latter half of 2008 that regardless of its merits and novelty, ATI was not financeable as a private company. In 2009, the decision was taken to wind up ATI and free the participating researchers and institutions from the disclosure obligations that could prevent them from pursuing individual commercial opportunities.

While ATI did not secure the financing originally anticipated, it demonstrated that it is possible to establish a virtual national stem cell incubator and the high value professional management could bring to both the basic research and product development plans of participating researchers. It would seem, however, that such a model would still require public investment to support product development until the point at which technologies could be spun out into individual start-ups or licensed to more established organizations. This approach is now being adopted in other areas of science and technology within Canada, funded through Industry Canada's Centres of Excellence for Commercialization and Research (CECR) program. As a for-profit entity, ATI was ineligible to apply to this program at its inception but, based on the positive experience of collaborating both through the SCN and ATI, a reconstituted and more focused subset of investigators is expected to apply for CECR funding in the near future.

The ATI experiment also proved that universities, hospitals, research funders, and economic development agencies can work together in a for-profit setting to commercialize research results. Several previous efforts to pool IP or consolidate commercialization activities in an area of science or technology within Canada had failed. Often and inappropriately, the policies and practices of TTOs and government funding agencies were cited as reasons for these failures. This assumption has now been shown to be untrue; furthermore, it is evident that if a compelling research partnership can be developed involving multiple institutions and corporate partners, the commercial agreements necessary to underpin the partnership can be negotiated and executed in a timely manner.

4.3 THE NEXT DECADE

Public investment in stem cell research will continue through SCN until at least 2015 and the strong federal and provincial support for an area of science that is recognized as a Canadian strength remains a given. As recently as 2008, a "state of the nation" report commissioned by the Science Technology and Innovation Council of Canada (2008) identified stem cells and regenerative medicine as key priority areas for federal investment. Notwithstanding this broad support, four emerging trends in Canada will shape the next decade of public investment:

- **Chemical biology**—Much of the interest and excitement in the stem cell field over the past 10 years has focused on the potential of cellular-based therapies for the replacement of damaged tissues or organs. While support for this strategy will remain strong, in Canada we expect the emergence of major initiatives examining the potential of growth factors and small molecules to promote endogenous repair. The two case studies cited earlier in this chapter involving clinical trials for stroke and neuroblastoma prove the effectiveness of this approach. The diversity of expertise within SCN offers the potential

TABLE 4.3
Stem Cell Therapy Developments by Canadian Companies

Company	Primary Indication	Preclinical	Phase I	Phase II	Phase III
Stem Cell Therapeutics	Stroke				■
	Multiple sclerosis	■			
Northern Therapeutics	Pulmonary hypertension			■	
	AMI	■			
Insception Biosciences	Perivascular disease		■		
	Diabetes	■			
Verio Therapeutics	DMD, AMI, diabetes	■			
Engine	Diabetes	■			

to develop extensive stem cell-based assays against which to screen for small molecules that promote differentiation and repair. Rapid advances in this area can be expected through partnerships with academic high-throughput screening facilities—such as those found in Toronto, Hamilton, Montréal, and Vancouver—and major pharmaceutical companies. It is an approach for which there is a well established business model, while private sector involvement will likely prove a catalyst for further public investment.

- **Commercial growth**—While only Organogenesis, a United States-based company, has brought a stem cell-based therapy to market and established it as standard of care to date, the next decade likely will see many more companies make that transition. As more successes emerge, private sector investment in the field that has largely stalled globally will begin to take off (see Chapter 5 by Prescott and Chapter 6 by Benjamin). This should catalyze further growth in public sector investment, where much of the new research funding is being directed toward applications with clear commercial potential *and* private sector partners. Table 4.3 summarizes the development status of stem cell-based therapies by major Canadian companies. In addition, Stem Cell Technologies, based in Vancouver, has established itself as one of the leading stem cell tool and reagent companies in the world. Ninety-five per cent of its revenues come from exports and the company is well positioned to benefit from the continuing global public investments in basic research.
- **Philanthropy and championing of stem cell research**—Over the past 5 years, the philanthropic community in Canada has made significant donations to establish new research centers across the country—the McEwen Centre for Regenerative Medicine in Toronto, the Sprott Stem Cell Centre in Ottawa, and the Hotchkiss Centre for Brain Repair in Calgary are only a few examples. These donations have leveraged substantial further infrastructure investments from federal and provincial governments. In the past, and in other areas of science, such centres have looked inward as they became operational, competing with each other for the same sources of funding. The recently established Canadian Stem Cell Foundation will change that paradigm by providing a forum for these visionary business

leaders to work together to excite and engage their peers to contribute to this important area of research. The foundation will also provide a vehicle for the general public to channel its support and will serve as a voice for patients in future public policy debates and for the silent majority that, opinion polls consistently show, supports stem cell research.

- **Internationalism**—It is evident from this publication that many jurisdictions are making public investments of hundreds of millions of dollars in stem cell research to gain a long-term scientific and commercial edge in this emerging field. As they have done so, most have quickly reinforced these research investments with the establishment of organizations with similar scope and mandate as the SCN. However, the past 5 years have also seen a parallel acknowledgment that catalyzing the translation of stem cell research into clinical applications will be an international pursuit. Canada has been an active participant in that movement and as early as 2004 led the creation of a coordinating group of these networks and national centers of excellence known as the International Consortium of Stem Cell Networks (ICSCN). Over the past 5 years, ICSCN has developed joint courses and workshops, paying particular attention to subjects in which the expertise in any single country may be fairly limited and large gains could be made through international collaboration. ICSCN has also created novel training opportunities for leading post-doctoral scientists from member countries to network with their peers around the world, and is working closely with the International Society of Stem Cell Research to combat stem cell "tourism."

The biggest opportunities and challenges lie in developing international research programs. Most funding agencies, networks, and centers of excellence have specific mandates for economic development and it will not be easy to align these interests within an IP and commercialization framework that bridges many jurisdictions and multiple institutions. Most public funding supports only those researchers working within that jurisdiction's borders, making the reallocation of resources according to changing project needs an additional challenge. Furthermore, the hands-on project management, inter-laboratory exchanges, and project team meetings that many networks have found crucial to the success of their projects become much more complex for an operation that spans four continents and multiple time zones. Nevertheless, rapid progress in the field will flow from breaking down these barriers. In 2009, Canada announced a partnership with the California Institute of Regenerative Medicine to fund two research teams committed to developing therapies that attack cancer stem cells. Each jurisdiction has committed $40 million to the teams, based on a single, unified proposal and evaluation process. If the projects succeed they will result in filings to conduct two Phase I trials within the next 4 years. This framework provides a model for developing other international research collaborations and Canada expects to be at the forefront of such international initiatives in the coming years.

In summary, stem cell research in Canada has come a long way since the first discoveries by Till and McCulloch in 1961. Over the past five decades it has built a strong, multidisciplinary and interconnected community of outstanding scientists, industries, non-profit partners, universities, and hospitals. It has demonstrated a willingness to undertake innovative approaches to commercialization and led the development of international networks. With sustained public investment, Canada will be well positioned to benefit from this paradigm-changing technology.

REFERENCES

Becker, A. J., A. E. McCulloch, and J. E. Till, 1963. Cytological demonstration of the clonal nature of spleen colonies derived from transplanted mouse marrow cells. *Nature* 197: 452–454.

Betting on HSC, Editorial (April 2002). *Nat Immunol* 3.

Brown, M. M., G. C. Brown, J. D. Stein, Z. Roth, J. Campanella, G. R. Beauchamp, 2005. Age-related macular degeneration: economic burden and value-based medicine analysis. *Can J Ophthalmol* 40: 277–287.

Canada Foundation for Innovation database of projects funded. Available from http://www.innovation.ca/en/projects-funded

Canadian Burden of Illness Study Group, 1998. Burden of illness of multiple sclerosis: Part I: Cost of illness. *Can J Neurol Sci* 25: 23–30.

Dawson, K. G., D. Gomes, H. Gerstein, J. F. Blanchard, K. H. Kahler, 2002. The economic cost of diabetes in Canada, 1998. *Diabetes Care* 25: 1303–1307.

Health Canada, 2002. The economic burden of illness in Canada, 1998 (catalogue#H21-136/1998E). Department: Ottawa: Public Works and Government Services Canada.

Health Canada and Parkinson's Society Canada, 2003. Parkinson's disease: Social and economic impact. Pages 1–6.

Heemstra, H. E., T. Zwaan, M. Hemels, B. M. Feldman, V. Blanchette, M. Kern, T. R. Einarson, 2005. Cost of severe haemophilia in Toronto. *Haemophilia* 11: 254–260.

Lapidot, T., C. Sirard, J. Vormoor et al., 1994. A cell initiating human acute myeloid leukaemia after transplantation into SCID mice. *Nature* 367 (6464): 645–648.

Nagy, A., J. Rossant, R. Nagy et al., 1993. Derivation of completely cell culture-derived mice from early-passage embryonic stem cells. *Proc Natl Acad Sci USA* 90: 8424–8428.

Reynolds, B. A. and S. Weiss, 1992. Generation of neurons and astrocytes from isolated cells of the adult mammalian central nervous system. *Science* 255: 1707–1710.

Rowley, E. and P. Martin. 2009. Barriers to the commercialisation and utilisation of regenerative medicine in the UK. EPSRC, University of Nottingham. http://www.nottingham.ac.uk/iss/research/Current-Research-Projects/Staff_projects/regenmed/reports_publications.htm

Science, Technology and Innovation Council of Canada, 2008. *Canada's Science Technology and Innovation System: State of the Nation.*

Shapiro, A. M., J. R. Lakey, E. A. Ryan et al., 2000. Islet transplantation in seven patients with type 1 diabetes mellitus using a glucocorticoid-free immunosuppressive regimen. *New Engl J Med* 343: 230–238.

Singh, S. K., C. Hawkins, I. D. Clarke et al., 2004. Identification of human brain tumour initiating cells. *Nature* 432(7015): 396–340.

Toma, J. G., M. Akhavan, K. J. Fernandes et al., 2001. Isolation of multipotent adult stem cells from the dermis of mammalian skin. *Nat Cell Biol* 3: 778–784.

Tropepe, V., B. L. Coles, B. J. Chiasson et al., 2000. Retinal stem cells in the adult mammalian eye. *Science* 287: 2032–2036.

U.K. Department of Health, 2005. U.K. Stem Cell Initiative, Section 4.1.2.2.

5 Investing in Regenerative Medicine
What Drives Private Investors?

Catherine D. Prescott

CONTENTS

5.1 Decisions within Context 59
    5.1.1 Demand 59
    5.1.2 Risk versus Reward 60
    5.1.3 Exit Options 61
    5.1.4 Acquisition Criteria 62
5.2 Business Models and Market Potential 62
5.3 Concluding Remarks 65
References 65

Regenerative medicines offer the prospect of treating medical conditions that require the replacement, repair, or regeneration of damaged cells and tissues and thereby have the potential to restore bodily functions. For the first time, we can look forward to tackling medical conditions that hitherto have been intractable to conventional approaches. Furthermore, by targeting the underlying causes of disorders, regenerative medicines offer the opportunity to better manage chronic conditions, possibly even leading to cures. It is therefore unsurprising that their potential has attracted a significant level of attention by a wide range of professionals, governments, and members of the general public. However, the question remains whether their support is sufficient to ensure the delivery of regenerative medicines to patients in a timely fashion and on a routine basis. Key to the delivery of any therapeutic is an appropriate level of financial commitment and this chapter sets out to review some of the trends and drivers that motivate both public and private investors to invest in the regenerative medicine industry.

5.1 DECISIONS WITHIN CONTEXT

5.1.1 DEMAND

Countries are challenged by the need to manage the increasing economic burden of healthcare. Data collated by the World Health Organization (WHO) showed that in 2006 the United States spent 15.3% of its gross domestic product (GDP) on health and the United Kingdom spent 8.2%. According to the World Health Statistics 2008 report, the leading causes of death today and predicted for the foreseeable future (data extrapolated until 2030) are non-communicable diseases. Ischemic heart disease, stroke, and chronic obstructive pulmonary disease are three major causes of mortality.

Chronic diseases are typically linked with associated complications and therefore incur co-morbidity costs. For example, over 20% of diabetic patients suffer from one or more complications such as kidney disease, heart disease, stroke, high blood pressure, blindness, and neuropathy (American Diabetes Association 2007). Diabetes is the leading cause of kidney failure and of new cases of blindness among adults aged 20 to 74 years. The risk of stroke and fatal heart attacks is two to four times higher in adults with diabetes. All of these conditions increase the economic burden of chronic diseases. Therefore, if a chronic disorder can be cured or at least the underlying cause of the disorder better managed so as to delay the onset of the associated conditions, healthcare expenditures would be reduced.

Chronic conditions not only cost a nation in terms of expenditures on healthcare but also in terms of loss in worker productivity. The Milken Institute (DeVol and Bedroussian 2007) calculated that the combined impact for seven of the most common chronic diseases costs the U.S. economy $1.3 trillion annually of which $1.1 trillion was attributable to losses of worker productivity. This represented nearly four times more than the costs of treatment (DeVol and Bedroussian 2007).

By taking into account both the economic and business health costs, it is in the clear interest of all stakeholders (including government, healthcare providers, and patients) to support the development of regenerative medicines that have the potential to lower the overall cost of healthcare and promote an active work force able to contribute to the GDP.

5.1.2 Risk versus Reward

The pharmaceutical and biotechnology industry sectors are actively engaged in pursuing varied and new approaches to treat chronic conditions (see Chapter 12 by Vertes). However, despite more than a doubling of expenditure on research and development, there has been an equivalent increase in the number of new drugs approved by the regulatory authorities (U.S. Food & Drug Administration [FDA]; PhRMA 2009). This "innovation gap" is in part a reflection of increasing demands for medicines that are effective and exhibit fewer side effects and better safety profiles. Such demands mean that more drug candidates to fail to meet the increasingly stringent clinical criteria.

The risk of failure raises concerns among stakeholders including investors whose decisions are often based on a managed-risk approach (venture not adventure!). For example, a venture capitalist will critically assess a company's strategy and the potential value of its pipeline and seek to mitigate any risk by tranching the investment against agreed milestones. This approach enables an investor to influence the future direction of a company and thereby to some extent manage the risk of the venture.

Investors in publicly traded companies who are unable to critically evaluate the long-term impact of an apparent "bad news" story may opt to mitigate their risks by withdrawing their investment and as a consequence undermine the financial viability of the company. There are all too many examples of companies whose potential was lost after a report of their failure to reach a single clinical milestone. An investor may also have a short-term perspective—seeking any increase (rather than maximal) in the value of an investment as a reflection of the performance of a single product rather than a greater realization of company value developed over a longer period.

Investors have finite funds and can choose where to invest subject to an agreed mandate. On the one hand, although the route to market for conventional therapeutics is by now well documented, the innovation gap reveals that this route is increasingly risky, the target patient population more narrowly defined, and therefore the market potential smaller than those of the former blockbuster markets. On the other hand, the route to market for cell-based regenerative medicines is ill defined and largely untested; the full market potential is unknown.

As of September 2009, only 5 cell-based products had reached the market and approximately 56 industry sponsored trials were in the clinic (ClinicalTrials.gov) (excluding bone marrow transplantation). If the regenerative medicine industry delivers its promise to revolutionize healthcare, the sector could offer a substantial investment opportunity. A recent estimate calculated that the global market potential for tissue engineering and regenerative medicine products would exceed $118 billion by 2013 (Life Science Intelligence 2009).

5.1.3 Exit Options

Investors take into account the investment risk, the potential for a return on their investment (ROI), and how they will be able to recover their money. The ROI is determined by how much investment a company received over the length of the investment period. A recent survey of U.S. and European biotechnology and specialty pharmaceutical companies between 2005 and 2008 showed that a trade sale was the most profitable exit route for private investment companies (Geilinger and Leo 2009). This was particularly true for companies acquired by large corporations.

To date, only a limited number of major corporations have made any formal commitment to the regenerative medicine sector and, therefore, the number of potential acquisition partners is relatively low. Pfizer took the lead in 2008 when it announced that it would launch two regenerative medicine research units based in the United Kingdom and United States (Pfizer Press Release 2008a). Other major corporations gained footholds in the sector by equity investment or through strategic alliances with regenerative medicine companies rather than commit substantial internal resources.

Roche and Novartis venture funds subscribed to two investment rounds in Cellerix (Cellerix Press Releases 2007, 2009), a company focused on the development of adult stem cell-derived therapies; Johnson & Johnson Development Capital (JJDC) led a $25 million series C investment in Novocell in 2007 (Novocell Press Release 2007) and also subscribed to successive rounds of investment in Tengion including a $21 million series C round (Tengion Press Release 2008). Pfizer invested $3 million in EyeCyte in 2008 (Pfizer Press Release 2008b)—the same year Genzyme announced a partnership deal with Osiris Therapeutics to develop and commercialize the mesenchymal stem cell products called Prochymal and Chondrogen (Osiris Therapeutics

Press Release 2008). The deal included a $130 million up-front payment by Genzyme plus additional milestone and royalty payments.

In July 2009, Johnson & Johnson became a founding member of the Alliance for Regenerative Medicine, a coalition of stakeholders with the common goal of advancing cell-based therapies (Alliance for Regenerative Medicine 2009; see Chapter 20 by Werner). Also in 2009, Sanofi-aventis announced that it had signed a memorandum of understanding with the Salk Institute for Biological Studies to launch a regenerative medicine program (Sanofi-aventis Press Release 2009) and Merck & Co., Inc. formed a Regenerative Medicine Oversight Committee (Merck 2009). Accordingly, there is evidence of an increasing trend by large corporations toward investment in regenerative medicines and increasing the prospect of a trade sale. This, in turn, will attract private investors.

5.1.4 ACQUISITION CRITERIA

To achieve a successful exit, it is important to understand what criteria influence the decision to acquire a company. According to an HBM survey (Geilinger and Leo 2009), at the time of a trade sale, 50% of the companies had products in the clinic and 11% of them had approved or marketed products, indicating a preference for a maturing and derisked product pipeline.

This trend was further supported by a more recent report that showed that the majority of the third quarter 2009 trade sale transactions were for companies with marketed products (Burrill 2009). Furthermore, partnering transactions for preclinical and Phase 1 products had decreased whereas later stage (Phase III and later) deals had remained strong. Since the regenerative industry is an emerging sector with relatively few products in the clinic or on the market, it is arguable that investors must pay particular attention to the probability of a regenerative medicine company achieving a successful trade sale. Few cell-based regenerative medicine companies currently fit the acquisition criteria. Those with market-approved products include Organogenesis (Apligraft), NuVasive (Osteocel), Intercytex (Valvelta), Genzyme (Carticel), Advanced BioHealing (Dermagraft), and TiGenix (ChondroCelect)—the first advanced therapy medicinal product (ATMP) to gain regulatory approval in Europe (TiGenix Press Release 2009). Osiris Therapeutics, Aldagen, Medipost, Aastrom, and Cellerix have one or more products in Phase III development (ClinicalTrials.gov).

The decision to invest in a private company is clearly influenced by the exit route and in the current economic climate this is determined by the appetite of the large corporations to acquire the private company. Based on the risk associated with the development (i.e., innovation gap) of both conventional therapeutics and regenerative medicines and the potential of regenerative medicines to revolutionize healthcare, why are major corporations reticent to fully engage in the sector? There are at least two reasons, relating to the market potential and the business model.

5.2 BUSINESS MODELS AND MARKET POTENTIAL

By comparison to conventional drugs, cell-based products will be expensive to manufacture, have a limited shelf life, and will typically be administered in a clinical setting. Therefore they must command a premium price relative to lower cost alternatives and as a consequence must demonstrate significant improvements in efficacy. Failure to do so could result at best in the relegation of cell-based therapies to the last line of resort or, at worst, failure to be eligible for reimbursement (see Chapter 21 by Meurgey and Wille and Chapter 22 by Faulkner). Alternatively, cell-based medicines must be targeted at indications for which there is a high unmet clinical need, similar to the strategy deployed by a specialty pharmaceutical business.

The extent to which cells are processed prior to their use as a therapy will exert a major impact on the market potential, business models, and therefore alignment with existing business practices. Autologous cells harvested from a patient, cultured, and subsequently returned to the patient (donor) could only be used in a chronic setting because of the time delay needed to expand the cells. Furthermore, two operational procedures are required (harvesting and subsequently administering the cell therapy), adding to the cost of the intervention over and above the manufacturing cost of the cells. If the procedures are invasive, then conventional surgery may represent a more attractive and lower cost alternative. The low number of cells available for a transplant would also restrict the number of indications that could be targeted. Despite the fact that autologous products are not aligned with traditional pharmaceutical product profiles (for example, mode of distribution and administration or potential market size), several large corporations have agreements in place with autologous cell therapy companies. Cordis (a Johnson & Johnson company) agreed to a co-development partnership with Mesoblast (Mesoblast Press Release 2005); Johnson & Johnson invested in Tengion (Tengion Press Release 2008); and Pfizer took an equity stake in EyeCyte (Pfizer Press Release 2008b).

An alternative approach to autologous therapies would be to harvest, process, and administer cells within a single operational procedure. Cytori's Celution technology permits the real-time processing of cells at the point of care. In 2009, the FDA announced that the Celution(R) 700 System would be regulated as a medical device (Cytori Press Release 2009), and indeed Cytori's business model is more aligned with that of a device company than a pharmaceutical business.

The pharmaceutical business model is principally based on the development of off-the-shelf products that can be targeted at large patient cohorts, are applicable in acute and/or chronic settings, and ideally are self-administered.

A number of regenerative medicine companies have adopted similar business models by developing soluble factors to either stimulate or mobilize a patient's own resident repair mechanisms. For example Genzyme's Mozobil (also known as perixafor, AMD3100, and JM 3100) mobilizes hematopoietic stem cells (Vose et al. 2009), Amgen's Neupogen (trade name for filgrastim) is an analog of human G-CSF (granulocyte colony-stimulating factor) and Medtronic's INFUSE Bone Graft contains recombinant human bone morphogenetic protein (rhBMP)-2 (McKay et al. 2007). Companies such as Fate Therapeutics (Fate Therapeutics Website) and Plasticell (Plasticell Website) have established discovery programs to identify small molecules that reprogram stem cells (see Chapter 11 by Choo).

Epogen (erythropoietin or EPO, a hematopoietic growth factor that stimulates erythroid progenitor cells) was arguably the first soluble factor-based regenerative medicine and was also the first blockbuster biotechnology drug (Harris 2009). Small molecules that act to modify disease are more in line with the traditional approach to drug development (compared to cell-based regenerative medicines that repair or replace damaged cells) and have the potential to address large markets. This alignment of the business model and market potential means that companies developing soluble regenerative medicines are anticipated to be attractive targets for investment and subsequent acquisition. This is in part borne out by the results of a preliminary survey of venture capital investment trends in regenerative medicine companies conducted in 2008 by students undertaking an MBA project at the University of Cambridge (personal communication). They concluded that of the 26 private companies studied, 66% focused on the discovery and development of soluble factors (chemicals and/or biologicals) and only a minority were focused on the development of cell-based products.

Allogeneic products are more aligned with the off-the-shelf product model permitting their use in both chronic and acute settings. The ability to expand and differentiate cells into a range of cell types extends their use for a broad range of clinical indications. It is therefore predictable that organizations developing allogeneic products would attract more attention from large corporations than those focused on autologous therapies: Genzyme/Osiris Therapeutics, Roche and Novartis/Cellerix, Pfizer/Novocell, Pfizer/University College London (University College Press Release 2009). However, although allogeneic products have greater potential than autologous products, the route to market remains ill defined and the regulatory hurdles high; thus such products pose a risk that challenges all stakeholders. This is especially true for companies developing products derived from pluripotent stem cell sources (e.g., embryonic stem cells) because the plasticity of these stem cells is a double-edged sword that requires a balance between their therapeutic potential and safety.

Geron worked alongside the FDA for 4 years prior to submitting 21,000 pages of documentation in support of the investigational new drug (IND) application to enter the clinic with the world's first human embryonic stem cell-derived therapy (GRNOPC1) for spinal cord injury. In May 2008, it was announced that the FDA had placed the IND on hold (Geron Press Release 2008). The FDA subsequently cleared GRNOPC1 to enter clinical trial in January 2009 (Geron Press Release 2009a) but in August the trial was again placed on hold due to evidence of non-proliferative cysts in animal studies that had been conducted in parallel with the clinical trial (Geron Press Release 2009b). In October, Geron and the FDA announced they had reached an agreement on the clinical hold and that Geron would aim to re-initiate the clinical trial in the third quarter of 2010 (Geron Press Release 2009c). Inevitably Geron's share price fluctuated according to the nature of the news announcements. Stringent regulatory demands that impede the rate of progress of a clinical product will likely deter all but the most confident long-term investors.

There is some indication that investors in the regenerative medicine sector view the correlation of market valuation, revenue, and pipeline potential differently from those who invest in conventional biotechnology. Comparing the operational expenses and revenues for 10 U.S. publicly held regenerative medicine companies with those for 10 biotechnology companies undertaking conventional approaches to drug discovery and development (matched within 5% according to market capitalization on a specified date) over 3 years (2006 to 2008) revealed that although the average operational expenses for both sectors were similar, regenerative medicine companies generated significantly lower revenues.

5.3 CONCLUDING REMARKS

Any investment decision must be evaluated within the context of all the available options and reflect a specific set of drivers that underpin the risk–reward profile. In terms of risk, Donald Rumsfeld noted that there are both "known unknowns" and "unknown unknowns." It is arguable that in gaining experience over time the pharmaceutical industry is better able to predict the challenges facing the development of conventional drugs ("known unknowns"). By contrast, the regenerative medicine industry is emerging, the route to market is poorly defined, and the market potential unrealized. Therefore the industry likely faces many unpredictable challenges ("unknown unknowns"). Investors recognize the risk inherent in any business and this is evaluated alongside the potential reward. If regenerative medicines are able to deliver on their promise to revolutionize healthcare, the rewards will be great. However, time is needed to develop both the science and business and interim support will require investors to commit substantial levels of funding over the long term.

REFERENCES

Alliance for Regenerative Medicine, 2009. Alliance for Regenerative Medicine Launches. www.alliancerm.org

American Diabetes Association, 2007. National Diabetics Fact Sheet http://www.diabetes.org/diabetes-basics/diabetes-statistics/

Burrill and Company, 2009. Merchant Banking News Letter Q3, 2009. http://merchant.burrillandco.com/newsletters.html

Cellerix Press Release 2007. Cellerix closes €27.2 m financing round to bring product Cx401 to market. http://www.cellerix.com

Cellerix Press Release 2009. Cellerix raises €27 million in a first closing. http://www.cellerix.com

ClinicalTrials.gov. A registry of federally and privately supported clinical trials conducted in the United States and around the world. http://clinicaltrials.gov/

Cytori Therapeutics Press Release 2009. Cytori's Celution(R) 700 system to be regulated as a medical device by US FDA. *ir.cytoritx.com*

DeVol, R. and Bedroussian, A., 2007. An unhealthy America: the economic burden of chronic disease: charting a new course to save lives and increase productivity and economic growth. http://www.milkeninstitute.org/publications/publications.taf?function=detail&ID=38801018&cat=ResRep

Fate Therapeutics Website: www.fatetherapeutics.com

Geilinger, U. and Leo, C., 2009. *HBM Partners: Trade Sales of Biotechnology and Speciality Pharma Companies 2005–2008.*

Geron Press Release 2008. FDA Places Geron's GRNOPC1 IND on clinical hold. www.geron.com/media

Geron Press Release 2009a. Geron receives FDA clearance to begin world's first human clinical trial of embryonic stem cell-based therapy. www.geron.com/media

Geron Press Release 2009b. Geron's IND for spinal cord injury placed on hold. www.geron.com/media

Geron Press Release 2009c. Geron and FDA reach agreement on clinical hold. www.geron.com/media

Harris, M., 2009. The billion-plus blockbusters: the top 25 biotech drugs (compiled from *BioWorld's Market-Leading Biotechnology Drugs 2009: Blockbuster Dynamics in an Ailing Economy*). www.bioworld.com

Life Science Intelligence, 2009. Worldwide markets and emerging technologies for tissue engineering and regenerative medicine. http://www.lifescienceintelligence.com/market-reports-page.php?id=IL600

McKay, W. F., Peckham, S. M., and Badura, J. M., 2007. A comprehensive clinical review of recombinant human bone morphogenetic protein-2 (INFUSE Bone Graft). *Int Orthop* 31: 729–734.

Merck & Co., Inc. Merck Research and Innovation Regenerative Medicine Research. http://www.merck.com/corporate-responsibility/research-medicines-vaccines/new-technologies/approach.html

Mesoblast Press Release 2005. Mesoblast and Cordis Corporation to join forces in adult stem cell heart trial. www.mesoblast.com

Novocell Press Release 2007. Novocell raises $25 million. Series C to support continued development of stem cell therapy. www.novocell.com

Osiris Therapeutics Press Release 2008. Genzyme and Osiris partner to develop and commercialize first-in-class adult stem cell products. http://investor.osiris.com

Pfizer Press Release 2008a. Pfizer launches global regenerative medicine research unit. www.pfizer.be/Media/Press+bulletins

Pfizer Press Release 2008b. EyeCyte Inc. secures series A funding from Pfizer. www.pfizer.be/Media/Press+bulletins

PhRMA, 2009. Pharmaceutical Industry Profile. http://www.phrma.org/profiles_%26_reports/

Plasticell Website: www.plasticell.co.uk

Sanofi-aventis Press Release 2009. Sanofi-aventis and the Salk Institute establish regenerative medicine research alliance. www.en.sanofi-aventis.com/press

Tengion Press Release 2008. Tengion announces second closing of series C financing. www.tengion.com

TiGenix Press Release 2009. TiGenix receives positive CHMP opinion on European MAA for ChondroCelect. www.tigenix.com

University College London Press Release 2009. UCL–Pfizer to develop pioneering stem cell sight therapies. www.ucl.ac.uk/news

U.S. Food & Drug Administration Centre for Drug Evaluation and Research Drug Approval Reports. http://www.accessdata.fda.gov/Scripts/cder/DrugsatFDA/index.cfm?fuseaction=Reports.ReportsMenu

Vose, J. M., Ho, A. D., and Coiffier, B. et al., 2009. Advances in mobilization for the optimization of autologous stem cell transplantation. *Leuk Lymphoma* 50: 1412–1421.

World Health Organization, 2006. Health statistics and health information systems. http://www.who.int/healthinfo/statistics/indhealthexpenditure/en/index.html

World Health Organization, 2008. World health statistics. http://www.who.int/whosis/whostat/2008/en/index.html

6 Public Investment Models
Coming out of the Closet and Going Public!

Reni Benjamin

CONTENTS

6.1	Introduction	67
6.2	Coming out of the Closet and Going Public!	68
	6.2.1 Initial Public Offering (IPO)	68
	6.2.2 Reverse Merger	69
6.3	You're Public: Now What?	70
	6.3.1 Fully Marketed Follow-On	71
	6.3.2 Public Investment in Private Equity (PIPE)	71
	6.3.3 Registered Direct Transaction	72
	6.3.4 Bought Deal	72
6.4	Market Reaction	72
6.5	Key Drivers of Valuation	74
	6.5.1 Cash Is King	74
	6.5.2 Execution, Execution, Execution	74
	6.5.3 It's All about Data	74
	6.5.4 Providing Hope, Not Hype	75
6.6	Valuation Metrics	76
	6.6.1 Multiples Analysis	77
	6.6.2 Cash Flow Analysis	78
	6.6.3 Comps Analysis	78
6.7	Case Study	79
6.8	Conclusions	79
References		82

6.1 INTRODUCTION

The pharmaceutical space represents one of the fastest growing sectors in the investment world given the steady and upward trajectory of healthcare spending. If left unchecked, spending could reach approximately 20% of the United States gross domestic product (GDP) by 2017 (Healthcare Costs, 2009). By way of reference, healthcare spending exceeded $2 trillion in 2006, representing 16% of the U.S. GDP (Healthcare Costs, 2009). Fundamentally contributing to this underlying growth in spending is that by 2030, 26% of the U.S. population will be 65 or older, as compared with 17% today (InfoWorld, 2009). In our opinion, due to the combination of increased spending in the healthcare space, a growing baby boom generation, and the countless billions of investment dollars waiting on the sidelines, the U.S. is positioned to usher in a new era of technological innovation that could transform the entire healthcare system as we know it.

Stem cell-based therapeutic innovations represent a revolutionary step forward in the healthcare space. Ever since the first successful bone marrow transplantation in 1968, scientists have been trying to harness the power of stem cells for therapeutic uses (Halme and Kessler, 2009). With the recent development of techniques to manipulate embryonic stem cells and an increased understanding of the pathways essential for cellular differentiation, an entire field seeking to create uses for embryonic (pluripotent) or adult (multipotent) stem cells has arisen. Today, the regenerative medicine sector encompasses specialized cells that either terminally or partially differentiated prior to transplant or contain varying mixtures of terminally or partially differentiated cell types (Halme and Kessler, 2009). Leading the charge in moving this field forward are more than 100 publicly traded and privately held biotechnology companies and a whole slew of universities conducting both the basic and translational science. While taxpayer dollars fund the majority of work conducted at the university level, both private and public biotechnology companies primarily generate funds through tapping the private or public capital markets. This chapter will focus on stem cell companies and the challenges and opportunities that exist for accessing capital through the public markets.

6.2 COMING OUT OF THE CLOSET AND GOING PUBLIC!

With venture capitalists increasing their demands on private companies and the prospects of multiple milestones coming up to potentially generate shareholder value, you (a company) have made the decision to go public and allow shareholders to take on risk and participate in the potential upside for the company. What are the options? In our view, two options are readily available to the average stem cell company as it makes the transformative decision to join the public markets.

6.2.1 INITIAL PUBLIC OFFERING (IPO)

The IPO market is the standard way companies enter the public marketplace. Typically armed with venture backing, relationships with multiple blue-chip investment banks, and hordes of lawyers, the stem cell company of interest files the appropriate documents with the Securities and Exchange Commission (SEC) and after approval begins marketing the initial offering to institutional investors—typically the managers of multi-billion dollar mutual funds and hedge funds, all the way down to funds managing over $1 million in capital. The challenges in completing an IPO are numerous and highly dependent on a variety of factors including:

- **Investor appetite**—Based on data from PricewaterhouseCoopers (PWC 2009) a total of two IPOs made it to market in the first quarter of 2009. In combination with the second (12 IPOs) and third quarters (20 IPOs), a

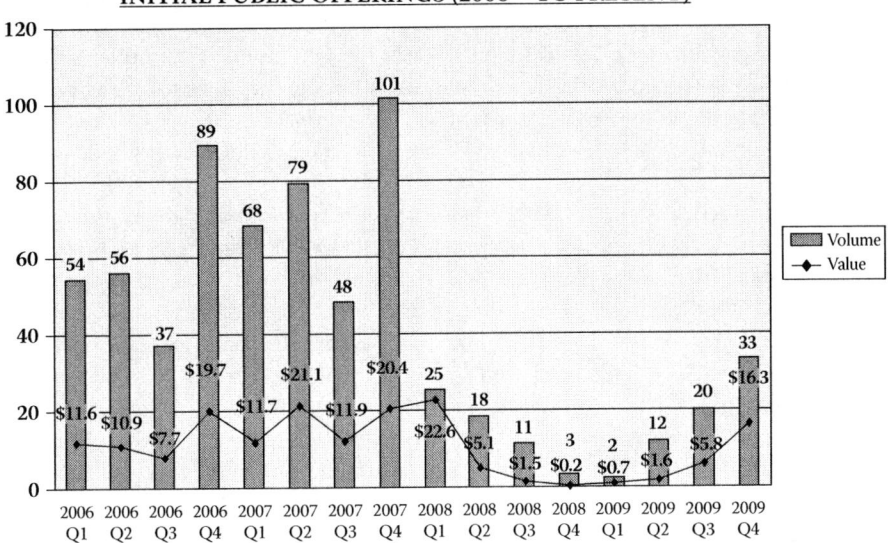

Public Investment Models 69

INITIAL PUBLIC OFFERINGS (2006 – TO PRESENT)

FIGURE 6.1 Initial public offerings (2006 to present). (Courtesy of PricewaterhouseCoopers Transaction Services.)

total of $8.1 billion in proceeds was raised. By way of comparison, 57 IPOs raising approximately $29 billion occurred in 2008, and 296 IPOs raising approximately $64 billion occurred in 2007 (Figure 6.1). To say that the IPO market slowed in 2009 is an understatement and represents a key challenge for companies pursuing this route of financing.

- **Company valuation**—A fair value for the company and an understanding of what investors are willing to pay for shares are essential aspects of a successful IPO. If the price is high, many investors may balk at participating in the IPO and the offering will need to be re-priced. Too low a price will cause extra shares to be issued in the offering, resulting in the dilution of holdings of existing shareholders unnecessarily.
- **Future drivers or milestones**—An understanding of milestones and other company-specific drivers is important for investors to gauge the potential enthusiasm for an IPO. Multiple opportunities to drive valuations higher could result in an over-subscribed deal with considerable excitement.

Clearly, if one is successful in overcoming the challenges of the public marketplace during the average IPO, the potential to successfully navigate future financings is in your favor, not to mention the distinction that such an event brings to the company and management.

6.2.2 Reverse Merger

Not all companies can obtain marque venture capital funding or tier one Big Pharma partners to fund development of their pipelines. In fact, many start-up stem cell companies were funded with monies from friends, family, and high net worth investors. But even these sources have their limits, and at some point a company must seek alternative means of financing its operations. When the IPO route is not an appropriate option, a second route to tap the equity markets is through the reverse merger process by which a private company merges with a public company. The three basic types of reverse mergers include (Rodman & Renshaw, LLC; privileged communications):

- **Virgin shell merger**—This type of reverse merger is free of operating history, assets, and liabilities and is created by filing the appropriate documents with the SEC.
- **Empty shell merger**—This is a merger of a private company into a public company that was once active, but for a variety of reasons, is now operationally inactive. This type presents some challenges in that the operating history, assets, and liabilities accrue to the new company.
- **Synergistic merger**—This merger involves two companies (one private and one public) with complementary or differing technologies; both seek to expand a platform.

The several advantages to the reverse merger route include (1) significantly lower costs compared to an IPO; (2) considerably less time to achieve public listing; and (3) no need for an underwriter. However, the path to liquidity in companies that use the reverse merger method may be considerably longer. For example, many companies that complete reverse mergers are listed on the bulletin board ("pink sheet") exchange listing functioning as "penny stocks" that very rarely trade significantly.

Also, such stocks are not commonly purchased by hedge funds or mutual funds because of internal rules that bar these funds from purchasing stocks under $5. As a result, shares are typically traded in very small blocks by retail investors. Company management teams are required to "hit the road" and market the company to all types of investors including retail brokers, venture funds that cross over and invest in publicly traded companies, and hedge funds.

Finally, management teams must secure the coverage of a sell-side analyst to help validate the company to investors. If management is successful in increasing the share price and trading volume, the company can then list its shares on either the AMEX or NASDAQ exchange. These exchanges carry the prestige and meet the regulatory requirements necessary to appear on the radar screens of a broader group of investors. In short, while both the IPO and reverse merger are commonly used to go public, access the capital markets, and require the same aftermarket support, an IPO tends to secure the investors, analysts, and bankers before a company goes public, whereas a reverse merger requires most of the work to be done after the company becomes public.

6.3 YOU'RE PUBLIC: NOW WHAT?

The public markets have created multiple opportunities for stem cell companies to access capital. Although the IPOs of some of the stalwarts in the regenerative medicine space have generated over $500 million in an IPO, what options do companies have after they become public and need to access the capital markets again? Several financing options exist for publicly traded companies:

- **Fully marketed follow-on**—This is an equity deal whereby the company sells registered shares to public investors under a previously filed shelf registration. A typical follow-on begins with a "road show" by which the company presents the deal to investors for several weeks—a process similar to an IPO.
- **Public investment in private equity (PIPE)**—A company sells unregistered common stock to a targeted group of institutional investors and files to register the securities within 30 days after the offering.
- **Registered direct transaction**—This is another equity-based deal whereby a company sells registered shares, but this time to a targeted group of institutional investors. This type of financing is typically handled on a confidential basis.
- **Bought deal**—This deal occurs when an underwriter agrees to purchase a block of stock from a company and then sells it to a large group of investors. The underwriter assumes the risk of placing the stock with investors.
- **Convertible debt deal**—This structure differs substantially from all the other four financing types mentioned above. In this case, a company sells unregistered convertible debt securities to institutional investors. The convertible debt pays out a fixed dividend and can be converted to common shares at an agreed-upon fixed stock price.

Naturally, there are pros and cons to each of the financing alternatives mentioned above. For our purposes, however, we will explore in more detail the four equity-based financing options.

6.3.1 Fully Marketed Follow-On

Like an IPO, a fully marketed follow-on is a labor-intensive effort that requires a registered shelf filing and targets a large group of investors in a period of 10 to 14 days. During the road show, management has the opportunity to communicate to a wide range of investor audiences, including large institutional investors. The key differences between this type of financing alternative and the other two mentioned below are the time required to complete the transaction and the potential impact on the share price as the road show proceeds.

6.3.2 Public Investment in Private Equity (PIPE)

The PIPE is a standard financing alternative commonly used by biotechnology companies to secure funds. A deal may be executed at a discount to the recent average closing price of the stock and typically requires the issuance of warrants and/or options to entice investors to invest in the company. This type of deal does not require up-front SEC registration. Typically, this type of investment is sold to a targeted group of 15 or fewer investors (mutual funds, public cross-over investors, private equity investors, financial institutions, and hedge funds) over a 3- to 5-day period, usually via a limited number of one-on-one meetings and conference calls.

This type of transaction is typically handled by a sole agent and transaction fees are negotiated up front. The benefits of such a transaction include (1) the ability to attract a broader investor base; (2) the ability to execute the financing quickly—typically within 2 weeks; and (3) the potential to not pressure the stock price when marketing the deal.

6.3.3 Registered Direct Transaction

The main difference between a registered direct and a common PIPE is that the equity being sold is already registered with the SEC in a registered direct deal. This key feature allows for the shares to be transferred to investors and, if the investor chooses, immediately be sold into the open market. As with the common PIPE, warrants and/or options are typically granted in an effort to entice investors to participate in the transaction. While many of the benefits are similar to the common PIPE, several key differences exist in a registered direct transaction including (1) the ability to attract a broader investor base; (2) execution of the transaction within one day; and (3) deal execution at slight discount to the current market price.

6.3.4 Bought Deal

The bought deal occurs when an underwriter agrees to purchase a block of stock from the company. It is different from the others mentioned because the underwriter must assume all the risk of purchasing the stock. The underwriter pays for the stock at a discount, under the belief that there will be a good measure of demand for the underlying equity so that the shares can be sold quickly to potential investors (*Financial Post* 2009). What is the benefit to the underwriter? The fee of course!

6.4 MARKET REACTION

The market reaction to any type of deal may be vastly different. Many factors ultimately influence the share price of an equity before and after a deal:

- **Fundamentals**—The fundamentals matter significantly when the behavior of an equity is analyzed both before and after a financing. For example, if the company is profitable, cash flow positive, and has a respectable compounded annual growth rate (CAGR), the equity will likely trade down briefly to the deal price after the deal is done, but then rebound over a short period as the fundamentals of the story take over. However, stem cell companies are typically cash burning machines that face considerable clinical and regulatory risk. The key fundamental value drivers include the data from clinical trials and the potential market size of the indications the therapies will target. Investors should expect that after a financing, the

stock will likely trade at or below the deal price. Due to the time required to generate clinical data and the risk involved, the stock will likely also take longer to rebound from the deal price.
- **Investors**—The players in the public markets range from fundamental long-term investors to technical investors and to hedge funds that employ a variety of techniques, including fundamentals and technical analyses, and have the ability to establish long or short positions on a stock. All these players are vital to fully functioning public and financing markets.
  - The ideal fundamental long-term investor employs an investment strategy driven by an analysis of the science, technological innovation, and the potential market size of the therapeutic. Based upon these metrics, the fundamental investor derives a value for the company. If a company trades at a market capitalization significantly lower than the perceived and/or calculated value, these investors will likely take a position in the company by participating in the deal. They tend to hold the stock until the fundamentals or valuation of the company significantly change.
  - The technical investor, for the most part, is not interested in the fundamentals of the company. On the contrary, this investor is keenly focused on momentum and volume associated with the underlying equity. These two key features allow the investor to gauge the directionality of the stock. If the volume is significantly higher than the average 30-day volume and allows the quick jettisoning of shares obtained from a financing, the technical investor will likely take a position in the company. This type of investor usually will sell shares of the company relatively quickly and hold on to the accompanying warrant and/or option positions obtained in the deal.
- **Dilution**—Dilution is the last thing in the world for a stem cell company to worry about. In reality, the choice is very simple: fund the company to be able to build shareholder value or declare bankruptcy. In many cases, the earnings per share or cash flow to be generated from the sales of therapeutics will not be recognized for many years. By that time, the average stem cell company could amend its corporate structure by implementing reverse splits to counter the effects of multiple financings. In essence, the more funds raised via financing and the more shares created and sold to investors, the more impact one might see on the stock price.
- **Signs of the times**—Credit markets freeze, IPO markets shut down, and equity markets slow down, but nevertheless funding opportunities always exist. Regardless of the equity price of a company, the only obstacle to securing financing is how onerous the terms of the deal are. In economically bad times when investors are skittish, they will be more than likely to demand much higher discounts regarding the price of the deal and significantly higher warrant coverage. On the other hand, in times of plenty, investors are willing to take a smaller discount and fewer warrants, just for the ability to participate in the deal and its potential upside.

6.5 KEY DRIVERS OF VALUATION

6.5.1 Cash Is King

The lifeblood of a stem cell company is the technology it possesses. The heart that pumps the lifeblood is the cash the company uses to fund the development and advancement of the technology. Without adequate technology and approximately 1 or 2 years of cash assets (as a worst case scenario) and ideally far more, the publicly traded stem cell company will be viewed by investors as being circled by vultures. What does that mean? After a company becomes a publicly traded entity, it faces new levels of disclosure and SEC requirements. Quarterly conference calls and statements allow the public to ascertain the burn rate associated with the ongoing programs and the extent to which its current cash position will fund operations.

However, life is less than perfect and many companies that have followed the IPO or reverse merger route are funded with only enough cash to achieve one or two valuation drivers—clinically validating events or proofs of principles of preclinical studies. If trials or data are positive (as long as the positive outcome has not already been fully anticipated by the market), a concomitant increase in share price should result and provide (1) opportunities for existing investors to sell their shares at a higher price; and (2) opportunities for new investors (should a financing be completed) to help fund the company to reach the next valuation driver (i.e., corporate partnership or positive data from more advanced clinical trials).

6.5.2 Execution, Execution, Execution

Investors reward management teams that do what they say! Execution of goals and milestones outlined at the beginning of a year are key to building shareholder value. The consistent advancement of products through the labyrinth of clinical development, the timely generation of meaningful news flows, and a straightforward communication of achievements are pivotal to establishing management credibility—an essential component of increasing a company's market capitalization and value.

A second component to execution involves the predictability of milestones. As institutional investors become comfortable in predicting future events, the fast money funds start to "play" events by buying options, buying public shares, or shorting shares prior to data or corporate announcements. All these activities, some of which are welcomed and others abhorred by management, serve one goal: to increase trading volume, which subsequently attracts bigger funds and allows them to buy shares. The increased volume also helps attract potential funding opportunities in the form of follow-ons, PIPEs, registered directs, and bought deals.

6.5.3 It's All about Data

For publicly traded stem cell companies, the fundamental driver of valuation is data. Whether preclinical or clinical, investors will ascribe a potential value to data. For example, while minimal value is given to companies in preclinical development, the moment a company advances a therapeutic to clinical development, a stepwise increase in market valuation occurs.

Public Investment Models 75

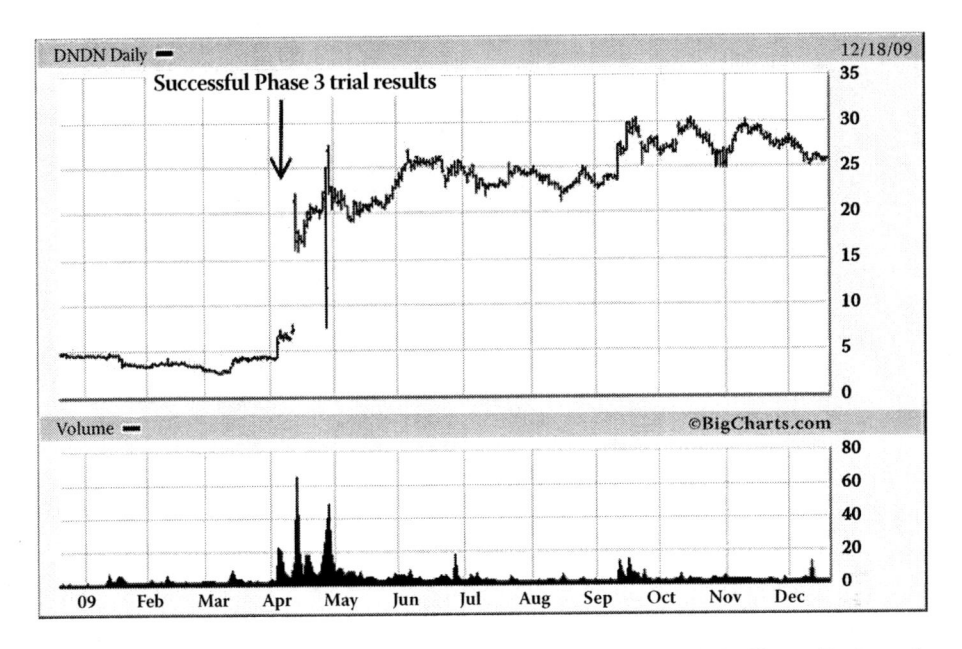

FIGURE 6.2 Positive clinical results drive valuation. (Courtesy of Big Charts; Rodman & Renshaw, LLC.)

Subsequently, if the company is able to advance from Phase I to Phase II studies, a further stepwise increase in valuation occurs. A more pronounced valuation increase may occur when a company achieves positive results from a pivotal Phase III study. The rewards (Figure 6.2) go to those few companies that successfully navigate the labyrinth of clinical trial designs and make it to the market on the other side. Based on prior expectations, the valuation increase can be quite sizeable.

However, this represents the best case scenario. While most investors wish for the linear timeline of clinical success, they should be prepared for the jagged saw-toothed reality. Biotechnology companies in general and stem cell companies in particular typically undergo many trial failures, many product iterations, and delays, all of which are reflected in their stock prices. For example, when Geron announced the FDA acceptance of its hESC stem cell investigational new drug application (IND), the stock significantly increased in value by approximately 50%. About 8 months later, the company announced that the clinical trial had been placed on hold pending additional review and the stock immediately sold off by about 10%. The whirlwind of valuation is a continual tug of war between investor expectations as to when a therapeutic will deliver promising clinical data and eventually hit the market—tempered by the realization that market presence could be delayed several years. The cost? Four hundred million dollars in valuation! See Figure 6.3.

6.5.4 Providing Hope, Not Hype

Stem cells stocks are largely driven by sentiment and hope, primarily because most lack revenues and even the semblance of collecting revenues in the near future (Miller

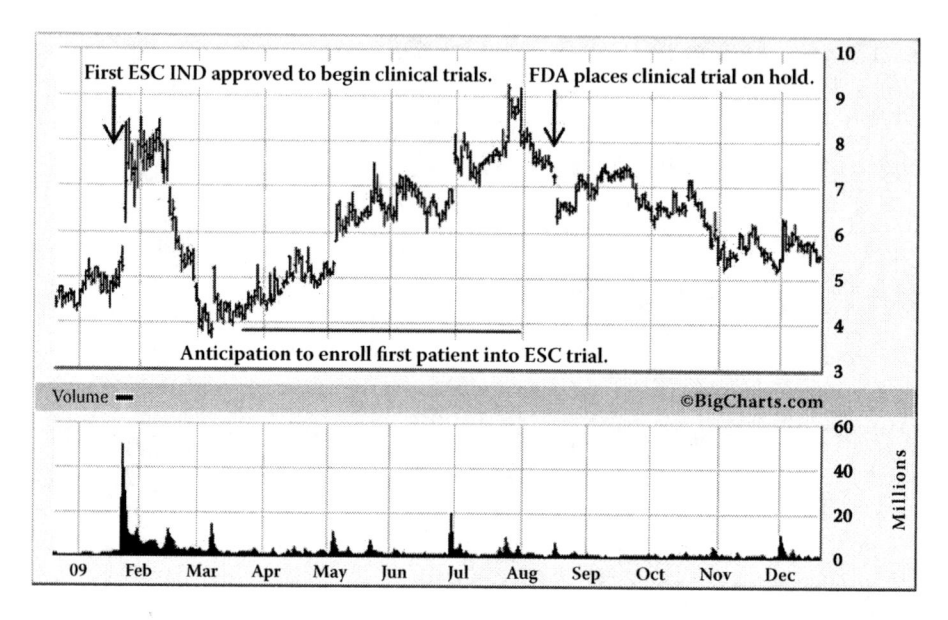

FIGURE 6.3 Fundamentals drive valuation. (Courtesy of Big Charts; Rodman & Renshaw, LLC.)

2006). Sentiment arises largely from news events that focus on human hardship and illness. For example, when Nancy Reagan speaks about Alzheimer's disease and Congress debates the pros and cons of repealing the ban on embryonic stem cell funding, one can often detect a concomitant increase in the interest of investors in stem cell stocks that tends to drive stem cell stock valuations higher, if only transiently.

Whether the story is Proposition 71 (California Stem Cell Research and Cures Initiative, 2004) or the approval of the first embryonic stem cell IND, the immediate investor reaction to such news is not what most outsiders would call rational. For example, when Proposition 71, the $10 billion embryonic stem cell funding initiative in California, appeared to be on its way to winning voter approval, one would reasonably predict that all embryonic stem cell companies located in California would increase in valuation given the potential for additional funding opportunities. In fact, however, all publicly traded stem cell players increased in valuation. Whether a company was developing technology that involved embryonic stem cells, adult stem cells, or umbilical cord stem cells was irrelevant in the eyes of investors.

What did this period of unbridled excitement, increased trading volumes, and higher share prices provide for these companies? Opportunities to access the public markets and raise capital without excessive dilution to existing shareholders (see Figure 6.4).

6.6 VALUATION METRICS

At the end of the day, potential investors need to place a value on a company in order to determine whether to participate in a financing or invest in the company

Public Investment Models

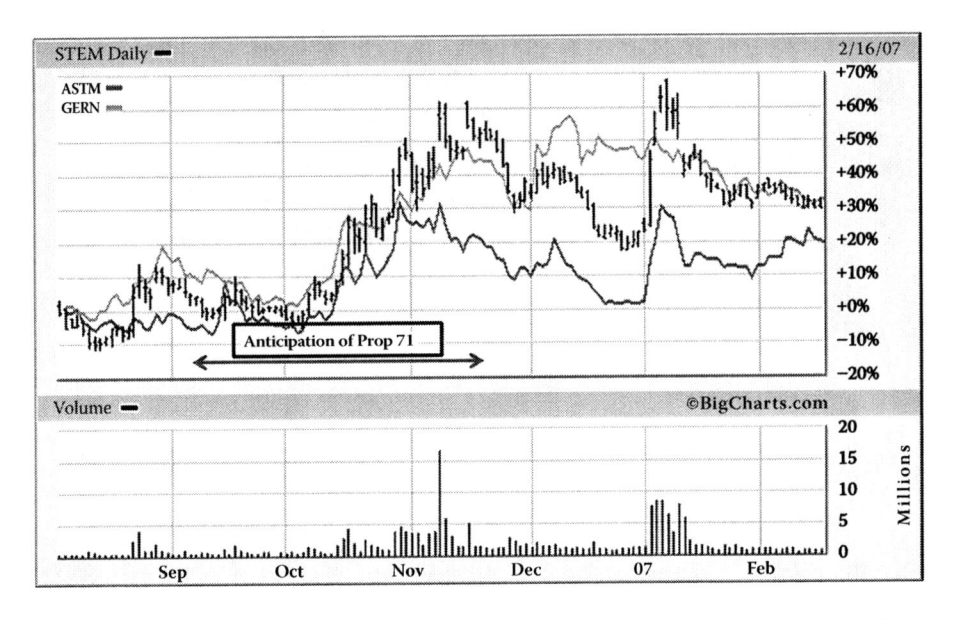

FIGURE 6.4 News flow drives group valuation. (Courtesy of Big Charts; Rodman & Renshaw, LLC.)

in the open market. Valuation methodologies vary greatly and range from simple analysis utilizing comparable companies ("comps analysis"), revenue and earnings multiple analysis, and present value or cash flow analysis. All of these valuation methodologies have their own pros and cons, and frequently more than one valuation method is warranted for investors to achieve an appropriate level of confidence in their assessment of the potential fair value of a company.

6.6.1 Multiples Analysis

Several key components are vital to a multiples analysis. The most important component is a model that incorporates revenue assumptions regarding the therapeutics under development or the potential for royalties under a partnership scenario, as well as future expenses including expenses for a potential in-house sales force and marketing. In short, considerable thought must be given to such assumptions, and an in-depth analysis of the one or more indications to be addressed by a therapeutic should allow an investor to calculate the respective market potential of each pipeline drug. Further analysis of the existing competitive landscape of marketed drugs and competing products in development should help an investor gauge the potential market penetration and further refine valuation estimates.

In addition, an investor should compile a list of comparable companies in order to determine the type of premium that investors are willing to pay for biotech companies at a similar stage of development or companies developing similar therapeutics. This premium is known as a multiple. This multiple, whether 45 times earnings for mature biotechs or a discounted multiple (e.g., 30 times) for biotech companies still in the development phase, is applied to the model projections. The resulting value divided by the shares outstanding is then discounted back to the present to determine the value of the future revenue or earnings stream in today's dollars. The discount rate used typically varies from 20% to 50% and reflects the clinical, regulatory, and commercial risks to achieving the projected revenues or earnings. Earlier stage stem cell companies with only preclinical studies or perhaps a Phase 1 asset would merit a discount rate at the upper end of the range. Companies with more advanced programs and therefore lower concomitant risks of eventually reaching the market would merit somewhat lower discount rates.

The multiples analysis is a common technique employed by sell-side analysts; it takes into account the current valuations of companies in the space in conjunction with their future revenue streams and projected operating expenses.

6.6.2 Cash Flow Analysis

Much like the previous analysis, a model must be constructed to reflect a company's future revenue potential and expenditures. The key difference is that a cash flow analysis model eliminates non-cash items that affect earnings. Since items like depreciation, amortization, and stock-based compensation can negatively affect the future earnings potential of a stem cell company, many investors will eliminate these items to ascertain the true cash impact of the potential future revenue stream.

6.6.3 Comps Analysis

Investors begin by compiling a list of comparable companies at a similar stage of development or companies developing similar therapeutics, much like the list compiled to determine the revenue or earnings multiple. However, for a comps analysis, the investor records the respective valuations, cash positions, debt positions, and enterprise values of the comparable companies. The enterprise valuation (market capitalization less cash on hand plus market value of debt) is a measure of value that investors place on a technology after taking into account the cash and outstanding debt of the company. Another way to appreciate the meaning of the enterprise value is to consider it the minimum dollar figure a potential acquirer would have to outlay to purchase the company outright.

For example, if a company trades at a market capitalization of $150 million and the company has $150 million in cash, one can assume that investors value the pipeline and technology at $0. In essence, investors believe that the company's underlying technology has absolutely no current value. Alternatively, an investor may have discovered an undervalued stock that is largely ignored—an oversight by the market that has nothing to do with a perceived absence of fundamental technology value for the company. Regardless of these nuances, one must compile a list of comparable companies to get an idea of what investors are willing to pay for the inherent technology of a similarly staged company. With that value in mind, one can calculate a potential "real" value for a company. For an investor, this is the quickest method to ascertain whether a particular company exhibits an equity upside.

6.7 CASE STUDY

While over 30 stem cell companies have gone public, one in particular is synonymous with the stem cell sector and has been able to capture the essence of the space—Geron Corporation. More importantly, management has been able to tap capital markets during good times and bad, to continue to fund development of the company's pipeline. In the company's illustrious near 20-year history (14 years as a public company), over $450 million in funds have been raised during its time as a public entity through a combination of convertible debt, PIPEs, registered directs, and follow-on offerings (Figure 6.5 and Table 6.1). The timeline, financing deals, and stock reactions to deals and news flow engender all the principles described in the previous pages.

6.8 CONCLUSIONS

The healthcare space represents one of the greatest investing opportunities of all time, but it is not for the faint of heart. Because of the underlying drivers such as an aging baby boomer generation, longer life expectancies, and the continuous need to develop therapeutics that substantially address afflictions of the elderly, many companies have taken up the challenge to change the healthcare space as we know it. However, this challenge comes at a great price. Tufts University publishes a yearly projection of the costs of developing a drug—the most recent cost to develop a drug exceeds $1.2 billion (Tufts, 2009). The primary sources to fund such an amount are investors. Publicly traded stem cell companies that can master the tools available to access the capital markets have the best chances to successfully navigate the eventual failures and potential successes on the road to regulatory approval and ultimately to commercialization.

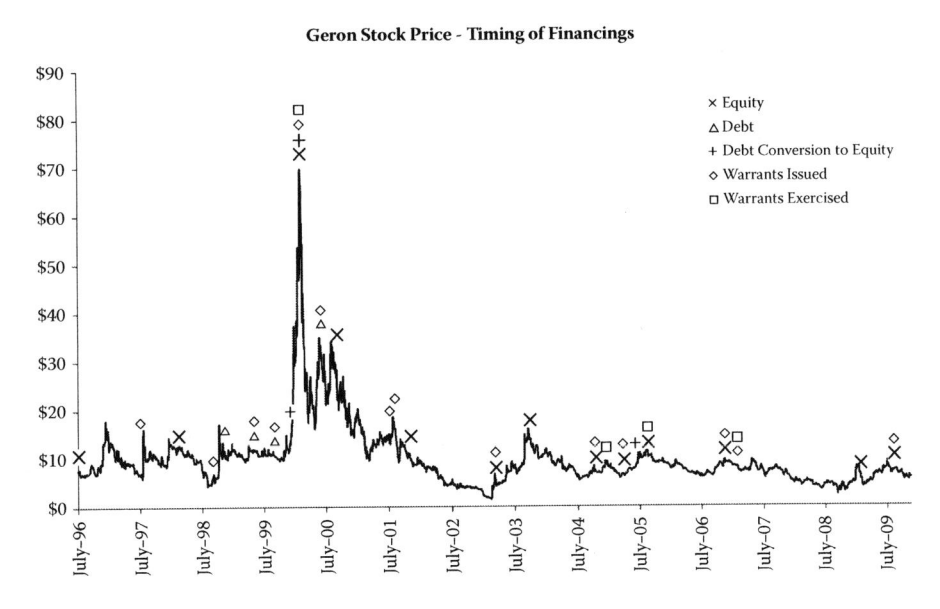

FIGURE 6.5 Smart companies always access the capital market. (Courtesy of Geron Corporation and Rodman & Renshaw, LLC.)

TABLE 6.1
Geron Corporation Financing 1996–2009

Date	Financing Vehicle		Share/Debenture Class	Share/Exercise/Conversion Price ($)	Shares Issued	Warrants Issued	Amount Raised (MM)
July 30, 1996	Equity—initial public offering	×	Common	8.00	2,312,500		$16.7
August 1997	Warrants issued—license agreement	◇	Common	5.78		25,000	
March 1998	Stock purchase agreement with Pharmacia and Upjohn	×	Common	At premium			$4.0
March 27, 1998	Equity—private placement	×	Series A convertible	1000.00	15,000		$15.0
October 1998	Warrants issued—license agreement	◇	Common	6.75		25,000	
December 10, 1998	Debt—institutional investors	◁	Series A convertible	10.00			$7.5
June 1999	Debt—institutional investors	◁	Series B convertible	10.00			$6.8
	Warrants issued	◇	Common	12.00		625,000	$0.7
	Debt—institutional investors	◁	Series C 2% coupon	10.25			$12.5
September 30, 1999	Warrants issued	◇	Common	12.50		1,000,000	
	Warrants issued	◇	Common	12.75		100,000	
1999	Debt conversion to equity	+	Series B convertible	10.00	450,000		
	Equity—institutional investor	×	Common		380,855		
	Warrants issued	◇	Common	67.09		200,000	$9.0
March 2000	Warrants issued	◇	Common	12.50		100,000	
	Debt conversion to equity	+	Series C 2% coupon	10.25	615,000		
	Warrants exercised	□	Common		1,100,000		$13.8
June 29, 2000	Debt—institutional investor	◁	Series D convertible	29.95			$25.0
	Warrants issued	◇	Common	37.46		834,836	

Public Investment Models

Date	Type		Security	Price	Shares	Warrants	Amount (M)
October 2000	Equity—financing facility	×	Common	~25	~100,000		$2.5
	Warrants issued—license agreement	◇	Common	14.60		100,000	
August 2001	Warrants issued—consulting agreement	◇	Common	22.56		9,000	
September 2001	Warrants issued—license agreement	◇	Common	9.07		5,000	
December 2001	Equity—financing facility	×	Common	~9	~810,000		$7.3
April 8, 2003	Equity—institutional investors	×	Common	4.60	4,000,000		$18.4
	Warrants issued	◇	Common	6.34		600,000	
October 29, 2003	Equity—public offering	×	Common	12.00	6,750,000		$69.0
	Equity—institutional investors	×	Common	6.10	6,500,000		$40.0
November 11, 2004	Warrants issued—institutional investors	◇	Common	6.10		2,000,000	
	Warrants issued—institutional investors	◇	Common	At premium		2,300,000	
January 12, 2005	Warrants exercised—institutional investors	☐	Common	6.10	2,000,000		$12.5
April 25, 2005	Equity—institutional investors	×	Common				$4.0
	Warrants issued—institutional investors	◇	Common				
June 30, 2005	Debt conversion to equity	+	Series D convertible	5.00	3,000,000		
	Equity—underwritten public offerings	×	Common	9.00	6,900,000		$62.1
September 16, 2005	Warrant exercise—Merck license agreement	☐	Common	9.00	2,000,000		$18.0
	Equity—institutional investors	×	Common	8.00	5,000,000		$40.0
December 14, 2006	Warrants issued—institutional investors	◇	Common	8.00		1,875,000	
	Warrants issued—institutional investors	◇	Common	At premium		3,000,000	
	Warrants exercised—institutional investors	☐	Common	8.00	1,875,000		$15.0
February 27, 2007	Warrants issued	◇	Common	At premium		1,125,000	
February 19, 2009	Equity—underwritten public offerings	×	Common	7.77	7,250,000		$56.3
	Equity—underwritten public offering	×	Common	6.55	550,000		$3.6
September 10, 2009	Warrants issued—institutional investors	◇	Common	At premium			

REFERENCES

California Stem Cell Research and Cures Initiative, 2004. http://www.cirm.ca.gov/pdf/prop71.pdf [accessed 12/1/2009].

Financial Post, 2009. http://www.financialpost.com/scripts/story.html?id=aa5689b0-c4b6-4a51-befc-e1ff5a9c90c&k=28670#ixzz0ZUYHJhSC [accessed 12/1/2009].

Halme, D. and Kessler, D., 2009. FDA regulation of stem cell based therapies. *New England Journal of Medicine* 355: 16.

Healthcare Costs, 2009. http://www.chcf.org/documents/insurance/Healthcare Costs09.pdf [accessed 12/1/2009].

InfoWorld, 2009. http://www.infoworld.com/t/business/aging-baby-boomers-will-drive-health-care-innovation-054 [accessed 12/1/2009].

Miller, D., 2006. The keys to biotech investing. In *Master Traders: Strategies for Superior Returns from Today's Top Traders*, Hamzei, F., Ed. New York: John Wiley & Sons, pp. 109–121.

PWC, 2009. http://www.pwc.com/US/en/press-releases/2009/PricewaterhouseCoopers-US-IPO-Watch.jhtml [accessed 12/19/2009].

Rodman & Renshaw., LLC. Privileged communication.

Tufts University, 2009. http://csdd.tufts/edu/NewsEvents/NewsArticle.asp?newsid=69 [accessed 12/1/2009].

Section 3

Business Models

7 Cell-Based Products
Allogeneic

Paul Kemp

CONTENTS

7.1 Introduction 85
7.2 Developing Business Models: Incorporating Manufacturing and
 Logistics Issues 86
    7.2.1 Obtaining Tissue: Sources, Donor Consents, and Safety Testing 87
    7.2.2 Cell Bank Manufacture 87
    7.2.3 Cell Bank Considerations 88
    7.2.4 Product Assembly and Manufacture 89
    7.2.5 Who Manufactures the Product? 89
    7.2.6 Manufacture: One Site or Several? 90
    7.2.7 Cryopreservation 90
    7.2.8 Shipping Choices: Hub and Spoke or Directly to Clinic? 91
    7.2.9 Shelf Life: Determining Minimum and Optimum 91
    7.2.10 Market Focus 92
    7.2.11 Defining Optimal Sales Channel 92
7.3 Regulatory Issues 93
7.4 Allogeneic Cell-Based Therapy Companies 93
7.5 Case History: Organogenesis 94
7.6 Analysis and Conclusions 95
References 96

7.1 INTRODUCTION

The implanted cells of an allogeneic transplant come from a human donor source other than the treated patient. Allogeneic transplants present certain advantages over allogeneic transplants in that:

- The cells can be tested and ensured to be genetically "normal."
- The cells can be tested for the absence of pathogens.
- The products can be sold off the shelf and do not require "bespoke" cell expansion as required with autologous transplants.
- The cells can be obtained from validated cell banks and expanded in culture prior to implantation.

Allogeneic transplants do, however, suffer from possible immunological rejections by patients if they are identified as non-self. Fortunately, this does not seem to be an issue for many allogeneic cell therapies, both approved and under development, although the reasons for this are poorly understood and for some products the therapeutic benefits seem to continue long after the implanted cells appear to have been cleared from the system (Wong et al. 2008; Phillips et al. 2002).

Allogeneic cellular transplants have a clinical and commercial history spanning more than 100 years, beginning with Eduardo Zim's first corneal transplant in 1905. Today over 20,000 cornea transplants are performed in the U.S. alone. Allogeneic cells are the bases for most organ transplantations and blood transfusions and they have now become very successful standard therapies for end stage organ failure of the kidney, liver, heart, lung, pancreas, and intestine. The access of patients to transplantation therapy, however, varies widely across the globe, partly determined by the underlying levels of economic development in different areas and partly by other factors such as the availability of organ, cell, and tissue donors and specialist provision of health services. For example, the number of kidney transplants performed in the U.S. and normalized for population is twice that of Europe and over 17 times that of Asia!

Entire organs consisting of complex systems of billions of different types of living human cells are transplanted from one person to another and, as long as the immunological issues are addressed, the organ continues to function in its new location for many years. Such transplants and transfusions have saved or improved the lives of several million patients. This combination of the functional success of organ transplantation combined with the frustrations caused by insufficient organ availability led to the development of strategies in which cells are first released from donated tissue and expanded in culture before implanting into recipients either directly (cell therapy) or after reassembly into a rudimentary tissue structure (tissue engineering). With such systems, potentially unlimited material could become available, depending upon the abilities of cells to be expanded in vitro. The current functional limitation to this process is the inability to reassemble a suspension of cultured cells back into a proto-organ system that can further develop upon implantation into a functional organ. Despite this, several allogeneic cell therapies are currently approved around the world and several business models have been developed.

7.2 DEVELOPING BUSINESS MODELS: INCORPORATING MANUFACTURING AND LOGISTICS ISSUES

Although this chapter focuses on the development of business models used in allogeneic cell-based products, those models are by necessity affected by outside factors such as the state of the art of current technology and the regulatory framework that acts as the critical gatekeeper between the technology and the marketplace. All of these factors change rapidly, and as a consequence, successful business models will by necessity also change.

The importance of a suitable and robust business model cannot be overemphasized. Several regenerative medicine companies have found themselves in severe difficulty, not because their products did not work but because their business models were not adequate. One example is Organogenesis Inc., a company that emerged

Cell-Based Products

from bankruptcy to be the first pure regenerative medicine producer to make a profit. This was achieved partly by changing from a license-based model to a fully integrated business model with its own sales force. Similarly, although Advanced Tissue Sciences did not emerge from bankruptcy, its flagship Dermagraft product became the basis of a new company, Advanced BioHealing, which also changed its business plan and created its own sales force rather than rely on a partner and a royalty stream. A company can adopt a number of options to commercialize its discoveries in allogeneic cell therapy and each company's approach in matching the various elements is unique. However, some best practices are emerging in the early 21st century and these will be discussed later in this chapter.

Several questions must be addressed *quantitatively* during development of a business model. Some relate to the original design of the product and must be considered during the research and development (R&D) phase and others are options dictated by cost and logistic issues. The true financial consequences of these decisions that are often made in isolation by scientists must be taken into account by managers early on in the life cycle of a product, although compromises and educated guesses are the best measures available in some cases without expending large amounts of research time and money to explore all the options. Time spent in planning, especially in the product design phase, may have critical and long-lasting effects.

Some of the major questions to consider when developing business models and business plans for an allogeneic cell-based product are described below. Many are interrelated and cannot be considered in isolation. When they are analyzed together, a picture emerges for each individual product. A single company may have different cell-based products and the solutions and drivers for each of the elements below are entirely different. The resulting overall business model of the company becomes somewhat hybrid to take into account the needs of the product portfolio.

7.2.1 Obtaining Tissue: Sources, Donor Consents, and Safety Testing

Tissue source is the first aspect of any allogeneic cell therapy to consider. Decisions about sources of tissues for products using differentiated cells are important. Foreskin tissue discarded as a result of neonatal circumcision has become the standard for various cell therapies involving skin-derived differentiated cells. Bone marrow aspirates are standard starting materials for mesenchymal stem cell-based therapies. Excess in vitro fertilization (IVF) embryos serve as sources of embryonic stem cells. However, the time and expense required to obtain such materials is not trivial and the regulators require extensive paper trails right back to the donors (and often the donors' parents) in the case of neonatal or embryonic material. In cases where a start-up business is the result of a university spin-out, the audit trail may be nonexistent although academic laboratories are becoming educated about the need to develop good manufacturing practice (GMP) cultures of cells under investigation.

7.2.2 Cell Bank Manufacture

Manufacture is the main differentiating aspect between autologous and allogeneic products and one of the largest defining aspects that directs a business plan. Using an allogeneic cell source requires and allows the production and testing of the cells. Testing such cell sources is extensive, costly, and time consuming. The testing starts with the donor and extends beyond culture expansion through end-of-production (EOP) testing. The tests required by the regulatory agencies are now fairly standardized (Guidance for Industry 1998) and the development of a working cell bank to be used in the manufacture of end product can take approximately two years and cost in excess of $100,000.

This significant investment needs to be spread over many finished items. Fortunately, through the mechanics of cell expansion, depending on the cell type, a very small starting tissue sample may eventually yield several billions of cells at the product stage that may translate into hundreds of thousands of production units. However, for an allogeneic product to be commercially feasible, the cell type in question must have the proliferative capacity to allow cell banks to be produced. For some types, such as dermal papillae cells used in Intercytex's hair regenerative program, this expansion potential is not currently sufficient to allow development of an allogeneic product and a commercially viable autologous approach is needed.

7.2.3 Cell Bank Considerations

Many enormously complicated issues surround the establishment of a cell bank and it would be outside of the scope of this chapter to fully address them. A thorough analysis needs to be developed for each product and program to fully determine the quantitative issues involved. The numbers of cells needed both for clinical studies and commercial use, the proliferative capacity of the cell type used, and the extent of ultimate demand for the product exert impacts on the sizes of the banks produced, the amount of tissue originally obtained, and the frequency at which cell supply needs change.

For example, many current regenerative medicine products are based, at least in part, on human dermal fibroblasts derived from neonatal foreskins. These fibroblasts have limited proliferative capacities, requiring manufacturers to change donor sources from time to time. Although extensive comparative testing must be carried out to demonstrate equivalence of the cells, such changes are generally acceptable by regulators. With essentially immortal embryonic stem cells, cell supply is not as much a problem, but the issue of cell source changes may well become one as it is generally established that each human stem cell line so far developed has its own unique characteristics, and the ability to differentiate and proliferate without chromosomal abnormalities differs from line to line (Catalina et al. 2008).

As well as the time and testing expense of developing cell banks, a business plan must also consider media cost and availability. Media compositions are initially developed during the R&D program, often at a university, and usually result from a long series of academic studies. Media components can become "historic" and whether their presence is critical to the performance of a product is not always determined. This can result in ongoing expenses during the development and manufacturing process and can also result in some justified and awkward questioning by the regulators when a company applies for a product license. One obvious, controversial, and highly expensive media component is serum, most commonly from bovine fetal or newborn calves. Serum can be an extremely expensive component of any product and

Cell-Based Products

its appropriate sourcing in these times of transmissible spongiform encephalopathy (TSE) can be a challenge for any purchasing department.

It is often a false perception, especially within the stem cell community, that the regulators will not approve any cell bank that has been exposed to bovine serum of any source. However, the use of this component is highly controlled and any company would be prudent from a cost and regulatory perspective to embark on studies aimed at reducing the use of this component to an absolute minimum or, if possible, removing it altogether.

7.2.4 Product Assembly and Manufacture

This stage is a unique element of each regenerative medicine operation and the point at which its intellectual property plays the largest role. The form in which the cells are presented to the patient defines the product and plays a major role in the development of the business plan. A prime example of this is also one of the first regenerative medicine products, Organogenesis' Apligraf, composed of allogeneic human dermal fibroblasts and keratinocytes. These cells are separately cultured, expanded, and tested, and then the fibroblasts are reconstituted into a "proto-dermis" by mixing into a gel of bovine type I collagen. After several days of culture, the keratinocytes are added to the upper surface; after more time in culture, the product is raised to an air–liquid interface that stimulates the keratinocytes to differentiate into a multi-layered epidermis. The whole process of assembly takes approximately 3 weeks and has been a huge focus of process development effort post-Chapter 11 to reduce the cost of goods and promote manufacturing efficiencies to levels that could better support the new business model.

7.2.5 Who Manufactures the Product?

The scale of cell-based manufacturing, even at the early commercial levels, is generally smaller than that of a classic non-cellular biological. Culture systems on a sub-liter scale are usually the norms for cell-based products rather than the massive, multi-liter fermentor processes developed over the years for the production of cell-derived acellular biologics such as recombinant proteins and vaccines. This means that increasing the output of cell-based products often occurs by scale-out, rather than scale-up, and also that a clinical scale and indeed an early commercial scale manufacturing facility could be built and made operational for costs ranging from $1 million to $5 million, depending upon the size.

This is very fortunate. Until the last few years, contract manufacture of cell-based products was not an option. The arrival of companies such as Lonza and Angel Biotech into the sector has made contract manufacturing an option and the economics of in-house versus contract manufacture can now be assessed for individual products (see Chapter 14 by Preti). It must be emphasized that contract manufacture can be brought in at various stages of the manufacturing process. For example, most companies contract for media manufacture and several now utilize contract manufacturers (such as Intercytex) to produce their cell banks even if the final assembly of the product is carried out in house.

7.2.6 MANUFACTURE: ONE SITE OR SEVERAL?

To a large extent, the answer to this question is based on market location, product shelf life, shipping costs, and logistics. The advantage of multiple sites is the redundancy in manufacture in case of a problem at one site, but this is offset by the cost of duplicating a very expensive quality system at each site and ensuring the outputs of all sites manufacturing complex cell-based products are equivalent (see Chapter 14 by Preti). To date, the general solution has been a single manufacturing site. Genzyme, Organogenesis, and Advanced Cell Therapy, for example, all produce commercial grade materials from a single manufacturing site and ship them around the U.S. and more recently to Europe. It is estimated that the absolute minimum requirements for a commercial (GMP) manufacturing and quality team is 10 people and the fixed costs of running such a team are in the region of £1 million to £2 million per year, depending upon the facility costs. This must be taken into account when developing a business plan that includes manufacturing.

One aspect of cGMP manufacturing that is often overlooked at the academic spin-out stage is illustrated in a quote from Mary Malarkey from the Center for Biologics Evaluation and Research (CBER) at the U.S. Food & Drug Administration (FDA): "The FDA considers the GMPs to be a 'sliding scale,' meaning that they must be more rigorously applied as products move forward in development.* In other words, during manufacture of Phase I material less detailed application of the regulations will be expected. By the time a product reaches Phase III clinical trials, GMPs are expected to be fully followed during manufacturing." Therefore, the ongoing costs incurred by a business to satisfy GMP will increase significantly as a product moves through clinical development and into commercial manufacture. These costs do not include on-and-off expenses of process validation required before any product can be licensed and these costs can be considerable.

7.2.7 CRYOPRESERVATION

Whether to utilize cryopreservation is one of the most critical decisions to be made about a product and the decision must be made early. Both approaches have been taken. For example, Dermagraft is cryopreserved; Apligraf is not. Pros and cons surround both approaches. The obvious advantage of a cryopreserved product is the huge increase in shelf life but it comes at a financial and logistical cost. Assuming a cryopreservation protocol can be developed (not a trivial task if a product has a large volume or is made of multiple cell types) certain considerations must be taken into account with a cryopreserved product:

- Need for and cost of storing product both at the manufacturing facility and more importantly at the clinic. Minus 80°C freezers are expensive, noisy, produce large amounts of heat, and are costly to maintain.
- Shipping product at −80°C is more expensive than shipping at near-freezing temperatures.

---

* http://www.gmpcompliance.net/CGMP_Defined/CGMP_Defined.htm

Cell-Based Products

- A suitable thawing protocol can be difficult to develop and monitor. Inappropriate thawing can easily kill the cells. This would not be known when the product was administered to a patient because there would be no time to test the product adequately before it was used.
- If product is stored at the clinic so that it is always available for use, a very expensive inventory must be maintained. The cost of a single unit of a cell-based product is typically at least $1,000.

7.2.8 Shipping Choices: Hub and Spoke or Directly to Clinic?

Shipping was one of the overlooked aspects of early regenerative medicine efforts. However, the technical issues and costs involved in transporting a living construct are far from trivial. When transatlantic shipping is required, one possible option is to ship in bulk first to a hub where the product is repackaged and then transported within the country or continent to the end-user clinic. The advantage of this approach (apart from the cost reduction in sending a single large package rather than many over large distances) is that during at least some of the shipping process the product is still within the manufacturer's control and therefore repackaging could be carried out while final product release testing is being carried out (see section on shelf life below). However this advantage may be offset by the cost of developing a GMP "repackaging center" able to receive the product and redistribute it to the various clinics.

7.2.9 Shelf Life: Determining Minimum and Optimum

Determining shelf life is a critical aspect of cell-based products and it drives many of the commercial decisions and costs. Protein-based biologics typically have shelf lives of months or years and this means that scale (bigger is cheaper) becomes a driving factor in any production system. Large sized batches and usable shelf lives make campaign manufacturing an attractive option. This is not the case with cell-based products where, without cryopreservation, a shelf life of only a few days or weeks is typical. This means that manufacturing needs to be constant and "just in time"; as a result, inventories of product are difficult to develop. An inventory can act as a buffer against manufacturing issues or batches that fail quality control testing. Without a buffer, a manufacturing operation is under constant pressure and the tendency is to develop a manufacturing safety margin that may lead to considerable waste of product. An issue that arises often during product development is the commercial need for extended shelf life—a difficult goal to communicate to R&D groups that also have other goals to meet. The total shelf life of a product is broken down into four elements:

- **Shelf life for manufacturing** is probably the least important of the shelf life elements. Any "left-over" material (after the other shelf life elements have been satisfied) would potentially allow for the rapid replacement of missed batches or product lost during processing. Accordingly, this would reduce the need to make back-up batches to ensure a constant supply to the clinic;

- **Shelf life for quality control** requires a minimum of 3 days for performing sterility assessment in a unit such as the BacT Alert (Biomerieux) but could be much longer. If a shipping system is such as a hub-and-spoke arrangement or direct shipment to hospital pharmacies is developed, it is possible to ensure that the product remains under control of the manufacturer during this phase and at least some of the quality control testing could take place during the shipping process.
- **Shelf life for shipping** may vary and depends upon where the product is shipped. Modern transport systems allow materials to move around the globe in a matter of days.
- **Shelf life for the clinic** may be the most important element of total shelf life. It should be at least a week to allow a patient to return for a missed clinic visit. Interestingly it has been argued that an extensive clinic shelf life actually devalues a product in the eyes of a clinician.

7.2.10 Market Focus

The key question of which territories to target (e.g., U.S., Europe, or Asia) must take into account the market size, regulatory requirements, and distribution logistics. Marketing is a very complex, multifaceted question, the answer to which relies on several, sometimes mutually exclusive, factors. Many regenerative medicine companies have, however, focused on their domestic markets which, for U.S. based companies, is perhaps not surprising but is less logical for European companies.

The U.S. is currently the obvious main market for regenerative medicine products. The FDA has more experience in regulating this area (although it has approved only one cell-based biologic as the early products were considered medical devices rather than biologics). The U.S. healthcare system arguably adopts new therapies more readily than most other countries and the reimbursement system although rigorous is used to paying high prices for therapies that work. How the Advanced Therapy Medicinal Product (ATMP) regulations will affect the regenerative medicine situation in Europe is too early to predict but will probably result in larger variations in reimbursement and product adoption between European states.

7.2.11 Defining Optimal Sales Channel

A company must consider multiple sales channels, for example, building an in-house sales force, license-out agreements, engaging a representative, and arrangements with contract distributors. Sales method was a critical issue in the success of early regenerative medicine companies such as Organogenesis and Advanced BioHealing. Sales forces are expensive to maintain and thus an obvious option is to license out a product to a larger organization that can add the cell-based therapy to the line of an existing sales force. However, experience indicates that selling a living product is not such an easy transition for companies more used to selling "white powders in bottles."

The experience of Novartis in selling Apligraf and the impact this had on Organogenesis is well documented and resulted in the Chapter 11 bankruptcy of

Cell-Based Products

Organogenesis in September 2002. Organogenesis emerged from bankruptcy as a privately funded company and focused primarily on Apligraf. Critically, it employed its own sales force, many of whom had experience selling Apligraf with Novartis. Organogenesis is now a profitable company with sales increasing year on year and is now beginning to look outside the U.S. (Section 7.5 below; Boston Bizjournals 2008; Business Week 2009).

7.3 REGULATORY ISSUES

Even after a 30-year history, the regulatory issues related to cell-based therapies are still rapidly changing. For example, 2009 saw the first approval (TiGenix's Chondroselect) in Europe under the new Advanced Therapy Medicinal Product (ATMP) regulations designed to cover both autologous and allogeneic cell-based therapies (see Chapter 19 by Bravery).

In the U.S., there is still some confusion over whether similar cell-based therapies developed for the same medical indication (e.g., venous stasis ulcers) are regarded as medical devices under the remit of the Center for Drug Evaluation and Research (CDER), as is the case for Apligraf (Organogenesis) and Dermagraft (Advanced BioHealing), or as biologics under CBER as is the case with Intercytex's Cyzact (see Chapter 22 by Faulkner). The regulatory situation in the rest of the world is even more unclear although the first approvals for cell-based products (autologous) have appeared in Japan. Europe may possibly take the regulatory lead with ATMPs and the first year has proven extremely promising.

7.4 ALLOGENEIC CELL-BASED THERAPY COMPANIES

Described below are some companies that specialize in producing allogeneic cell therapies and have products in clinical trials or on the market.

Advanced BioHealing (U.S.)—Like Organogenesis, Advanced Tissue Sciences filed for Chapter 11 bankruptcy protection in 2002. In this case, however, Advanced BioHealing bought its technology and resuscitated its Dermagraft product for treating diabetic ulcers and also adopted the fully integrated model by recruiting its own sales force. Also like Organogenesis, Advanced BioHealing shelved other products that were not approved, greatly reducing the R&D burden. Advanced BioHealing's fortunes have completely turned around and the company now looks to increase the indications of Dermagraft by running additional clinical studies.*

Cellerix (Spain)—Founded in 2004, Cellerix is a biopharmaceutical company that develops treatments based on adipose-derived adult stem cells (see Chapter 8 by Bravo and Blanco-Molina). Cellerix's pipeline covers three main therapeutic areas: gastroenterology, dermatology, and immune-based diseases. Its product portfolio includes Ontaril®, an injectable autologous suspension that contains stem cells for the treatment of complex perianal fistulas of cryptoglandular origin and associated with Crohn's disease; Cx501, a chimeric skin equivalent containing autologous and allogeneic components for treating epidermolysis bullosa, the allogeneic Cx601

---

* http://www.abh.com/

products for treating perianal fistulas associated with Crohn's disease; Cx611 for the treatment of autoimmune and inflammatory diseases; and a Treg platform (Cx911) targeting autoimmune diseases.

Intercytex (U.K.)—Intercytex became operational in 2000 and had several products in preclinical and clinical development. It floated on the London Stock Market's AIM in 2006 as Intercytex Group Plc. and purchased Axordia, a Sheffield-based embryonic stem cell company, in 2008. Unfortunately Intercytex was unable to raise funds in the 2009 recession and had to sell many of its assets. There was a management buy-out of Vavelta, an injectable allogeneic fibroblast asset used to treat dystrophic epidermolysis bullosa and scar contractures; Intercytex Ltd. began operations in 2010.

Organogenesis (U.S.)—This company's remarkable history makes it the subject of a brief case history. See Section 7.5 below.

Osiris Therapeutics (U.S.)—Osiris was founded in 1992 to develop and commercialize cellular therapies based on stem cells isolated from readily available adult bone marrow. The primary thrust of the near-term business strategy revolves around the clinical development of product candidates for a variety of indications. The inflammatory program focuses on providing support for bone marrow transplantation after chemotherapy or radiation therapy. The goal is to treat graft-versus-host disease (GvHD), a life-threatening complication following the receipt of bone marrow from another individual. The cardiac program focuses on MSCs to improve heart function following myocardial infarction (MI) and to prevent progression to congestive heart failure (CHF). Osiris entered into a strategic alliance with Boston Scientific Corporation for this product area (Osiris Press Release 2003). In orthopedic applications, MSCs may be useful in regenerating meniscal tissue in the weeks following surgery for knee injuries, thereby preventing arthritis. In November 2008, Osiris and Genzyme announced a strategic alliance for the development and commercialization of Prochymal and Chondrogen. Under the terms of the agreement, Osiris retains commercialization rights to both products in the United States and Canada, with Genzyme having these rights in all other countries. In 2008, Osiris entered into a contract with Lonza for Lonza to manufacture Prochymal for the treatment of GvHD (Osiris Press Release 2008).

ReNeuron (U.K.)—This company is developing genetically stable neural stem cells for the treatment of stroke. Also, through the acquisition of Amcyte, it is looking at micro-encapsulated islet cells. The company was founded in 1997, went public on AIM, was re-privatized in 2003, and refloated in 2006. It has signed an agreement with Angel Biotech to manufacture its product and has gained regulatory and conditional ethical approvals to commence a Phase I clinical trial in the U.K. with its lead ReN001 stem cell therapy for disabled stroke patients (ReNeuron 2010).

7.5 CASE HISTORY: ORGANOGENESIS

Organogenesis Inc., one of the first regenerative medicine companies, was the first to become profitable. Its history is a reflection of the field and its times and also illustrates the importance of a business model that complements the product. The company was founded in 1986 based on technology invented by developmental biologists at Massachusetts Institute of Technology. Its lead products at the start were a so-called living skin equivalent (LSE) and a small calibre living blood vessel equivalent (LVE). See Weinberg and Bell 1986.

The company went public on a small exchange very early on and was funded through most of its life by so-called "high net worths." Because absolutely no contract manufacturing options were available then, Organogenesis developed and built its own GMP manufacturing facility in the U.S. Other products were added to the portfolio and two strategic arrangements were developed, one with Eli Lilly to fund development of the LVE, and a licensed marketing deal with Novartis to sell the LSE, then known as Apligraf (TechAgreements 1996), and approved by the FDA in 1998 (U.S. Food & Drug Administration 1998). At this stage, Organogenesis had a market cap around $700 million to $1 billion and a staff of about 200. Novartis was selling Apligraf for the treatment of venous stasis ulcers.

However, the business model revealed severe problems relating to the amount of material sold, the transfer price, and the large fixed costs of the manufacturing facility borne by Organogenesis. This led to an ultimate collapse and Organogenesis filed for bankruptcy protection in October 2002 (Boston Bizjournals 2003). But all was not lost. Three private investors bought the company and invested further money to re-start it (Boston Bizjournals 2003). New management was brought in from the Novartis sales force and the product was resupplied to clinics. The company was now fully integrated, manufacturing and selling products, and within a few short years was able to turn profitable.

During that time, the essential Apligraf product did not change, but the company worked hard to reduce the cost of its manufacture and introduced automation to further reduce costs. It also lobbied the reimbursement industry to increase the price that medical insurance companies would pay for the product. Another step was modifying shipping containers to reduce shipping costs significantly but more importantly also allowed a shelf life increase from 5 to 15 days (10 days at clinic), greatly improving the logistics. The company is now highly successful, with steadily increasing sales, expanding manufacturing facilities, and a broadening product portfolio. The success was due to changes in business plan and management rather than changes in the underlying science.

7.6 ANALYSIS AND CONCLUSIONS

Current regenerative medicine companies have all addressed the issues of how to turn cell-based technologies into viable businesses in different ways. A feature that probably differentiates this group from other biotechnology companies is that regenerative medicine companies tend to be more fully integrated than similar sized biotechnology companies—although this may reflect the relative maturities of these two sectors. The support infrastructure is not yet in place for regenerative medicine and outsourcing is beginning. As an example, Osiris Therapeutics outsources mesenchymal cell manufacturing to Lonza; ReNeuron outsources to Angel Biotech. Although some of these companies are over 20 years old, the field is still in its infancy and business plans are steadily changing to take account of the changing landscape. What is absolute, however, is that an appropriate business strategy is crucial from day one in order to bring these exciting potential therapies to the market and to use them to develop sustainable businesses.

REFERENCES

Boston Bizjournals, 2008. http://boston.bizjournals.com/boston/stories/2003/09/08/daily27.html

Boston Bizjournals, 2003. http://www.bizjournals.com/boston/stories/2003/08/25/story5.html

Business Week, 2009. http://www.businessweek.com/magazine/content/09_ 24/b4135000953288.htm?chan=top+news_top+news+index+-+temp_news+%2B+analysis

Catalina, P., Montes, R., and Ligero, G. et al. Human ESCs predisposition to karyotypic instability: is a matter of culture adaptation or differential vulnerability among hESC lines due to inherent properties? *Molec. Cancer* 7: 76, 2008.

Osiris Press Release, 2003. Boston Scientific and Osiris Therapeutics announce stem cell alliance to develop and commercialize cardiovascular therapies using adult mesenchymal stem cells. http://investor.osiris.com/releasedetail.cfm?releaseid=202618

Osiris Press Release, 2008. Securities & Exchange Commission filing. http://investor.osiris.com/secfiling.cfm?filingID=1047469-09-2710

Phillips, T. J., Manzoor, J., and Rojas, A. et al. The longevity of a bilayered skin substitute after application to venous ulcers. *Arch. Dermatol.* 138: 1079–1081, 2002.

ReNeuron, 2010. Regulatory Update. http://www.reneuron.com/news__events/news/document_225_237.php

TechAgreements, 1996. http://www.techagreements.com/agreementpreview.aspx?num=64504&title=Novartis+Pharma+Ag+/+Organogenesis+Global+Manufacturing+And+Supply+Agreement

U.S. Food & Drug Administration, 1998. Guidance for Industry: Human Somatic Cell Therapy and Gene Therapy. http://www.fda.gov/biologicsbloodvaccines/guidancecompliance-regulatoryinformation/guidances/cellularandgenetherapy/ucm072987.htm

Weinberg, C. B. and Bell, E.A., 1986. Blood vessel model constructed from collagen and cultured vascular cells. *Science* 231: 397–400, 1986.

Wong, T., Gammon, L., and Liu, L. et al., Potential of fibroblast cell therapy for recessive dystrophic epidermolysis bullosa. *J. Invest. Dermatol.* 128: 2179–2189, 2008.

8 Autologous Cell-Based Products

Fulfilling the Promise of Cell Therapy

Eduardo Bravo and Magdalena Blanco-Molina

CONTENTS

8.1 Introduction 97
8.2 Business Model 98
    8.2.1 Choosing a Viable Business Model 98
    8.2.2 Autologous Cell Therapy Business Model 98
8.3 Regulatory Issues 100
8.4 Process and Logistics of Manufacturing Autologous Cell Therapies 101
    8.4.1 Production Process 101
    8.4.2 Quality Control 102
    8.4.3 Logistics 103
8.5 Autologous Cell Therapy Market 103
    8.5.1 Clinical Trials 103
    8.5.2 Current Treatments and Future Prospects 103
    8.5.3 Autologous Cell Companies 104
8.6 Case Study: Cellerix 108
    8.6.1 Product Pipeline 109
    8.6.2 Ontaril Production 111
    8.6.3 Ontaril Regulatory Strategy 112
    8.6.4 Ontaril Commercial Strategy 112
8.7 Analysis and Conclusions 112
    8.7.1 Challenges for Development of Autologous Therapies 113
References 114

8.1 INTRODUCTION

Autologous cell therapy (ACT) is the use of a cellular medicine that contains the patient's own cells. The advantages of such approach include the minimization of risks from systemic immunological reactions, bioincompatibility, and disease transmission. To date, this form of therapy has been used successfully to bioengineer skin substitutes, counteract chronic inflammation, treat burns, and improve postoperative healing (Kazmi et al. 2009) and exhibits exciting potential as a treatment for heart failure and other cardiovascular diseases (Fine et al. 2008).

Since the bioengineering revolution over the past decades, ACT has become a rapidly evolving field. ACT is involves various technologies and disciplines ranging from cellular and molecular biology to virology, although the main interest in this chapter focuses on cell-based therapies. The use of ACT was reported as early as 1985 when autologous conjunctival transplants were used for the restoration of damaged ocular surfaces (Weise et al. 1985). The majority of products currently in clinical trial stages are autologous (Markson and Hill 2009).

In this chapter we will review the main subjects and challenges regarding autologous cell therapies, including the manufacturing process, the regulatory issues, and the market potential.

8.2 BUSINESS MODEL

8.2.1 Choosing a Viable Business Model

Different business models are required for the successful commercialization of cell-based therapies because of the complexity and nature of these products. When deciding on an appropriate model, important distinctions (Sleigh 2008) include:

- Type of cell therapy (autologous versus allogeneic)
- Complexity of product (cells only versus combination cell)
- Duration of therapy required (chronic treatment versus single treatment)

Autologous and allogeneic products should follow different business models due to their different characteristics such as different ways of manufacturing, harvesting of donor tissues, scalability and manipulation time required before implanting cells back into patients. In addition, it is important to consider that autologous therapy products are suitable only for a small patient base and allogeneic products are applicable for larger patient groups. Table 8.1 illustrates four possible business models for commercializing cell-based therapies.

Model A is not an achievable business model today because autologous therapies are not scalable or low cost (reagents needed for cell manipulation are expensive and qualified personnel are needed). Model B is the only feasible business model for an autologous therapy nowadays. Given the high cost of these therapies, these products must target areas of high unmet need and demonstrate higher efficacy compared to standard treatments to achieve reimbursement.

Model C is the large pharma model: low costs due to scalable manufacturing and "off the shelf" products that are easily delivered to a patient. No cell therapies exist in this space yet. Model D is the biotech company model with products targeting high unmet needs that can tolerate higher costs because they show good efficacy and so, they are reimbursable.

8.2.2 Autologous Cell Therapy Business Model

Table 8.2 illustrates a typical flowchart for the manufacturing of an autologous cellular medicine-based model.

TABLE 8.1
Four Possible Business Models for Commercializing Cell-Based Therapies

	Large Indication (High Incidence)	**Niche Indication (Small Incidence)**
Autologous (Personalized medicine)	Not commercially viable (A) • Extraordinary therapeutic benefit needed to compete with low cost therapies. • Manufacturing not scalable due to high costs. • High risk of substitution. • Low barriers to competitive entry.	Autologous model (B) • Orphan indications with no alternative therapies. • Manufacturing not scalable, but profitable therapy if superior efficacy demonstrated. • Key opinion leaders really important.
Allogeneic (Off-the-shelf)	Big Pharma model (C) • Low costs, scalable manufacturing. • Off-the-shelf cell therapies → potential competitors to biological drugs. • Possible premium prices due to efficacy (not palliative therapies).	Biotech company model (D) • Efficacious therapy that targets populations with high unmet needs. • Moderation of costs as scalability is reached. • Products targeting several indications.

TABLE 8.2
Autologous Cellular Medicine Manufacturing Process

MANUFACTURING PROCESS OF AN AUTOLOGOUS CELLULAR MEDICINE

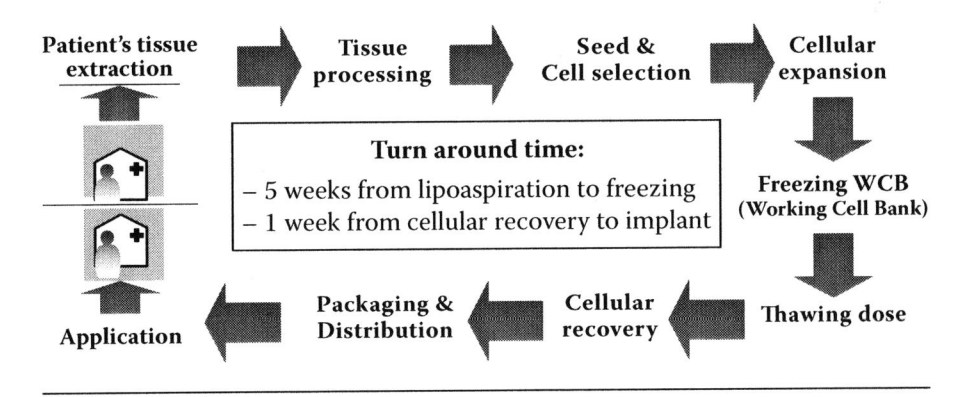

Patient's tissue extraction → Tissue processing → Seed & Cell selection → Cellular expansion → Freezing WCB (Working Cell Bank) → Thawing dose → Cellular recovery → Packaging & Distribution → Application

Turn around time:
− 5 weeks from lipoaspiration to freezing
− 1 week from cellular recovery to implant

8.3 REGULATORY ISSUES

Somatic cell therapy falls within the requirements for Advanced Therapy Medicinal Products (ATMPs; see Chapter 19 by Bravery). In order to efficiently develop ATMPs and assure product accessibility to patients, a consistent regulatory framework is an essential prerequisite along with available guidance documents (technical requirements and guidelines). It is also desirable to have tools for earlier patient access as well as regulatory assistance for applicants from proper authorities (Garcia-Olmo et al. 2007a and b). A harmonized regulatory framework guarantees a high level of health protection for the patients and is crucial to (1) harmonizing market access and ensuring the free movement of these products by establishing a tailored and comprehensive regulatory framework for their authorization, supervision, and post-authorization vigilance and (2) fostering the competitiveness of undertaking operations in this field.

Advanced therapies are based on new and complex manufacturing processes. The heterogeneity and particularities of ATMPs must be taken into account for their preclinical and clinical development. Clinical development of cell-based products might be associated with some special requisites such as proof of concept, cell kinetics, and issues related to appropriate study design and unique therapeutic goals (Garcia-Olmo et al. 2007a and b).

The regulatory framework for ATMPs was established by Regulation 1394/2007 covering advanced therapy medicinal products. The regulation was designed to ensure the free movement of these medicines within the European Union (EU), to facilitate their access to the EU market, and to foster the competitiveness of European pharmaceutical companies in the field, while guaranteeing the highest level of health protection for patients. This regulation also establishes the new expert Committee on Advanced Therapies (CAT; Garcia-Olmo et al. 2007a and b).

Some of the key topics identified for cell-based products by the EMA (European Medicines Agency) in its guidelines for human cell-based medicinal products (EMA/CHMP/CPWP/323774/2005) are the following:

- Development, manufacturing, and quality control: manufacturing process, in-process controls, and release and stability tests
- Non-clinical aspects: need for biodistribution studies, cell viability, proliferation, level and rate of differentiation, and duration of in vivo function
- Clinical aspects: peculiarities in dose finding, risk stratification, and special pharmacovigilance issues (including traceability)

Gene and cell therapies have been classified as medicinal products and are regulated as such in the EU. The current legislation covering autologous cell therapy in Europe is mainly based on the existence of three directives: 2003/63/EC, 2001/20/EC, and 2004/23/EC regarding the European Community's code relating to medicinal products for human use, clinical trials, and cells and tissues, respectively.

Additionally, the principle of a compulsory community marketing authorization has been established for gene therapy medicinal products and somatic cell therapy medicinal products resulting from any biotechnology processes referred to in the Annex to Regulation 726/2004/EC. The application for marketing authorization (MA) of these products must be made to the centralized procedure submitted to EMA. At present, the presentation of the MA application dossier for cell therapy products must fulfill the same administrative and scientific requirements as for any other medicinal product, as laid down in the legislation.

Directive 2001/83/EC further requires that medicinal products must be prepared according to Good Manufacturing Practice (GMP) requirements. Requirements for production and testing of cell-based products should be as strict as those for traditional medicinal products, but production specifications and test methods should be tailored specifically to these products.

In the EU guidelines, these directives apply to both autologous and allogeneic products. In the U.S., the Food & Drug Administration (FDA) issued two specific guidance documents for autologous cell products, indicating the importance of such therapies (U.S. FDA 1996, 1997).

8.4 PROCESS AND LOGISTICS OF MANUFACTURING AUTOLOGOUS CELL THERAPIES

Cell therapy products must match the safety, purity, potency, and identity criteria that apply to conventional drugs (FACT-JACIE). Even if current technologies allow for the complete testing of all of these parameters, two additional and important features of cell therapy products exert major impacts on the production process:

- Short shelf life (sometimes even shorter than the time required to obtain final results of release testing, for example, for sterility).
- Inability to survive a terminal sterilization step prior to patient infusion. As a consequence, the manufacturing process must be carried out aseptically and special emphasis must be given to the in-process quality controls imposed on the production environment, raw materials, and manipulation to ensure product safety.

8.4.1 PRODUCTION PROCESS

The production process should be robust enough to be reproducible and maximize the chances of success regardless of the initial tissue quality, especially in the case of autologous processes in which patient co-morbidities can alter the quality of the tissue (see Chapter 14 by Preti). The process should be validated and reproducible to ensure that the quality of the final product is always within pre-established specifications (Garcia-Olmo et al. 2007a and b).

Novel methodologies applicable to cell therapy product manufacturing have been developed and validated to ensure standardized and controlled processes. Examples include cell selection or separation devices (e.g., magnetic beads, density centrifugation) and culture systems (e.g., flasks, cell factories, bioreactors, etc.); see Garcia-Olmo et al. 2007a and b.

8.4.2 Quality Control

Cell products are manufactured in accordance with GMP principles and guidelines that require the implementation of a pharmaceutical quality assurance (QA) system GMP guidelines.* Prior to the manufacture of clinical batches, procedures must be in place to ensure proper production supervision as described in the GMP guidelines, including programs for product manufacturing, quality control (QC), and quality assurance (QA). Manufacturing and distribution sites must be audited by local regulatory authorities and demonstrate their compliance with these guidelines (see Chapter 14 by Preti).

Each batch of a biological product requires testing prior to release, in accordance with official compendia of standards such as the European Pharmacopeia (Eur. Ph) or the United States Pharmacopeia (USP). However, the methods described in these compendia are not necessarily adaptable to the special nature of cell therapy products. Novel and alternative methods are required to deal with the special characteristics of cell products under development.

Facility characteristics—Facilities should be designed and built so as to prevent any kind of contamination or mixing of products, facilitate the flow of materials, and minimize unnecessary traffic of personnel. A cell therapy manufacturing plant should comply with the following requirements: a good ventilation scheme, pressure control, and maintenance of strict relative humidity and temperature in processing areas to eliminate particles in suspension and minimize the chances of cross contamination via airbone microbial agents. An appropriate cleaning regimen with validated disinfectants to remove most surface microbial contamination and HEPA filtration of the air pulsed into clean rooms is essential.

Personnel—Cell therapy processes are labor intensive because of the characteristics of the product manipulation and the potential risk of product alteration. Therefore, the operators should be specifically trained in adequate gowning and compliance with proper procedures in clean rooms to reduce human-generated particle contamination.

Autologous sourcing of raw material—Segregation of raw material is a key element in autologous treatment manufacture. All efforts must be made to dedicate raw materials to one single patient and avoid their re-use for another. (Products available solely to the patient should be labeled "For autologous use only.") The segregation strategy also needs to be applied during manufacturing. The manipulation of materials belonging to different patients can occur in the same aseptic cabinet only if separated by a strict line clearance (cleaning included). When possible, the option should be incubation of cells within closed systems (Garcia-Olmo et al. 2007a and b). Among all of these requirements, it is crucial that the identity of the donor should be clearly indicated by clerical and tracking procedures.

---

* http://www.ich.org/cache/compo/363-272-1.html.

8.4.3 Logistics

Tissue source—Patient-derived cells or tissues from different sources (blood, bone marrow, adipose tissue) should be shipped urgently to the manufacturing facility to be processed as soon as possible. Cell origin traceability is essential in autologous therapies.

Distribution and Transportation—The delivery of cell therapies is complex and requires extensive product control and monitoring. In order to succeed, the commercialization of cell therapies will need to overcome several weaknesses held by the major carriers, including the lack of a true point-to-point chain of control; non-controlled x-rays and inspection; no guarantee of package orientation, handling, or storage conditions; and in many cases, no standard operating procedures. A successful transportation network for the cell therapy industry would require several key components, to handle the cellular materials from collection through manufacturing and delivery back to the patient, at each transfer point:

- Secure chain of control and custody
- Standard operating procedures in all phases of transit
- A highly specialized and trained air and ground courier network
- Quality assurance

Shelf life—One of the main issues of working with cell therapies is their short cell life making logistics an important issue. In the case of an autologous therapy, it is essential to ensure communication between company–hospital–patient so that the patient is ready for the cell therapy when it arrives at the hospital.

Storage—The storage of cell therapy products is another challenging issue. Cell therapies are living products whose efficacy depends upon viability. Cryopreservation is usually the method of choice for long-term storage. With the rapid emergence of cell therapy and tissue engineering, the successful application of cryopreservation for storage is becoming increasingly critical. Conventional cryopreservation protocols often result in significant cell loss (greater than 50% in many cases). This loss is detrimental, as is the associated loss in quality of the cell product. A loss of this scale translates to loss in product value and in the case of a cell therapy product, uncertainty about the therapeutic dose.

8.5 AUTOLOGOUS CELL THERAPY MARKET

8.5.1 Clinical Trials

There are currently 605 ongoing clinical trials on autologous cell therapies at different stages of completion worldwide. Table 8.3 summarizes these ongoing clinical trials.

8.5.2 Current Treatments and Future Prospects

The scope of autologous cell therapy is vast and has immense therapeutic potential. This includes treatment for limb ischemia (Napoli et al. 2005), bone marrow

TABLE 8.3
Cellerix's Ongoing Clinical Trials of Autologous Cell Therapies as of December 2009

NUMBER OF CLINICAL TRIALS ONGOING (DECEMBER 2009) ON AUTOLOGOUS CELL THERAPIES

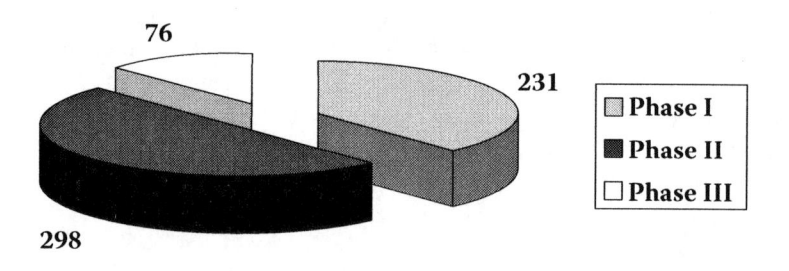

- Phase I: 76
- Phase II: 231
- Phase III: 298

transplant for haematological disorders (Locasciulli et al. 2007), ischemic stroke (Dharmasaroja 2009), ischemic heart disease (Perin 2004), autologous hematopoietic cells as targets for gene transfer to treat various blood disorders (Sadelain et al. 2004), and autoimmune diseases (Tyndall and Daileker 2005); the use of autologous macrophages to treat spinal cord injury (Assina et al. 2008), cell therapy with autologous lymphocytes to treat various cancers (Gervais et al. 2005), or autologous adipose-derived stem cells to treat fistulas (Garcia-Olmo et al. 2009).

Advances in the potential of stem cells to induce tissue regeneration and revascularization have increased interest for its use in the treatment of cardiovascular diseases (Orlic et al. 2003). The injection of autologous mononuclear bone marrow cells into areas of ischemic myocardium has been shown to improve the condition of a patient during Phase I clinical trials (Tayyareci et al. 2008). The treatment of neurodegenerative diseases has also been an avenue of research for autologous therapies. Alzheimer's disease (Sugaya 2005), multiple sclerosis (Mancardi 2009), and Parkinson's disease are among the most studied indications for these therapies. Different types of autologous treatments exist that seek to improve wound healing, counteract chronic inflammation, and aid wound closure. Various medical conditions have benefited from such modes of therapy including the management of burns, pressure ulcers, postoperative healing, and various skin disorders (Edmonds et al. 2009).

8.5.3 Autologous Cell Companies

A variety of different approaches are being used to develop autologous stem cell therapies. In this context, we can find a number of companies developing autologous cell therapies using different cell sources, harvesting, and expansion methods (www.medtrack.com).

Aastrom Biosciences Inc., U.S. (www.aastrom.com)—Aastrom is engaged in the development of autologous cell products for the regeneration of human tissue.

Its proprietary tissue repair cell (TRC) technology involves the use of a patient's own cells to manufacture products to treat a range of chronic diseases and serious injuries affecting vascular, bone, cardiac, and neural tissues.

Advanced Cell Technology (ACT), U.S. (www.advancedcell.com)—ACT focuses on applying stem cell technology in the field of regenerative medicine to bring effective, patient-specific therapies to the bedside. Its product folio includes the Myoblast Program, an autologous adult stem cell therapy for the treatment of heart disease; the Retinal Pigmented Epithelial Cell Program for the treatment of eye diseases including macular degeneration; and the Hemangioblast (HG) Cell Program for cardiovascular disease, stroke, and cancer.

Azellon Ltd, U.K. (www.azellon-ltd.com)—Azellon is engaged in R&D and commercialization of an adult autologous stem cell technology that has shown great promise in vitro for healing meniscal tears. The company's technology is based on harvesting adult stem cells from the iliac crest, expanding them under GMP conditions, and seeding the increased population of these autologous cells into a membrane. Once implanted, these new cells lead to a strong and long-term repair without removing any tissue.

BioHeart Inc., U.S. (www.bioheartinc.com)—This company focuses on the discovery, development, and commercialization of autologous cell therapies to treat chronic and acute heart damage. Its lead clinical candidate, MyoCell, is a cell therapy product in which myoblasts isolated from a patient's thigh are expanded using a proprietary process and injected into the scar tissue of the heart (infarct area) via a needle-injection catheter. The myoblasts then engraft, form new muscle, and enhance pumping of the heart.

Brainstorm Cell Therapeutics Inc., U.S. (www.brainstorm-cell.com)—This biotech company is developing stem cell technologies to provide treatments for incurable neurodegenerative diseases. The company is focused on developing neurotrophic factor cells (NTFs) from a patient's bone marrow to treat Parkinson's disease, amyotrophic lateral sclerosis (ALS), and spinal cord injuries. Its core technology is NurOwn™ for autologous transplantation to treat the three diseases cited above.

Capricor Inc., U.S. (www.capricor.com)—Capricor is engaged in the discovery and development of human cardiac stem cell therapeutics to support or replace cells damaged or lost due to chronic and acute heart disease. Capricor's proprietary technology is based upon harvesting heart cells from adult human biopsies and clinically manufacturing an autologous heart stem cell therapy. The company is specializing in the development of human cardiac stem cell technology using cells derived from adult heart tissue. It uses a proprietary process to isolate, purify, and expand rare candidate stem cells found in adult human tissue.

Cardio3 Biosciences SA, Belgium (www.cardio3bio.com)—Cardio3's focus is on regenerative therapies for the treatment of chronic class II or III heart failure. Its C-Cure™ is a bio-therapeutic product consisting of autologous adult stem cells guided in vitro to the cardiac lineage (cardiopoietic cells) before implantation in a failing heart.

Cellerix SA, Spain (www.cellerix.com)—Cellerix is a biopharmaceutical company developing treatments based on adipose-derived adult stem cells. Its pipeline covers three main therapeutic areas: gastroenterology, dermatology, and immune-based diseases. Cellerix's product portfolio includes Ontaril®, an injectable autologous suspension that contains stem cells for treating complex perianal fistulas of cryptoglandular origin and associated with Crohn's disease; Cx501, a chimeric skin equivalent containing autologous and allogeneic components for epidermolysis bullosa; allogeneic Cx601 for perianal fistula associated with Crohn's disease; Cx611 for the treatment of autoimmune and inflammatory diseases; and a Treg platform (Cx911) targeting autoimmune diseases.

Entest Biomedical Inc., U.S. (www.entestbio.com)—This biotech company conducts research and development on therapeutic solutions for traumatic brain injury using autologous adipose-derived stem cells. Its R&D activities include stem cell therapies, diabetes testing, and medical devices. The company also focuses on isolating hematopoietic stem cells from peripheral blood and bone marrow for differentiation into T cells to generate new populations of immune cells from a patient's own stem cells to rejuvenate weak or injured immune systems.

Gamida Cell Ltd., Israel (www.gamida-cell.com)—Gamida specializes in the clinical development of hematopoietic stem cell therapeutics for the treatment of cancer as well as future regenerative cell-based medicines to treat cancer, cardiac disease, and peripheral vascular disease. The company is developing an autologous bone marrow product, CardioCure, expanded ex vivo to treat patients after acute myocardial infarction.

Genzyme Corporation, U.S. (www.genzyme.com)—This is a biopharma operation engaged in discovering and developing products and services to improve the lives of patients with debilitating diseases. Genzyme's research platforms include protein and antibody therapies, polymers and small molecule therapies, cell and gene therapies, and biomaterials. Its lead product in the stem cell field is Carticel containing autologous cultured chondrocytes and indicated for the repair of symptomatic cartilage defects of the femoral condyle caused by acute or repetitive trauma.

Innovacell Biotechnologie, Austria (www.innovacell.at)—Its mission is to restore the natural function of damaged and defective muscle tissue using the body's own healing mechanisms. The company specializes in the research and production of autologous human muscle stem cells (myoblasts) and their regeneration potential and is examining further uses of regeneration using autologous muscle cells in several projects, particularly in the fields of urology and orthopedics.

Karocell Tissue Engineering AB, Sweden (www.karocell.com)—Karocell is developing methods to repair and regenerate autologous tissue. It integrates engineered tissue with a patient's own tissue. This can be done either by culturing cells in vitro for autotransplant or by guiding cells in vivo to regeneration. The company's R&D areas include breast tissue, fat, cartilage, chronic ulcers, matrices, culturing methods, and cell applications. Its core operations are production of autologous cells and management of skin and cell banks.

MedCell Bioscience Ltd., U.K. (www.medcell.eu)—This biotech focuses on stem cell technologies to provide musculoskeletal regenerative therapeutics via cell therapy, cell expansion, and tissue scaffolds that are the cornerstones of regenerative medicine. Its technologies are: MS-ten (autologous stem cell treatments for tendon and ligament injuries); NovaPod (cell expansion bioprocessing systems);

and TheraGlass (binary sol–gel glass technologies for the treatment of tendon and ligament injuries using adult stem cells).

NuPotential Inc., U.S. (www.nupotentialinc.com)—It specializes in producing de-differentiated cell lines suitable for re-differentiation into multiple lineages and somatic cell nuclear transfers. NuPotential has developed a cell reprogramming platform that utilizes small molecule effectors to produce renewable sources of autologous reprogrammed pluripotent stem cells (RePSCs). Its epigenetics-based small molecule approach to regulate cell reprogramming and restore cell differentiation potential does not rely on nuclear transfer. The pathways and molecules identified are being used to create new cell lines and a unique family of antibodies and assays.

Opexa Therapeutics Inc., U.S. (www.opexatherapeutics.com)—This company is engaged in the development of patient-specific cellular therapies to treat autoimmune diseases such as multiple sclerosis (MS) and diabetes. Opexa has two proprietary therapeutic platforms: T cell therapy and autologous stem cell therapy. It has developed a proprietary adult stem cell technology to produce monocyte-derived stem cells (MDSCs) from blood. These MDSCs are derived from a patient's monocytes, expanded in laboratories, and then administered to the same patient. The company's product pipeline includes Tovaxin, a novel T cell vaccine indicated for the treatment of multiple sclerosis that is specifically tailored to a patient's disease profile, and an autologous T cell vaccine based on T cell vaccination technology to treat rheumatoid arthritis.

Stempeutics Research Pvt. Ltd., India (www.stempeutics.com)—Stempeutics engages in research in the areas of biology of mesenchymal stem cells, human embryonic stem cells, a cardiovascular research program, diabetes research, and neurobiology. Its focus is to carry out research in bone marrow derived mesenchymal stem cells to be used for both autologous and allogeneic therapeutic treatment of various degenerative diseases.

T2 Cure GmbH, Germany (www.t2cure.com)—T2 is developing regenerative therapeutics from a patient's own stem cells. It is also engaged in the development of novel progenitor cell-based regenerative therapeutics to provide new treatment options to patients suffering from cardiac and peripheral vascular diseases such as myocardial infarction and peripheral artery occlusive disease. T2's therapeutics consists of autologous progenitor cells prepared from patient bone marrow. These progenitor cells have the potential to induce repair processes in ischemic cardiac or peripheral tissues.

TCA Cellular Therapy LLC, U.S. (www.tcacellulartherapy.com)—Its specialty is R&D and marketing of adult stem cell therapies for serious and life-threatening diseases. TCA's therapeutic areas include spinal cord injuries, Parkinson's disease, ALS, and cardiac diseases. The company focuses on regenerative medicine that has the potential to revolutionize medicine by being able to produce specialized cells obtained from the same patient for use in a wide-array of novel and important therapies.

Tengion Inc., U.S. (www.tengion.com)—Focused on developing neo-organs and tissues derived from a patient's own cells, its mission is to transform the lives of patients in need of organ transplants or augmentations by developing autologous neo-organs and tissues that harness the regenerative power of a patient's own healthy tissue, enabling it to restore vital functions. The company's platform supports a broad potential product portfolio that addresses multiple tissues and organs in the genitourinary and cardiovascular systems.

TiGenix NV, Belgium (www.tigenix.com)—ToGenix develops local treatments for damaged and osteoarthritic joints. Its product portfolio is focused on treatment of articular cartilage defects; development of simpler surgical procedures to correct articular cartilage defects; repair of cartilage defects in patients with early osteoarthritis; and repair of traumatic lesions of the meniscus. The company's lead product, ChondroCelect, is an improved autologous chondrocyte implantation product for the structural and functional repair of cartilage defects of the knee. ChondroCelect was approved by the European Commission in October 2009 as the first advanced therapy medicinal product.

Tristem Corporation, U.K. (www.tristemcorp.com)—Tristem is engaged in R&D of its retro-differentiation technology used to create stem cells from mature adult cells. Tristem's technology offers the potential to prepare large quantities of stem cells from a patient's own blood by the process of retro-differentiation. Its technology provides large quantities of stem cells which are by definition DNA matched and therefore bypasses the political and ethical issues related to therapeutic cloning. Tristem is evaluating the clinical potential of its retro-differentiation stem cell technology for aplastic anemia, thalassemia, and leukemia—diseases that ordinarily require bone marrow transplants (BMTs).

TxCell, France (www.txcell.com)—The development of new therapies for inflammatory and autoimmune diseases is the focus of TxCell. Its technology platform centers on Tr1 cells that are considered immunosuppressive tools for delivering cytokines to inflammatory tissues. The process begins with the collection of a patient's whole blood followed by isolation, growth, cloning, and high-scale amplification of antigen-specific Tr1 cells. After injection of the cells to the same patient, a specific antigen is administered to activate the cells. This antigen is designed for a specific delivery at the targeted site of inflammation. The encounter of Tr1 cells with their specific antigen will induce an activation of the Tr1 cells followed by a local secretion of immunosuppressive cytokines.

8.6 CASE STUDY: CELLERIX

Cellerix is a product-focused biopharmaceutical company developing innovative medicines based on cell therapy from adult origin. Cellerix has generated a portfolio of autologous (patient-derived), allogeneic (donor-derived), and chimeric (patient- and donor-derived) products for well-characterized orphan indications of unmet clinical needs. Its vision is to become a world leading biopharmaceutical company in the area of cell therapy; to achieve this, Cellerix has designed its strategy upon four pillars:

- **Commercialize products in Europe and distribute beyond Europe via license partners**—Cellerix plans to become a fully integrated biopharmaceutical company with R&D, manufacturing, and sales and marketing capabilities for its products in Europe, while ensuring development and commercialization beyond Europe through licensing partners.

TABLE 8.4
Cellerix's Acquisition Strategy

SUMMARY OF CELLERIX'S STRATEGY

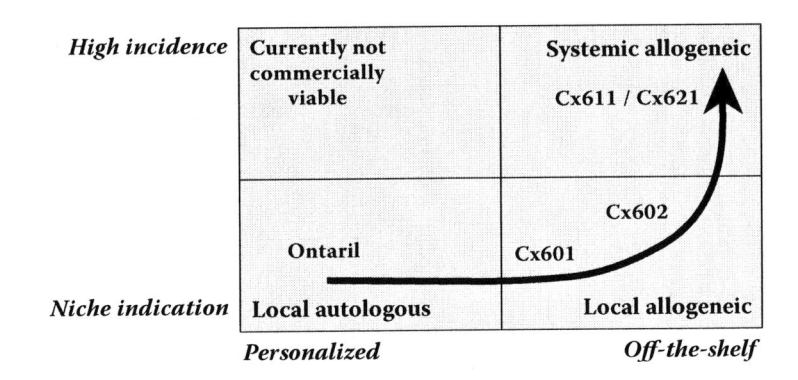

	Currently not commercially viable	Systemic allogeneic Cx611 / Cx621
High incidence		
	Ontaril	Cx602 Cx601
Niche indication	Local autologous	Local allogeneic
	Personalized	*Off-the-shelf*

- **Use revenues from autologous and chimeric products as near-term drivers of growth.** Cellerix's lead products (Ontaril and Cx501) are positioned to receive regulatory approval in 2010 or 2011 and reach the European market shortly thereafter, which will put the company in a position to profit from near-term revenues.
- **Catalyze transformation into a global biopharmaceutical company through allogeneic products**—The allogeneic products currently in development will allow Cellerix to target much larger markets in the autoimmune disease area. Products have already been tested in animal models with promising results for three indications.
- **Pursue strategic acquisitions of other cell therapy assets to build a dominant presence**—Cellerix is actively evaluating cell therapy assets with solid intellectual property positions and proofs of concept that can complement its existing pipeline. Its strategy is outlined in Table 8.4.

8.6.1 Product Pipeline

Cellerix is focusing its near-term development on products that address well characterized markets with sufficient unmet medical needs. In the longer term, Cellerix will aim to use its allogeneic technology to address much larger markets and thereby catalyze the company into a global biopharmaceutical business. Table 8.5 presents an overview of its proprietary product pipeline. It has two product candidates that are well advanced along clinical development paths.

Cx401—an autologous product based on expanded adult stem cells derived from human adipose tissue (eASCs)—showed excellent efficacy and safety results in Phase I and II clinical trials and has already finalized a Phase III clinical trial investigating its potential for treating complex perianal fistulae of cryptoglandular origin

TABLE 8.5
Overview of Cellerix's Proprietary Product Pipeline

AN OVERVIEW OF THE COMPANY'S PROPRIETARY PRODUCT PIPELINE

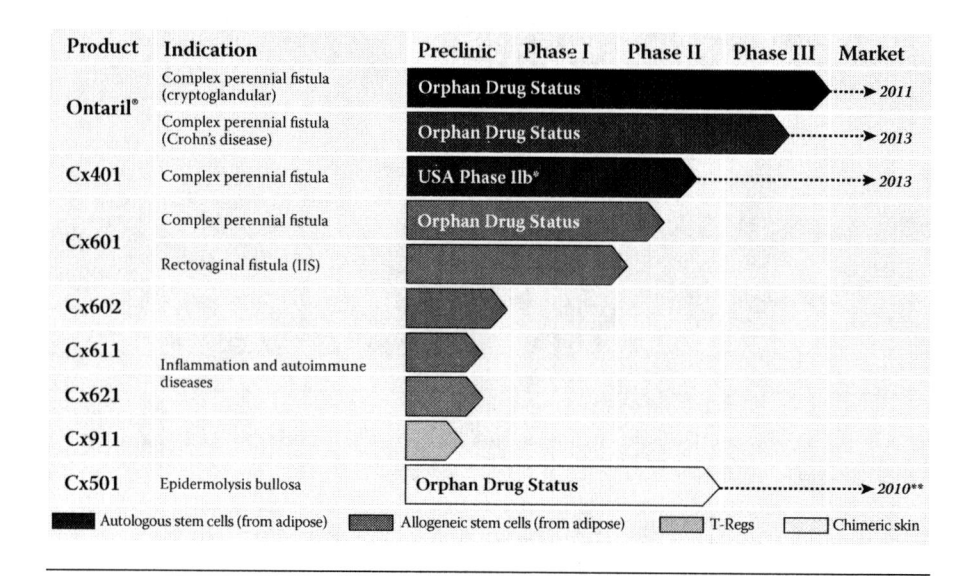

Product	Indication	Preclinic	Phase I	Phase II	Phase III	Market
Ontaril®	Complex perennial fistula (cryptoglandular)		Orphan Drug Status			▶ 2011
	Complex perennial fistula (Crohn's disease)		Orphan Drug Status			▶ 2013
Cx401	Complex perennial fistula		USA Phase IIb*			▶ 2013
Cx601	Complex perennial fistula		Orphan Drug Status			
	Rectovaginal fistula (IIS)					
Cx602						
Cx611	Inflammation and autoimmune diseases					
Cx621						
Cx911						
Cx501	Epidermolysis bullosa	Orphan Drug Status				▶ 2010**

■ Autologous stem cells (from adipose)   ▨ Allogeneic stem cells (from adipose)   ▢ T-Regs   ▢ Chimeric skin

(FATT 1). In 2008, Cellerix started a second Phase III trial to study the treatment potential of Cx401 in patients with complex perianal fistulae suffering from Crohn's disease (FATT 2). Cx401 will be marketed in Europe under the Ontaril trade name. The North American rights to Cx401 have been licensed to Axcan Pharma, a multinational pharmaceutical company focused on gastroenterology.

Cx501 is an innovative chimeric skin product that utilizes expanded autologous keratinocytes and allogeneic fibroblasts to create a skin-equivalent that is less susceptible to patient rejection and represents a long-term skinreplacement therapy for patients suffering from recessive dystrophic epidermolysis bullosa (RDEB). Cx501 is in Phase II trials and could potentially gain approval under the regulatory regimen of "exceptional circumstances" in 2012.

Based on the relatively rare prevalence, severe nature, and lack of effective treatments of the therapeutic indications for which Cellerix's clinical stage product candidates are being developed, Ontaril, Cx501, and Cx601 have received orphan drug status from the EMA.

Cellerix is focusing its long-term strategy on the development of a new generation of products based on expanded allogeneic (donor-derived) stem cells supported by the results of Cellerix' research, which showed eASCs from adipose tissue not to cause rejection or have any other adverse effects. In this area, the company's pipeline is currently composed of four programs.

Autologous Cell-Based Products

Cx601 and Cx602 involve expanded allogeneic adipose-derived stem cells for local delivery. Cx601 has already been administered to humans (with no related side effects) under compassionate use regulation in Spain. Based on this data, Cellerix advanced this program into the clinic in two Phase I/II clinical trials: (1) a Phase I/IIa open label trial in 24 perianal fistula–Crohn's disease patients, scheduled for completion by the end of 2009 and (2) a Phase I/IIa investigator-initiated study (IIS) of 10 rectovaginal fistulasis patients, approved by the Spanish Medical Regulatory Agency (AEMPS). Orphan drug designation for Cx601 was achieved in record time.

Cx611 and Cx621 involve expanded allogeneic adipose-derived stem cells for systemic delivery. These products had already been validated in various inflammatory and autoimmune animal models during their preclinical development. These allogeneic programs will allow Cellerix to target much larger markets in the inflammatory and autoimmune disease areas with off-the-shelf products.

Cx911 is based on selective induction of regulatory T cells (Tregs) for adoptive immunotherapy. It has been demonstrated that hASCs exert their immunosuppressive activity via the selective induction of Tregs with suppressor capacity over self peripheral blood mononuclear cells (PBMCs) in a dose-dependent manner (Gonzalez-Rey 2009). Based on this, Cellerix initiated the development of an immunotherapy platform to target autoimmune diseases. The development of this immunotherapy platform will allow Cellerix to target larger markets in the autoimmune disease area of personalized medicine.

With these products, Cellerix aims to broaden the applicability of its current product candidates, increase manufacturing efficiencies, and target larger autoimmune and inflammatory diseases.

8.6.2 Ontaril Production

Cellerix's proprietary eASC preparation and application technology begins with liposuction of a patient. The adipose tissue (up to 300 ml) collected in the procedure must be delivered to the company at a controlled temperature within 48 hours of the liposuction.

During production, the company performs the required quality control procedures related to microbiological sterility, cell viability, potency, identity, purity, and morphology. Batches are identified by codes and traceability is assured throughout the process. Upon completion of the production process the cells are thawed, recovered in culture, trypsinized, washed, and resuspended in the final formulation. They are then placed into a vial and packaged at Cellerix's facilities to be delivered to the hospital in a special temperature-controlled unit that maintains product temperature at a constant 18 to 25°C to maintain stability. The finished product must be delivered and administered to the patient within 48 hours from the time of packaging. Shelf life is defined as the period during which cell viability is maintained at a value greater than 70%. The shelf life of Ontaril is currently 48 hours. Cellerix is working on several internal and external projects to increase shelf life viability and is confident that shelf life will be substantially increased before Ontaril reaches the market.

8.6.3 ONTARIL REGULATORY STRATEGY

Cellerix plans to submit a filing to the European Medicines Agency (EMA) based on the results of Fistula Advanced Therapy Trial (FATT)-1. The submission will be reviewed via the centralized procedure (mandatory for advanced therapies) and lead to approval of the drug for the treatment of complex anal fistula of predominant cryptoglandular origin (non-Crohn's indication) throughout Europe. If approved, Ontaril will be the first stem cell therapy product to be reviewed via this procedure. Ontaril's orphan drug designation has allowed the company to take extensive scientific advice from the EMA on the design and conduct of preclinical and clinical studies as well as chemistry, manufacturing, and control (CMC), allowing the company to be confident that it is well positioned to gain regulatory approval by mid-2011.

Following approval in non-Crohn's patients, Cellerix expects to seek a label expansion to include patients with Crohn's disease, based on results of FATT-2, a Phase III trial for the indication of complex perianal fistulae derived from Crohn's disease.

8.6.4 ONTARIL COMMERCIAL STRATEGY

Cellerix's strategy is to market Ontaril directly to European hospitals and out-license the rights to the product in other regions. In line with this, Cellerix has granted North American rights for Cx401 to Axcan Pharma.

Given that the therapeutic indication for which Ontaril is being developed has a relatively low incidence rate, with a significant unmet medical need, that the specialists are readily identifiable (GI specialists), and that the number of competing products is low, the company believes that a relatively small central marketing structure with a sales force of specialist hospital representatives should be sufficient to successfully introduce Ontaril to gastroenterologists in European hospitals.

8.7 ANALYSIS AND CONCLUSIONS

It is known that the enormous potential of stem cells allows them to treat diseases and to cure them. However, the challenges and costs associated with achieving those aims are extremely high. The challenges and opportunities for stem cell science are (Sleigh 2008):

- Ethics
- Funding
- Intellectual property
- Manufacturing
- Regulation
- Reimbursement
- Delivery

Stem cell transplant has been used to treat hematological malignancies for the last four decades and now serves as a routine and life saving procedure, although problems may occur with finding a suitable donor. The challenge facing cell therapy today arises

Autologous Cell-Based Products

from the fact that when cells are modified in any way (e.g., expanded) or are intended for heterologous use, they are considered by regulatory bodies as pharmaceutical products (drugs) and the cell therapy industry must face the challenge of developing these cells in line with the regulatory requirements for pharmaceutical products.

Stem cells from different sources (adult, fetal, placental tissue, umbilical cord, or embryonic stem cells) are being developed for a huge range of conditions and several autologous and allogeneic treatments are under clinical development. The use of a patient's own tissue introduces logistical challenges and additional cost, making these therapies less attractive from a commercial view, although safer for patients. However, the use of allogeneic products may introduce additional complications due to rejection of the cells by the immune system. Progress in the development of stem cell-based therapies has been slower than expected because of the many challenges in ensuring high quality manufacturing procedures, undertaking complex clinical trials, and handling ethical issues.

The market for stem cell-based therapies is in its infancy, with few therapies derived from stem-cells having reached the market. It has been anticipated to be a massive growth sector, dependent on products currently in development proving they are safe and effective and barring any major adverse events that put a hold on the development of the field.

The reality of autologous therapies today is that they hold little promise to expand regenerative medicine into large markets in which current standards of care are simple, albeit not curative solutions (e.g., diabetes, congestive heart failure) (Smith 2008).

Exciting times in the healthcare industry remain ahead as regenerative medicine products provide the promise of a cure for diseases that have long been thought as "incurable" (e.g., Parkinson's disease, multiple sclerosis, chronic obstructive pulmonary disease, etc.).

8.7.1 Challenges for Development of Autologous Therapies

Challenges differ depending on whether the therapy is autologous or allogeneic as they have different characteristics as shown in Table 8.6. The major challenges to be addressed by companies developing autologous therapies include:

- Manufacturing: improving expansion and manipulation methods
- Logistics: improving shelf life
- Demonstrating safety and efficacy

Furthermore, several questions must be answered in order to successfully develop an autologous cell therapy:

- Competitive barriers to entry: Do I have a novel technology protected by intellectual property rights?
- Barrier to exit: Will potential entrants see limited opportunity for multiple competitors?

TABLE 8.6
Characteristics of Autologous and Allogeneic Therapies

Autologous	Allogeneic
Pros:	**Pros:**
• No adventitious agents or karyotype QC needed	• Patient benefits because no invasive harvesting required; logistics issues simplified; medicine ready to use when needed
• Safety, fast track to clinic	
• Eventual rejection minimized	
Cons:	• Universal (standardized off-the-shelf) product for treating many patients
• Single batch production per patient	• Easier development of robust, high quality manufacturing procedures; reduction of costs
• Not suitable for acute conditions (in which treatment delays cannot be tolerated); several weeks are needed to have product ready to be implanted in patient	
	Cons:
• Stem cell harvesting methods invasive to patients	• Potential issues with immune rejection, depending on cell type and source
• Difficult logistics	• Unresolved regulatory issues
• High manufacturing costs	• More QC needed
• Difficulty in designing preclinical experimental package	

- Costs: Can the company scale up the manufacturing process to reduce the costs?
- Indication: Does the target indication have a sufficient unmet need for which costly therapeutics appear viable?

REFERENCES

Assina, R., Sankar, T., and Theodore N. et al., 2008. Activated autologous macrophage implantation in a large-animal model of spinal cord injury. *Neurosurg Focus* 25: E3.

Dharmasaroja, P. 2009. Bone marrow-derived mesenchymal stem cells for the treatment of ischemic stroke. *J Clin Neurosci* 16: 12–20.

Edmonds, M., Bates, M., and Doxford, M. et al., 2000. New treatments in ulcer healing and wound infection. *Diabetes Metab Res Rev* 16: 51–54.

European Pharmacopeia (Eur.Ph). http://www.edqm.eu/site/page_584.php

FACT-JACIE, 2007. *International Standards for Cellular Therapy, Product Collection, Processing and Administration*, 3rd ed. Omaha, NE: Foundation for the Accreditation of Cellular Therapy.

Fine, G. C., Liao, R., and Sohn, R. L., 2008. Cell therapy for cardiac repair. *Panminerva Med* 50: 129–137.

Garcia-Olmo, D., Garcia-Verdugo, J. M., and Alemany, J. et al., 2007. In *Cell Therapy: Regulatory Aspects of Human Somatic Cell Therapy*, Pascual, M., Ed. New York: McGraw Hill, pp. 71–79.

Garcia-Olmo, D., Garcia-Verdugo, J. M., and Alemany, J. et al., 2007. In *Cell Therapy: Production of Cell-Based Medicine for Somatic Cell Therapy*, Fernández-Miguel, G. and Bovy, T., Eds. New York: McGraw Hill, pp. 34–41.

Garcia-Olmo, D., Herreros, D., and Pascual, I. et al., 2009. Expanded adipose-derived stem cells for the treatment of complex perianal fistula: a phase II clinical trial. *Dis. Colon Rectum* 51: 79–86.

Gervais, A., Bouet-Toussaint, F., and Toutirais, O. et al., 2005. Ex vivo expansion of antitumor cytotoxic lymphocytes with tumor-associated antigen-loaded dendritic cells. *Anticancer Res* 25: 2177–2185.

González-Rey, E., González, M. A., and Varela, N. et al., 2009. Human adipose-derived mesenchymal stem cells reduce inflammatory and T cell responses and induce regulatory T cells in vitro in rheumatoid arthritis. *Ann Rheum Dis* 69: 241–248.

Good Manufacturing Practice Guidelines. 6a. cGMP (U.S., FDA): http//www.fda.gov/cdrh/devadvice/32.html; 6b. EU GMP (Europe, EMA): http://ec.europa.eu/enterprise/pharmaceuticals/eudralex/homev4.html.

Kazmi, B., Inglefield, C. J., and Lewis, M. P., 2009. Autologous cell therapy: current treatments and future prospects. *Wound Res* 9.

Locasciulli, A., Oneto, R., and Bacigalupo, A. et al., 2007. Outcome of patients with acquired aplastic anemia given first line bone marrow transplantation or immunosuppressive treatment in the last decade: a report from the European Group for Blood and Marrow Transplantation (EBMT). *Haematologica* 92: 11–18.

Mancardi, G. 2009. Further data on autologous haemopoietic stem cell transplantation in multiple sclerosis. *Lancet Neuro* 8: 219–221.

Markson, G. and Hill, E., 2009. How to commercialise cell therapy. Presentation from Cambridge Consultants, 17th February 2009.

Medtrack database: www.medtrack.com

Napoli, C., Williams-Ignarro, S., and de Nigris, F. et al., 2005. Beneficial effects of concurrent autologous bone marrow cell therapy and metabolic intervention in ischemia-induced angiogenesis in the mouse hindlimb. *Proc Natl Acad Sci USA* 102: 17202–17206.

Orlic, D., Kajstura, J., and Chimenti, S. et al. 2003. Bone marrow stem cells regenerate infarcted myocardium. *Pediatr Transplant* 7: 86–88.

Perin, E. 2004. Transendocardial injection of autologous mononuclear bone marrow cells in end-stage ischemic heart failure patients: one-year follow-up. *Int J Cardiol* 95: 45–46.

Sadelain, M., Rivella, S., and Lisowski, L. et al. 2004. Globin gene transfer for treatment of the beta-thalassemias and sickle cell disease. *Best Pract Res Clin Haematol* 17: 517–534.

Sleigh, S., 2008. Stem cell-based therapeutic delivery: challenges and opportunities. *Pharma-Vision.*

Smith, D. M., 2008. Successful business models for cell-based therapies. *World Stem Cell Report.*

Sugaya, K. 2005. Possible use of autologous stem cell therapies for Alzheimer's disease. *Curr Alzheimer Res* 2: 367–376.

Tayyareci, Y., Umman, B., and Sezer, M. et al., 2008. Intracoronary autologous bone marrow-derived stem cell transplantation in patients with ischemic cardiomyopathy: results of 18-month follow-up. *Turk Kardiyol Dern Ars* 36: 519–529.

Tyndall, A. and Daikeler, T., 2005. Autologous hematopoietic stem cell transplantation for autoimmune diseases. *Acta Haematol* 114: 239–247.

U.S. Food & Drug Administration, Guidance on Applications for Products Comprised of Living Autologous Cells Manipulated Ex Vivo and Intended for Structural Repair or Reconstruction. (FDA Notice, 61 *Fed Reg* 26523, May 28, 1996).

U.S. Food & Drug Administration, Guidance for the Submission of Chemistry, Manufacturing and Controls Information, and Establishment Description for Autologous Somatic Cell Therapy Products, 1997. (http://www.fda.gov/downloads/BiologicsBloodVaccines/GuidanceComplianceRegulatoryInformation/Guidances/CellularandGeneTherapy/UCM169060.txt.4.)

U.S. Pharmacopeia (USP). http://www.usp.org

Weise, R. A., Mannis, M. J., and Vastine, D. W. et al., 1985. Conjunctival transplantation: autologous and homologous grafts. *Arch. Ophthalmol* 103: 1736–1740.

9 Business Models for Cord Blood

Suzanne M. Watt

CONTENTS

9.1 Introduction 117
9.2 Current Uses of Cord Blood 118
    9.2.1 Umbilical Cord Blood for Allogeneic Hematopoietic Stem Cell Transplantation and for Treating Genetic Disorders 118
    9.2.2 Umbilical Cord Blood for Autologous Transplants: Research and Developing Future Therapies 121
9.3 Business Models of Cord Blood Banks 122
    9.3.1 Public Model 122
    9.3.2 Private Model 123
    9.3.3 Hybrid Public–Private Model 125
9.4 Ethical Issues 126
References 128

9.1 INTRODUCTION

Hematopoietic stem cell transplantation has been carried out for over four decades, yet its optimal clinical use in patient treatments is still in development. Three sources of hematopoietic stem cells are used for transplantation. These originate from (1) bone marrow, (2 peripheral blood following mobilization from the bone marrow with granulocyte colony-stimulating factor (G-CSF) with or without other mobilizing factors such as the CXCR4 antagonist, Perixafor, and (3) umbilical cord blood (Fesler et al. 2009, Donahue et al. 2009, Fruehauf et al. 2009, Watt and Forde 2008). All three sources of hematopoietic stem cells may be used for autologous and related or unrelated allogeneic transplants and, with some differences between cell sources, are used routinely for treating malignant and non-malignant disorders of the blood. Their use is dependent on a number of factors including individualized patient need, the hematological or other disease being treated, patient age, disease status, conditioning, availability of a donation or donor, and the country where the transplant takes place.

In the United Kingdom, autologous hematopoietic stem cell transplants constitute approximately 60% of all hematopoietic stem cell transplants. The remaining 40% account for allogeneic transplants, the cells for which may be sourced from related or unrelated donors. Worldwide and depending on ethnicity, on average, around 30% of patients requiring matched allografts have human leukocyte antigen (HLA) identical sibling donors available. The remainder will receive transplants of bone marrow, mobilized peripheral blood, or cord blood sourced from unrelated donors, with the choice of personalization of a donor for each patient, depending on such issues as those listed above. Umbilical cord blood is becoming increasingly important in these unrelated donor transplant settings, principally for treating hematological malignancies and certain inherited genetic diseases. Such cord blood units are most often sourced from public not-for-profit cord blood banks for these purposes.

Over the past few years, there has been an explosion in private (for-profit) cord blood banking. Such banks are for-profit and, in most cases, bank cord blood units for future autologous transplants or for future allogeneic transplants for a donor's immediate family. These cord blood units are often procured and stored on the basis of future promises for their use in regenerative medicine or for future insurance in the event that the child will develop a hematological or non-hematological illness that can be treated by cord blood transplantation. More recently, a hybrid public–private cord blood bank model has been developed. The cord blood units are made available in part for public use and in part stored for future autologous transplants or transplants for a donor's immediate family, but a proportion of the proceeds from the donated cord blood may be made available for research on cord blood stem cells. These three business models of cord blood banking are described below, along with the ethical issues that collecting and banking cord blood and the business models raise.

9.2 CURRENT USES OF CORD BLOOD

9.2.1 Umbilical Cord Blood for Allogeneic Hematopoietic Stem Cell Transplantation and for Treating Genetic Disorders

While there are over 14 million bone marrow volunteer donors listed on registries worldwide, it has been estimated that a significant number of patients requiring unrelated hematopoietic stem cell transplants (15 to 40%) will not have suitable HLA-matched donors as many tissue types may be under-represented on donor registries. This, of course, varies with ethnic origin and it has been estimated (reviewed by Sullivan 2008, Smith and Wagner 2009) that unrelated bone marrow donors will be found for 75% or more of patients of Western European origin and only about 25% from ethnic minority groups within a particular country such as the U.K. Altruistic donations of cord blood have been advocated as appropriate sources of cells to make up this shortfall.

Table 9.1 summarizes indications for allogeneic hematopoietic stem cell transplants that also encompass the use of allogeneic cord blood. The table is based on a more detailed summary presented by Sullivan (2008) and also described by Watt (2010) in more detail. Among allogeneic donations, directed donations of cord blood represent a special case because they are sourced from sibling donors of a family member who may require a cord blood transplant (Reed et al. 2003, Smythe et al. 2007). These have been primarily used to treat hemoglobinopathies such as severe sickle cell disease and beta-thalassemia major, particularly where there is an identical HLA-match (Smythe et al. 2007, Locatelli et al. 2003).

Diseases treated with directed donations are generally at the request of a transplant clinician and/or referring obstetrician and may be selected using prenatal or preimplantation genetic testing and HLA typing (Smythe et al. 2007). The directed

TABLE 9.1
Allogeneic Hematopoietic Stem Cell Transplant Indications for Cord Blood*

Leukemias and Lymphomas
Very high risk acute lymphoblastic leukemia and acute myeloid leukemia
Acute lymphoblastic leukemia and acute myeloid leukemia relapses
Secondary treatment-related acute myeloid leukemia
Juvenile myelomonocytic leukemia
Chronic myeloid leukemia relapse
Chronic lymphocytic leukemia
Advanced state Hodgkin's disease
Hodgkin's disease relapse
Non-Hodgkin's lymphoma relapse

Myeloproliferative Diseases
Myelodysplastic syndrome
Multiple myeloma

Lymphohistiocytic Disorders
Langerhans cell histiocytosis relapse
Hemophagocytic lymphohistiocytosis

Bone Marrow Failure and Genetic Disorders
Congenital aplastic anemia
Severe aplastic anemia
Fanconi's anemia
Congenital anemias
Genetic metabolic disorders
Hemoglobinopathies
Immune deficiency syndromes

* See also Watt (2010).

cord blood units are often stored without volume reduction to reduce loss of cells and without the minimum volume criteria established for unrelated cord blood donations. The resulting matched directed sibling donor transplants generally demonstrate less graft-versus-host disease than matched unrelated donations (Smythe et al. 2007, Locatelli et al. 2003, Watt et al. 2009, Watt 2009, Watt and Contreras 2005, Watt (2010), and references therein). Compared to unrelated donations, the number of directed donations is unknown worldwide, but thought to be small. Donations collected and banked with the support of NHS Blood and Transplant in the U.K. represent 5 to 10% of allogeneic cord blood units collected and banked annually (Smythe et al. 2007, Watt et al. 2009).

The first successful allogeneic cord blood transplant was a direct donation conducted in 1988 for a child with Fanconi's anemia. The cells were obtained from an HLA-identical sibling (Gluckman et al. 1989) and the transplant recipient remains alive and well, having returned to full blood counts and full donor hematopoietic chimerism (reviewed in Gluckman 2009). Further transplant advances have developed since then (reviewed in Gluckman et al. 1989, Gluckman 2009). For unrelated transplants, key advances have included:

1. The collection from volunteer donors and the storage of unrelated cord blood units for transplantation. This technique commenced with the establishment of the New York Cord Blood Bank in 1993 (Rubenstein et al. 1993, Rubenstein et al. 1995).
2. The first unrelated cord blood transplant in a child took place in 1993 (Kurtzberg et al. 1996) followed by transplantation in an adult (Gluckman et al. 1997).
3. Double unrelated cord blood transplants. Smith and Wagner (2009) have reviewed double unrelated cord blood transplants in adults including those receiving non-myeloablative or reduced intensity conditioning (i.e., not completely destroying stem cells in the bone marrow with drug treatments prior to transplant) (Delaney et al. 2009). This innovation extended the age for such transplant recipients up to 75 years (Brunstein et al. 2007).
4. One or two cord blood units for transplantation were supplemented with ex vivo expanded or subset-selected cord blood units or third party cells, such as mesenchymal stem cells or CD34+ (CD34 is a marker of human blood cells that includes blood stem cells and their immediate progeny that generate all blood cells) cells from mobilized peripheral blood (reviewed in Fernandez 2009).
5. Gradually increasing improvements in defining the quality of the cord blood units stored for clinical transplantation, e.g., nucleated cell dose, CD34+ cell dose, cell viability, patient conditioning prior to transplant, time of administration of cord blood units (both manipulated and unmanipulated), HLA matching, disease type and status, and other factors (reviewed in Smith and Wagner 2009, Brunstein et al. 2007, Fernandez 2009, Rocha and Gluckman 2009, Querol et al. 2010, Watt 2010).

Cord blood donations have a number of advantages over bone marrow donors including (1) ready availability when collected and banked as an HLA-typed and viral pathogen-free or -tested hematopoietic stem cell source, compared to the 3 or 4 months required to secure a donation from a volunteer bone marrow donor, (2) the absence of donor attrition found with some bone marrow donors, and (3) the greater tolerance of one or two HLA loci mismatches without adversely affecting survival or increasing graft-versus-host disease in recipients (reviewed in Eapen et al. 2010, Smith and Wagner 2009, Watt and Contreras 2005). Disadvantages can include (1) restricted total nucleated and CD34+ cell numbers in a single cord blood unit, (2) delays in hematological reconstitution (the most significant predictors for successful cord blood transplant outcome are the cell dose infused and HLA match), and (3) the lack of availability of a back-up stem cell harvest for tandem transplants or of donor lymphocytes for subsequent infusions to treat infections or enhance graft-versus-tumor response (reviewed in Smith and Wagner 2009).

Smith and Wagner recently made recommendations for the use of unrelated single cord blood units in unrelated allogeneic hematopoietic stem cell transplantation after

Business Models for Cord Blood

reviewing the outcomes of a variety of retrospective studies (see Smith and Wagner 2009 and articles therein). These were:

1. Selection of hematopoietic stem cell source (cord blood, bone marrow, or mobilized peripheral blood) based on choice of the transplant physician to best suit individual patient need.
2. For malignant disease, 6/6 HLA-matched cord blood is the preferred choice for both children and adults provided cell dose is adequate.* The second choice is 8/8 HLA-matched unrelated bone marrow or peripheral blood stem cell donation or a 5/6 or 4/6 HLA-matched cord blood unit, since survival rates appear comparable. However, selection would also be based on cell dose, the future need for tandem transplants or donor lymphocyte infusions, and the urgency of the transplant.
3. For non-malignant diseases, Smith and Wagner (2009) consider the information available for making a solid recommendation is insufficient, although they suggest 8/8 that HLA bone marrow in children and adults or peripheral blood stem cell harvest in adults is the gold standard, with an adequate dose of 6/6 or 5/6 HLA-matched cord blood as the next choice should the former be unavailable.

9.2.2 Umbilical Cord Blood for Autologous Transplants: Research and Developing Future Therapies

Umbilical cord blood is often cited as a source of hematopoietic stem cells for autologous transplants or for use in hematological and regenerative medicine research in order to develop or assess its potential for future therapies (Harris 2009). As discussed above, stem cells from umbilical cord blood are currently used in the regeneration of blood supplies from diseased bone marrow most often in an allograft setting and hence may be considered a first generation regenerative medicine product.

Sullivan (2008) reviewed the potential use of autologous cord blood units for transplantation and cites no value for them in treating leukemias, bone marrow failure syndromes (except possibly acquired aplastic anemia), or genetic and metabolic disorders (except possibly as sources of HLA-matched cells for gene therapy procedures); and limited value for treating diseases that affect the bone marrow. Furthermore, he predicts their clinical use in the first 20 years of life based on known indications would be no greater than 1 in 15,000.

Others predict lower rates (Samuel et al. 2008). This compares with usage of c.≥2% of unrelated cord blood units banked in many established public cord blood banks such as the NHS Cord Blood Bank in the U.K. (Watt et al. 2009), and a much higher relative usage for directed donations where they are supported via not-for-profit organizations such as the Stem Cell and Immunotherapy Department of NHS Blood and Transplant (Smythe et al. 2007).

Harris described the increased storage of autologous cord blood units (estimates are around 500,000 to 1,000,000 worldwide (Harris 2009, Samuel et al. 2008)) and

---

* Cell dose adequacy is defined as total nucleated cell count/kg recipient body weight of >3 × 10^7 for 6/6, >4 × 10^7 for 5/6, and >5 × 10^7 for 4/6 HLA matched cord blood units (Smith and Wagner 2009).

their increasing use in regenerative medicine clinical trials (Harris 2009). Areas of use described by Harris and colleagues include their therapeutic use for stroke, bone and joint repair, cornea and skin wound repair, treating juvenile Type I diabetes, cerebral palsy, traumatic brain injury, and hearing loss (Harris 2009), as well as cardiac repair (Copeland et al. 2009). Despite the promise of cord blood in treating these disorders, the composition of cord blood in terms of stem cells or their progeny and the mechanisms by which these treatments might demonstrate therapeutic effects remain matters of some debate. It is well established that the umbilical cord and its blood contain hematopoietic stem cells; that at least a proportion of cord blood donations contains mesenchymal stem cells (MSCs) capable at least of adipogenic and osteogenic differentiation; and that almost all cord blood units contain endothelial precursor cells (Zhang et al. 2009, Watt 2010). Other types such as unrestricted somatic stem cells (USSCs) that have fibroblastoid appearances similar to MSCs and which lack adipogenic potential, but are multipotent (Kögler et al. 2006). Ratajczak's group identified very small embryonic-like (VSEL) stem cells in umbilical cord blood (Zuba-Surma et al. 2009). Although cells expressing ectodermal, endodermal, and mesodermal lineage markers appear to be derived in vitro from these cells, the generation of such cells in vivo into functional lineages and in sufficient numbers in the clinical setting of regenerative medicine is unclear.

In other studies, infused cells from hematopoietic tissues appear to mediate their effects principally through paracrine or anti-inflammatory mechanisms by which they may enhance revascularization, promote the proliferation of endogenous stem cells, limit scarring or apoptosis, or modulate tissue remodeling without contributing substantial numbers of cells to the replacement of damaged tissue such as cardiac muscle, neuronal cells, and others (Martin-Rendon et al. 2008a, Martin-Rendon et al. 2009, Brunskill et al. 2009, Martin-Rendon et al. 2008b, Hill et al. 2009, Watt 2010). The use of autologous cord blood for regenerative medicine is currently speculative and is the focus of a great deal of basic research.

9.3 BUSINESS MODELS OF CORD BLOOD BANKS

There are essentially three business models of cord blood banks established on the basis of differentially treating known indications of disease or for more speculative regenerative medicine treatments. These are the public, the private, and the hybrid public–private models. They all play different roles in cord blood banking and provision and are described and compared below.

9.3.1 Public Model

The public model for cord blood banking is based on the findings that unrelated or directed related donations (for defined use) of cord blood can be successfully collected, stored, and used as alternatives to bone marrow or mobilized peripheral blood stem cells to treat a series of hematological disorders (described in Table 9.1) and some non-hematological disorders. These are not-for-profit cord blood banks (Sullivan 2008, Watt 2010, and references therein). Under this model, unrelated or directed related donations are collected altruistically (no payment by the donor family). The unrelated altruistic donations are made available usually worldwide for public use through international registries. They are of particular importance for patients requiring hematopoietic stem cell transplants and who lack matched related or unrelated bone marrow donors. This often occurs for ethnic minority recipients within particular countries, especially in relation to indigenous, mixed race, and immigrant populations. Therefore, a proportion of public cord blood banks at least will aim to preferentially collect from ethnic minority donors for this reason, given the cost associated with collecting and long-term banking of cord blood units. Directed donations are in contrast collected and banked where there is a sibling or family member with a disease that may be treated by cord blood transplantation (Smythe et al. 2007). These directed cord blood donations are not made available on national or international registries but are banked for future use by the donor family for a particular accepted disease indication.

Public banks may be funded through government grants in aid supplemented with funding related to the supply of cord blood units for transplantation because they are unlikely to ever achieve self-sufficiency in terms of full cost recovery by supplying cord blood units to transplant centers. Examples of government support for public cord blood banking are the funding by the U.S. Congress to increase the number of cord blood units collected and banked in the U.S. to a maximum of 150,000 units, and in the U.K. where there is current government support in England to increase the cord blood units stored to 20,000, and with black and ethnic minority donors being particularly targeted (see Sullivan 2008, Watt et al. 2009, Howard et al. 2008, Howard et al. 2005).

Cord blood donations for public use may also be collected via banks supported through charitable funding (Querol et al. 2009a and b). A model for cord blood banking in the U.K. suggested that the target for cord blood units banked publicly should be on the order of 50,000 high quality units to meet local demand for around 36% of ethnic minority patients and 80% of total patients if unbiased collections are made, or around 70% for both patient groups if collections are biased toward ethnic minority groups (Querol et al. 2009a and b). A cost benefit analysis by Howard for increasing the U.S. public cord blood units banked from 50,000 to 150,000 estimated the cost per life year gained was about $55,873 and this was considered by the U.S. Congress to be cost effective (Howard et al. 2005).

There are now around 400,000 cryopreserved cord blood units from unrelated donors stored worldwide in over 50 public cord blood banks. Over 14,000 of these units have been used for transplantation (http://www.bmdw.org; http://wmda.org; Kurtzberg 2009, Gluckman and Rocha 2009). As indicated above, the numbers of cord blood units sourced from related and/or sibling donors (often referred to as directed donations) and collected by not-for-profit organizations for specific clinical purposes and for use of family members are unknown. There are thus well defined and accepted ethical, scientific, and clinical based evidence or arguments for maintaining such public cord blood banks nationally.

9.3.2 PRIVATE MODEL

Private cord blood banks have been established worldwide, except in Spain, Dubai, and Italy, to store cord blood units collected from and for use by the donor or their family members (see (Sullivan 2008) and reference therein for exceptions in Australia and New Zealand). These are for-profit companies. Although there is and has been a great deal of hope and hype for the use of cord blood stem cell therapies to treat a variety of degenerative diseases, cord blood units are currently in routine use to treat both children and adults principally with hematological malignancies, and for children to treat certain cancers (e.g., neuroblastoma) and certain inherited diseases (e.g., inherited anemias, metabolic storage diseases). Sullivan (2008) reported 150 private cord blood banks listed worldwide on the Parent's Guide to Cord Blood Website (http://parentsguidecordblood.org) and further details which transplants might benefit from allogeneic or autologous sources of cord blood. Autologous cord blood transplants for example were potentially indicated only for advanced stage or relapsed Hodgkin's disease, relapsed non-Hodgkin's lymphoma (NHL), multiple myeloma, severe aplastic anemia, and stages 4 and 5 neuroblastomas, although these could also be treated with autografts of bone marrow or mobilized peripheral blood hematopoietic stem cells (Sullivan 2008). Sullivan (2008) also predicted that no more than and most likely substantially fewer than 1 in every 150,000 private cord blood donations are likely to be used for personal or family use for established treatments for diseases at present, although other estimates have been significantly lower (Samuel et al. 2008).

The current cost for private storage of cord blood units ranges upward from about $900 in the U.S. or £1120 in the U.K. for 20 years' storage (http://parentsguidecordblood.org). Using the assumptions that the cost of procurement, banking, and storage for 20 years is $3,620 and that there is a 0.04 and 0.07% chance, respectively, of a child requiring an autograft or allograft, Kaimal et al. (2009) estimated that private cord blood banking costs an additional $1,374,246 per life year gained. They further conclude that cost effectiveness for private donations would only be reached if the cost is less than $262 or the likelihood of using cord blood in pediatric transplants is around 1 in 110. The challenges faced in terms of private cord blood banking have been reviewed by Fisk and Atun (2008) and include:

1. The very small chance that a cord blood unit will be used for present indications during its 20-year storage time and the speculative nature of future therapies for degenerative diseases that develop often in older age
2. Insufficient information about the type and quality of and storage conditions for cord blood stem cells required for future non-hematological therapies
3. The logistics of collecting high quality cord blood units from maternity units across the country and the regulatory issues that must be met in procuring cells for human use (Watt 2010)
4. The risk of corporate failure

Thornley et al. (2009) audited 93 pediatric transplant physicians (72% of those approached) on their usage of privately banked cord blood units in the U.S. and Canada. Of only 50 cord blood units transplanted, 18% were for autologous use, while the remainder were allografts for related family members. The audit concluded that the majority of pediatric transplant physicians would not advocate private cord blood banking at present in the absence of a known indication requiring hematopoietic stem cell transplantation. Further conclusions on the usage of cord blood for routine transplants in this audit included:

1. Pre-emptive transplants outnumbered prophylactic transplants.
2. Allografts exceeded autografts.
3. Allografts were preferable to autografts for acute lymphoblastic leukemia (ALL) because of their potential graft-versus-leukemia effect and the possible presence of the leukemic clone in the cord blood sample originating from the patient.
4. The probability of using an autograft of cord blood for relapsed ALL in the absence of a suitable allograft ranges from 12 to 17 per million during the first 20 years of life in the U.S. and Canada.
5. More than half (55%) of responding transplant physicians would use autologous cord blood for treating severe aplastic anemia in place of an unrelated donor if a sibling donor were unavailable, although the estimated use was predicted as fewer than 3 per million for all age groups in the U.S.
6. More than half (55%) of responding transplant physicians would use autologous cord blood for high risk neuroblastoma provided cell dosage and viability met the appropriate standard of quality, although the estimated usage was predicted around 3 to 5 per million in U.S. children below 15 years of age.

Another anonymous survey by Fox et al. (2007) of 325 pregnant women indicates that many women are ill informed about banking cord blood units for private or family use. Harris (2008 and 2009) has, in contrast, argued for the use of autologous cord blood in transplantation for a spectrum of non-hematological diseases. In 2008, he cited two clinical trials in children using autologous cord blood, one for treating Type I diabetes led by the University of Florida and one for treating cerebral palsy led by Duke University, with 23 and 50 respective transplants recorded. To date there are only anecdotal reports of improvement in some patients in non-randomized clinical trials (Harris 2009) and details of outcomes are awaited. Thus, if outcomes appear positive, a great deal of information and future double-blinded randomized clinical trials will be required to demonstrate the efficacy of these treatments.

As discussed below, there is generally a lack of support by professional organizations for private for-profit autologous cord blood banking at the present time at least in the U.K. and U.S., particularly in the provision of "biological insurance" for treating low-risk individuals and for speculative future treatments. This may change with future research.

9.3.3 Hybrid Public–Private Model

In 2007, the Virgin Health Group announced the opening of a hybrid public–private cord blood bank (Fisk and Atun 2008, Silversides 2007, Mayor 2007a and b, Polymenidis and Patrinos 2008). In this model, 80% of the cord blood is stored for public use and 20% is stored and available for private use. The publicly available cord blood remains available to the donor and his or her family if not used or subsequently required (Virgin's predicted chance of use by an unrelated recipient is less than 1%). The costs for collection and storage are similar to those incurred by private cord blood banks (about £1500 for 20 years' storage). However 50% of the proceeds may be donated for research on cord blood stem cells. Key elements to this approach which differ from many private banks include advising donors of the very low chance of use of their cord blood units and the unlikely use of autologous cord blood to treat such blood diseases as leukemia. The challenges faced by such hybrid cord blood banks have been addressed in detail by Fisk and Atun (2008) and may be summarized as follows:

1. The potential risk of compromising the quality of a cord blood unit in terms of reduced total nucleated cell numbers when only 80% of the cord blood unit is made publicly available. The selection of a cord blood unit from a public registry by transplant physicians is in part based on sufficient cell numbers for transplantation to achieve the best outcomes (Querol et al. 2010, Watt 2010).
2. Insufficient cells in a private donation sample to provide stem cells (hematopoietic, mesenchymal, and endothelial stem and/or progenitor cells) for current and future potential therapies.
3. The speculative nature of future therapies and the unknown or undefined storage requirements for cord blood stem cells other than hematopoietic.
4. The logistics of collecting from many maternity units across the country and the regulatory issues that must be met, for example, in Europe, at least in procuring cells for human use (Watt 2010).
5. Storage for 20 years, when regenerative therapies may be sought beyond the fifth and sixth decades of life.
6. The risk of corporate failure.

Evidence of the sustainability and viability of such hybrid banks is awaited due to their recent establishment, but the current indication is that uptake by the public may be slow, at least in the U.K. Recently, a not-for-profit public cord blood bank in the U.S., CORD:USE, has opened a cord blood bank for private family donations of cord blood so that these are of the same quality as the altruistic cord blood donations banked.

9.4 ETHICAL ISSUES

A variety of ethical issues relating to cord blood banking need consideration. They were reviewed by the Institute of Medicine of the National Academies of Science in the U.S. at the request of Congress some years ago (Meyer et al. 2005). Particular ethical issues and recommendations have included issues of fully informed consent, testing and follow-up of donations, traceabilty, security, and confidentiality.

These ethical considerations are, among other issues, also covered by standards for cord blood banking provided by international accreditation agencies such as FACT (Foundation for the Accreditation of Cellular Therapies)-Netcord, the U.S. Food & Drug Administration (FDA), and the European Commission Directives and Legislation. The European Commission requirements are enacted into law in the U.K. and cord blood banks are licensed via the Human Tissue Authority (Austin et al. 2008, Watt 2010). In England and Wales, the Human Tissue Act 2004 makes consent the key element for the legal removal, storage, and use of tissue or cells from living donors.

The European Union Tissues and Cells Directives 2004/23/EC, 2006/17/EC, and 2006/86/EC benchmarked European standards that must be met by activities involving cells and tissues for human use. The directives were implemented into U.K. law in 2007 through the Human Tissue (Quality and Safety for Human Application) Regulations 2007 and now cover the procurement, testing, processing, storage, distribution, and import or export of cells and tissues for human application, whether through public, private, or hybrid cord blood bank models.

Robust international standards are important for cord blood banking because as many as 40 to 75% of cord blood units may cross international borders. Errors in identification and tracking of cord blood units can potentially lead to serious outcomes for patients (Querol 2009, McCullough et al. 2009, Watt 2010), and a cord blood bank's key objective must be to offer patients the best chances of cures.

International FACT-Netcord standards have sought to provide a regulatory framework that crosses international boundaries. More recently, these have been supplemented with FDA standards (Watt 2010). Research and clinical experience developed over the past four decades have concentrated principally on optimizing cord blood transplants for hematological disorders. However, the types of cells and the quality standards for banking cord blood units required for potential future applications in regenerative medicine or future cell therapies, if these do become available, are unfortunately ill-defined at the present time.

Other ethical issues relate principally to concerns related to banking autologous cord blood units in cases where (1) payment is made by the donor family for cord blood storage essentially as future insurance against diseases that have a low risk of developing or for treating a variety of disorders for which treatments are not developed or are far from being optimized and (2) where the consenting donor mother may be ill informed or lack adequate understanding of the scientific and clinical issues (Sullivan 2008, Fisk and Atun 2008, Fisk et al. 2005, Ballen et al. 2008).

Many professional organizations, such as the Royal College of Obstetrics and Gynecologists and the Royal College of Midwives in the U.K., the American Society for Blood and Marrow Transplant (ASBMT), the World Marrow Donor Association (WMDA), the American Academy of Pediatrics, and the European Commission's Ethics in Science and New Technologies group have not at the present time supported autologous cord blood banking for individuals considered at low risk of developing diseases that may be treatable by autologous cord blood transplantation and their concerns have been extensively reviewed, although these organizations will review the evidence base for use at regular intervals (Sullivan 2008, Fisk and Atun 2008, Fisk et al. 2005, Ballen et al. 2008). It is, however, paramount that the best treatment is offered to the patient to provide the best chance of a cure and this must be based on robust scientific and clinical research and evidence-based medicine.

REFERENCES

Austin, E. B., M. Guttridge, D. Pamphilon, and S. M. Watt 2008. The role of blood services and regulatory bodies in stem cell transplantation. *Vox Sang* 94: 6–0017.

Ballen, K. K., J. N. Barker, S. K. Stewart et al. 2008 American Society of Blood and Marrow Transplantation. Collection and preservation of cord blood for personal use. *Bone Marrow Transplant* 14: 356–363.

Brunskill, S. J., C. J. Hyde, C. J. Doree et al. 2009. Stem cell treatment for acute myocardial infarction. *Eur J Heart Fail* 11: 887–896.

Brunstein, C. G., J. N. Barker, D. J. Weisdorf et al. 2007. Umbilical cord blood transplantation after non-myeloablative conditioning: impact on transplantation outcomes in 110 adults with hematologic disease. *Blood* 110: 3064–3070.

Copeland, N., D. Harris, and M. A. Gaballa 2009. Human umbilical cord blood stem cells, myocardial infarction and stroke. *Clin Med* 9: 342–345.

Delaney, C., J. A. Gutman, and F. R. Appelbaum 2009. Cord blood transplantation for haematological malignancies: conditioning regimens, double cord transplant and infectious complications. *Br J Haematol* 147: 207–216.

Donahue, R. E., P. Jin, A. C. Bonifacino et al. 2009. Perixafor (AMD3100) and granulocyte colony-stimulating factor (G-CSF) mobilize different CD34+ cell populations based on global gene and microRNA expression signatures. *Blood* 114: 2530–2541.

Eapen, M., V. Rocha, G. Sanz, et al. 2010. Effect of graft source on unrelated donor haemopoietic stem-cell transplantation in adults with acute leukaemia: a retrospective analysis. *Lancet Oncol* Jun 15. [Epub ahead of print]

Fernandez, M. N. 2009. Improving the outcome of cord blood transplantation: use of mobilized HSC and other cells from third party donors. *Br J Haematol* 147: 161–176.

Fesler, M. J., G. Kudva, and P. J. Petruska. 2009. Plerixafor and pegylated filgrastim: a case of safe and effective hematopoietic stem cell mobilization. *Bone Marrow Transplant* Dec 7. [Epub ahead of print]

Fisk, N. M., I. A. Roberts, R. Markwald, and V. Mironov 2005. Can routine commercial cord blood banking be scientifically and ethically justified? *PloS Med* 2: e44.

Fisk, N. and R. Atun 2008. Public–private partnership in cord blood banking. *Br Med. J* 366: 642–644.

Fox, N. S., C. Stevens, R. Ciubotariu et al. 2007. Umbilical cord blood collection: do patients really understand? *J Perinat Med* 35: 314–321.

Fruehauf, S., M. R. Veldwijk, and T. Seeger. 2009. A combination of granulocyte-colony-stimulating factor (G-CSF) and Plerixafor mobilizes more primitive peripheral blood progenitor cells than G-CSF alone: results of a European phase II study. *Cytotherapy* 11: 992–1001.

Gluckman, E. 2009. Ten years of cord blood transplantations: from bench to bedside. *Br J Haematol* 147: 192–199.

Gluckman, E., H. E. Broxmeyer, A. D. Auerbach et al. 1989. Hematopoietic reconstitution in a patient with Fanconi's anemia by means by umbilical cord blood from an HLA-identical sibling. *New Engl J Med* 321: 1174–1178.

Gluckman, E. and V. Rocha 2009. Cord blood transplantation: state of the art. *Haematologica* 94: 451–4.

Gluckman, E., V. Rocha, A. Boyer Chammard et al. 1997. Outcome of cord blood transplantation form related and unrelated donors. *New Engl J Med* 337: 373–381.

Harris, D. T. 2008. Cord blood stem cells: worth the investment. *Nat Rev Cancer* 8: 823–828.

Harris, D. T. 2009. Non-haematological uses of cord blood stem cells. *Br J Haematol* 147: 177–184.

Hill, A. J., I. Zwart, H. H. Tam et al. 2009. Human umbilical cord blood-derived mesenchymal stem cells do not differentiate into neural cell types or integrate into the retina after intravitreal grafting in neonatal rats. *Stem Cells Dev* 18: 399–409.

Howard, D. H., M. Maiers, C. Kollman et al. 2005. A cost-benefit analysis of increasing cord blood inventory levels. In *Cord Blood: Establishing a National Haematopoietic Stem Cell Bank Program*, Meyer, E. A. et al., Eds. Washington: National Academies Press, pp. 221–241.

Howard, D. H., D. Meltzer, C. Kollman, et al. 2008. Use of cost-effectiveness analysis to determine inventory size for a national cord blood bank. *Med Decis Making* 28: 243–253.

Kaimal, A. J., C. C. Smith, R. K. Laros et al. 2009. Cost-effectiveness of private umbilical cord blood banking. *Obstet Gynecol* 114: 848–855.

Kögler, G., S. Sensken, and P. Wernet 2006. Comparative generation and characterization of pluripotent unrestricted somatic stem cells with mesenchymal stem cells from human cord blood. *Exp Hematol* 34: 1589–1595.

Kurtzberg, J. 2009. Update on umbilical cord blood transplantation. *Curr Opin Pediatr* 21: 22–29.

Kurtzberg, J., M. Laughlin, M. L. Graham et al. 1996. Placental blood as a source of hematopoietic stem cells for transplantation into unrelated recipients. *New Engl J Med* 335: 157–66.

Locatelli, F., V. Rocha, W. Reed et al. 2003. Related umbilical cord blood transplantation in patients with thalassemia and sickle cell disease. *Blood* 101: 2137–2143.

Martin-Rendon, E., S. J. Brunskill, C. J. Hyde et al. 2008a. Autologous bone marrow stem cells to treat acute myocardial infarction: a systematic review. *Eur Heart J* 29: 1807–1818.

Martin-Rendon, E., J. A. Snowden, and S. M. Watt 2009. Stem cell-related therapies for vascular diseases. *Transfus Med* 19: 159–171.

Martin-Rendon, E., D. Sweeney, F. Lu et al. 2008b. 5-Azacytidine-treated human mesenchymal stem/progenitor cells derived from umbilical cord, cord blood and bone marrow do not generate cardiomyocytes in vitro at high frequencies. *Vox Sang* 95: 137–148.

Mayor, S. 2007a. World's first public-private cord blood bank launched in UK. *Br Med J* 334: 277.

Mayor, S. 2007b. World's first public-private cord blood bank launched in United Kingdom. *Br Med J* 334: 229.

McCullough, J., D. McKenna, D. Kadidlo et al. 2009. Mislabeled units of umbilical cord blood detected by a quality assurance program at the transplantation center. *Blood* 114: 1684–1688.

Meyer, E. A., K. Hanna, and K. Gebbie 2005. Ethical and legal issues. In *Cord Blood: Establishing a National Haematopoietic Stem Cell Bank Program*, Meyer, E. A. et al., Eds. Washington: National Academies Press, pp. 106–119.

Polymenidis, Z. and G. P. Patrinos 2008. Toward a hybrid model for the cryopreservation of umbilical cord blood stem cells. *Nat Rev Cancer* 8: 823.

Querol, S. 2009. A case of mistaken identity. *Blood* 114: 1459–1460.

Querol, S., P. Rubinstein, S. G. Marsh et al. 2009a. Cord blood banking: providing cord blood banking for a nation. *Br J Haematol* 147: 227–235.

Querol, S., G. J. Mufti, S. G. Marsh et al. 2009b. Cord blood stem cells for hematopoietic stem cell transplantation in the UK: how big should the bank be? *Haematologica* 94: 536–541.

Querol, S., S. G. Gomez, A. Pagliuca, et al. 2010. Quality rather than quantity: the cord blood bank dilemma. *Bone Marrow Transplant* 45: 970–978.

Reed, W., R. Smith, F. Dekovic et al. 2003. Comprehensive banking of sibling cord blood for children with malignant and non-malignant disease. *Blood* 101: 351–357.

Rocha, V. and E. Gluckman 2009. Improving outcomes of cord blood transplantation: HLA matching, cell dose and other graft- and transplantation-related factors. *Br J Haematol* 147: 262–274.

Rubenstein, P., L. Dobrila, L., R. D. Rosenfield et al. 1995. Processing and cryopreservation of placental/umbilical cord blood for unrelated bone marrow reconstitution. *Proc Natl Acad Sci USA* 92: 10119–10122.

Rubenstein, P., R. D. Rosenfield, J. W. Adamson et al. 1993. Stored placental blood for unrelated bone marrow reconstitution. *Blood* 81: 1679–1690.

Samuel, G. N., I. H. Kerridge, and T. A. O'Brien 2008. Umbilical cord blood banking: public good or private benefit? *Med J Aust* 188: 533–535.

Silversides, A. 2007. Interface of private and public faces proposed cord blood bank. *CMAJ* 177: 705–706.

Smith, A. R., J. E. Wagner 2009. Alternative haematopoietic stem cell sources for transplantation: place of umbilical cord blood. *Br J Haematol* 147: 246–261.

Smythe, J., S. Armitage, D. McDonald, et al. 2007. Directed sibling cord blood banking for transplantation: the ten-year experience in the National Blood Service in England. *Stem Cells* 25: 2087–2093.

Sullivan, M. J. 2008. Banking on cord blood stem cells. *Nat Rev Cancer* 8: 555–563.

Thornley, I., M. Eapen, L. Sung, et al. 2009. Private cord blood banking: experiences and views of pediatric hematopoietic cell transplantation physicians. *Pediatrics* 123: 1011–1017.

Watt, S. M. 2009. Realising the potential of cellular transplant therapies for NHS patients: the role of NHS blood and transplant. *NAPC Rev* Spring: 70–71.

Watt, S. M. 2010. Umbilical cord blood stem cell banking. In: *Comprehensive Biotechnology. 2nd Edition. Vol. 5. Medical Biotechnology and Healthcare.* Moo-Young, M. et al. Eds. Amsterdam: Elsevier Academic Press, (in press).

Watt, S. M., K. Coldwell, and J. Smythe 2009. Comparisons between related and unrelated cord blood collection and/or banking for transplantation or research: the UK NHS blood and transplant experience. In: *Clinical Use of Placenta, Amniotic Fluid, Umbilical Cord and its Contents: Current Medical Perspective.* (in press).

Watt, S. M. and M. Contreras 2005. Stem cell medicine: umbilical cord blood and its stem cell potential. *Semin Fetal Neonatal Med* 10: 209–220.

Watt, S. M. and S. P. Forde 2008. The central role of the chemokine receptor, CXCR4, in haemopoietic stem cell transplantation: will CXCR4 antagonists contribute to the treatment of blood disorders? *Vox Sang* 94: 18–32.

Watt, S. M., C-C. Su, and J. H-Y. Chan. 2010. The therapeutic potential of stem cells in umbilical cord and umbilical cord blood. *Taiwan Med J* (in press).

Zhang, Y., N. Fisher, S. E. Newey et al. 2009. The impact of proliferative potential of umbilical cord-derived endothelial progenitor cells and hypoxia on vascular tubule formation in vitro. *Stem Cells Dev* 18: 359–375.

Zuba-Surma, E. K., I. Klich, N. Greco et al. 2010. Optimization of isolation and further characterization of umbilical cord blood-derived very small embryonic/epiblast-like stem cells (VSELs). *Eur J Haematol* 84: 34–36.

10 Changing the Game of Drug Discovery

John Walker

CONTENTS

10.1 Groundbreaking Technology: The Promise and the Challenge 131
10.2 Answering a Productivity Crisis in Drug Discovery 132
10.3 Cell Therapy and Its Caveats 133
10.4 Foundations of a New Platform Technology 134
10.5 A New/Old Model for Pharmaceutical Partnerships 135
10.6 Proprietary Programs Driven by Dire Needs 136
10.7 Walking the Risk/Reward Tightrope 137
10.8 A Solid Intellectual Property Portfolio 138
10.9 Finance beyond Term Sheets 139
References 140

10.1 GROUNDBREAKING TECHNOLOGY: THE PROMISE AND THE CHALLENGE

Being on the ground floor of a scientific breakthrough—a truly game-changing technology—presents a challenge for those lucky enough to be involved. The enormous potential of cellular reprogramming, particularly human induced pluripotent stem (iPS) cells from tools for studying basic biology to toxicology predictors and from drug discovery to cell therapy is exciting, but the breadth of that potential presents so many opportunities that there is a temptation to try to address them all. Therefore, one strategy is to focus on the application of cellular reprogramming to drug discovery and development, so that real therapies can be brought to patients as quickly as possible.

Drug discovery has traditionally been highly dependent on surrogate systems, engineered animal models, and affinity assays that eventually led to human clinical studies—and much disappointment when the experimental results from these rather imperfect substitutes did not translate into reliable clinical practice. Now, with the promise of cellular reprogramming, we have the ability to put the patients at the front of the discovery cycle and use patient-derived assays throughout the discovery and preclinical development processes.

Shinya Yamanaka's 2006 paper on the creation of induced pluripotent stem cells from adult mouse cells was a watershed event for biology (Takahashi and Yamanaka 2006). The production of human iPS cells a short time later by Kazuhiro Sakurada of

Bayer Schering and subsequently by Yamanaka and others opened a new era in drug discovery. Like embryonic stem cells, iPS cells can be programmed to replicate most cells of the body and truly represent a new platform for the screening of drug candidates.

To accelerate the development of drug discovery opportunities, one strategy and business model is to achieve a set of partnerships relatively early in the evolution of a company in conjunction with the development of internal and focused proprietary programs. For example, joining forces with an established tool company to industrialize the iPS technology rapidly and broadly and collaborating with leading pharmaceutical companies to address both specific diseases and broad therapeutic categories constitute viable partnership models. In today's uncertain financial environment, such a partnering strategy should provide adequate capital for a company to advance iPS cells as a platform technology and bring drugs to the clinic without multiple rounds of dilutive equity offerings. Furthermore, sharing technology will allow a company to benefit from the expertise of the broader scientific community, bringing this groundbreaking technology more rapidly to market and new therapies more quickly to patients with unmet medical needs.

10.2 ANSWERING A PRODUCTIVITY CRISIS IN DRUG DISCOVERY

The common wisdom is that bringing a new drug to market currently takes on average about 10 years and $1 billion and it is estimated that 9 of 10 drug candidates fail in clinical studies. As bad as those numbers sound, those in the business know the real odds are much worse. The reality is probably closer to 1 candidate in 25 or 30 if preclinical failures are included, putting the actual cost at over $2.5 billion per approved new chemical entity (NCE). Clearly those odds are not sustainable (see Chapter 12 by Vertes).

Why are the numbers so bad? The answer is that we are witnessing the fundamental limitations of a discovery paradigm based on serendipitous finds like aspirin and opium, whose derivatives are still among the most commonly prescribed drugs, or based on the screening of vast chemical libraries against targets of unknown and unquantifiable relevance. The weaknesses of these random approaches are now painfully visible in the declining productivity of the pharmaceutical industry.

The early promise of biotechnology was to take drug discovery away from such shots in the dark to a model premised upon expanded knowledge of the molecular pathways of disease and the production of proteins, peptides, and antibodies that would bind with receptors of known structure. The success of Genentech and its kin is testimony to the validity of this approach, but it is limited to diseases with well documented causes. Too many diseases are the results of still unknown pathologies in inaccessible cells.

The truly surprising statistic should be the number of effective drugs developed despite the inability to harvest relevant cells from living patients and despite the inadequacy of animal models to predict the safety or efficacy of potential drugs in humans. Lacking a supply of cells from diseased human organs, research scientists perform in vitro screens against mouse fibroblasts or the ubiquitous Chinese hamster ovary (CHO) cells. While such assays can demonstrate activity in a cell, they are only marginally predictive of a compound's likely benefit to a patient. Similarly, toxicology studies performed in laboratory animals or rodent liver cells often fail to demonstrate the human body's ability or inability to tolerate a potential drug.

With the successful industrialization of human iPS cells, medical researchers will have access to a virtually inexhaustible supply of the most powerfully predictive drug discovery target possible: living cells from patients afflicted with a disease of interest. Access to such cells reduces the randomness of small molecule screening because the relevance of the target is established a priori and it also obviates the need for known receptor sites of well characterized structures, permitting the assaying of biologics without exhaustive foreknowledge of the disease pathway. The process is like an in vitro clinical trial, with the patient's disease in a Petri dish.

10.3 CELL THERAPY AND ITS CAVEATS

Why not use iPS cells as drugs? It seems obvious that a patient whose disease resulted in the deaths of neurons or heart muscle tissue would benefit from the administration of cells reprogrammed to fill the deficit. After all, the first successful products of biotechnology were all replacement therapies, from human growth hormone to erythropoietin. Furthermore, iPS cells avoid two major obstacles with human embryonic stem cells: the immune response to and rejection of cells from another individual and the ethical and political objections to harvesting stem cells from discarded embryos.

We believe that iPS cell therapies will eventually be available but, as with nearly every new discovery in biology, the path will not be as easy as it appears and the process will take time. Phase I clinical studies of a human embryonic stem cell-based product for the treatment of spinal cord injury were recently put on hold after continuing animal studies raised concerns that the drug caused a proliferation of cysts (Geron, 2009). The past history of promising modalities like gene therapy and antisense suggests that rushing preclinical work and driving new technologies into the clinic is likely to be counterproductive.

The clinical application of iPS cells faces significant hurdles, some shared with embryonic stem cell therapy, some unique. Current techniques for generating specific types of differentiated cells of therapeutic interest (e.g., neurons, cardiomyoctyes, and pancreatic cells) are inefficient. They yield mixes of differentiated cell types containing relatively small percentages of desired cells. Thus, in order to obtain clinically relevant quantities of a specific cell type, technically challenging and cumbersome enrichment techniques are required after differentiation of embryonic stem (ES) or iPS cells. An important concern is that even a highly purified population of differentiated cells may still contain a tiny number of undifferentiated cells or a small number of differentiated cells that spontaneously dedifferentiate. In both scenarios, these cells could give rise to tumors after administration to a patient.

Among the unique obstacles to iPS cell therapies is that most iPS cells are created using retroviruses or lentiviruses to insert the reprogramming genes into the host cell genome. As cells become reprogrammed, the reprogramming genes are gradually "turned off" over time and generally stay turned off in iPS cells. However, in rare cases, the introduced reprogramming genes may spontaneously "turn on" again and drastically alter the properties of iPS cells or differentiated cells derived from them, possibly leading to tumor formation. A further concern is that, as insertion of the reprogramming genes by retroviruses and lentiviruses occurs somewhat randomly within the genome, it is possible for such insertions to inactivate genes that regulate cell growth, thereby stimulating tumor growth. Thus, it is generally believed that iPS-based cell therapy will require iPS cells generated by non-viral-based methods, which, as of yet are extremely inefficient for the production of iPS cells (Yamanaka 2009).

These obstacles can assuredly be overcome with money and time, but the advantage of using iPS cells as a starting point to develop predictive assays is that the current limitations of the technology do not hinder their predictive power in drug discovery. Using iPS cells for in vitro drug or toxicology screens is viable today.

10.4 FOUNDATIONS OF A NEW PLATFORM TECHNOLOGY

The unsung heroes of the biotech revolution were the makers of tools and enabling technologies. Although Kary Mullis of Cetus ultimately received a Nobel Prize for the development of the polymerase chain reaction (PCR), a technique for generating thousands to millions of copies of a particular DNA sequence, the industrialization of recombinant proteins and monoclonal antibodies required many advances in the hardware and software of biological processes. This sequence of steps is likely to apply also to iPS cells.

Biotech pioneers like Cetus often had to invent and produce their own tools, but today we can draw upon a community of sophisticated vendors that grew up along with the industry. These toolmakers present the first broad partnering opportunities for iPierian, because as we build our own drug discovery program we will inevitably develop a number of innovations and tools that are key to the technology but not core to our business model (see Chapter 13 by Goswami). Accordingly, we plan to strike a partnership with a single leading company for an exclusive long-term relationship by which the company can turn our innovations into product opportunities, and we can access upfront capital as well as an ongoing stream of royalty payments. Our partner will enable us to monetize the offshoots of industrializing iPS technology that are essential for the platform but not to our therapeutic discovery programs.

It's worth reiterating here that although "industrializing" is a tidy sounding gerund, the actual task of scaling up iPS cell processes from laboratory quantities to commercial production is far from trivial. Yields must improve by at least an order of magnitude, if not two, coupled with a concomitant improvement in purity. Working closely with a company in the business of turning innovation into ready-to-use products will benefit companies such as iPierian directly, by cutting the time required to reduce science to practice. Collaborations will also benefit the broader life sciences community by making products more widely available.

For iPierian this kind of partnering should be a win–win situation. Trying to keep a genuine breakthrough technology in house limits learning and hinders the ability for the science to advance quickly. Certainly, some companies have opted not to share the wealth and have hoarded their intellectual property, but our observation is they have not moved as quickly or effectively as they might have advanced with partners. We believe that we can benefit not only our company and our investors, but also patients much more efficiently by allowing other firms and academic researchers to develop this technology with us.

10.5 A NEW/OLD MODEL FOR PHARMACEUTICAL PARTNERSHIPS

The same philosophy applies to our partnering strategy with pharmaceutical and biotechnology companies. Here we have two distinct models, one for specific diseases that iPierian is focusing on internally, for example, spinal muscular atrophy, a neurodegenerative disease that is the leading genetic cause of death in infants and toddlers. A second model will serve companies working on broad therapeutic areas like metabolic disease, cancer, and cardiology.

In both cases, however, we will seek to partner much earlier than is common practice today, where collaboration is commonly deferred until proof of concept, which may not be achieved until a molecule is in Phase II trials. That risk-averse approach makes sense to small companies seeking to maximize the financing of a core asset and to large companies looking for product candidates with near-term potential. But with a breakthrough technology like cellular reprogramming, it is possible to create real value and receive proper compensation for partnering far earlier in the process, much the way the first biotech companies did in the late 1970s, throughout the 1980s, and into the early 1990s.

The value proposition rests on the remarkable predictive power of the iPS platform compared to other available assays. The ability to show a response to a drug candidate in an in vitro study based on living cells from a diseased patient will enable demonstration of proof of concept long before the commencement of clinical trials.

Consider amyotrophic lateral sclerosis (ALS), commonly known as Lou Gehrig's disease, a progressive, fatal disease caused by the degeneration of motor neurons, the nerve cells in the central nervous system that control voluntary muscle movement. The causes of ALS are unknown, but a small percentage of patients with an inherited form of the disease have a mutation in the SOD1 gene that produces superoxide dismutase, a powerful enzyme and antioxidant that protects the body from superoxide, a toxic free radical (Rosen et al. 1993). Because transgenic mice can be bred with this human mutation, these animals are commonly used in ALS studies, even though results may be relevant only to 1 to 2% of all ALS patients.

In 2008, a team at the Harvard Stem Cell Institute led by John Dimos derived iPS cell lines from an 82-old woman suffering from ALS, and were than able to direct the cells to form motor neurons and glia, the cells that are most affected by the disease (Dimos et al. 2008). These cells now have the potential to be used to observe the effects of potential drugs on human motor neurons without the need to solely rely on the possible therapeutic relevance of mouse data. The peer-reviewed journal *Science* called the Harvard team's work the Breakthrough of the Year for 2008.

One partnership strategy is to collaborate on ALS with a pharmaceutical partner by collecting a set of tissue samples that represents a diverse group of patients with the disease. An alternative strategy would to take the research and development program downstream, developing assays, and performing the in vitro clinical trial and toxicology studies through preclinical work and into clinical trials as a proprietary program. iPierian's initial focus will be on small molecule candidates, because their ability to cross the blood–brain barrier is crucial in neurodegenerative diseases. The screening platform can be extended to investigate other indications as well as to exploit proteins, monoclonal antibodies, and other biologics as methods for modulating the disease process.

The partnering model can be much broader, tackling an entire therapeutic area representing a constellation of drug development opportunities. In these cases, the option would be to join with a large established company from the outset, before starting to build assays or models internally, as the partner would provide its knowledge and expertise in the disease category and in determining the appropriate patient cohort from which people would be drawn at the outset of the discovery program.

iPS cells represent a product-enabling platform, as opposed to a pure technology play, because it constitutes an entirely new way of approaching drug discovery. The platform comprises an enabling set of methods and technologies that allows researchers to take an entirely different approach, to change fundamentally the way we think of drug discovery, because it starts with patients and with the way the disease manifests itself in humans, not in a surrogate model system.

10.6 PROPRIETARY PROGRAMS DRIVEN BY DIRE NEEDS

With such a broadly applicable platform, iPierian could have built a business as a pure tools company. There are worthy examples of firms that have prospered this way, but greater value for shareholders can be created through the delivery of much-needed new therapies to patients by employing iPS cell technology in the discovery and development of drugs. However, from a pragmatic perspective, a small company is limited in the medical markets it can address. Even after the science is proven, after a candidate drug successfully passes through three phases of clinical trials and approval by the Food and Drug Administration, the company faces immense logistical and financial hurdles in building adequate manufacturing, marketing, and sales capabilities to bring that drug to market.

Choosing the right partner eases some of the burden, but it is also important to choose the right disease category. A critical part of our strategic planning was to perform a set of rule-in/rule-out exercises to see whether we could identify a set of diseases that (1) were truly grievous, (2) presented a dire need for new treatments, and (3) were not tractable with current drug discovery technologies. The category that emerged from this process was neurodegenerative disease.

Anyone who has seen a child struck down by spinal muscular atrophy (SMA) or a loved one ravaged by ALS, Parkinson's, or Huntington's disease knows how truly devastating these illnesses are. Our search of the literature for available laboratory systems, both in vitro and in vivo, found a dearth of models that could validate targets or predict the efficacies of candidate therapies. No products on the market today provide any more than temporary palliative treatment, and a veritable graveyard of large and small molecules were tested against these diseases without success. It is estimated that SMA strikes 1 of 6,000 persons; the most severe form occurs in the first 6 months of life, with a mean survival time of 3 to 4 years. Patients with teenage onset can have normal lifespans but are typically confined to wheelchairs. The disease has a relatively straightforward genetic profile in that it is monogenic

Changing the Game of Drug Discovery

recessive; the most common form of SMA is caused by mutation of the SMN gene that encodes for survival motor neuron protein (Kostova et al. 2007). SMA appears "solvable," having been termed closest to treatment among more than 600 such diseases by the National Institute of Neurological Disorders and Stroke.

SMA is particularly suited to the iPS discovery process. It is clear that deletion or mutation of a single gene leads to the disease but how the mutation leads to the disease is not known. That means there is no clear target for a potential drug compound and screening must be pursued phenotypically, which is how iPS cells can play their unique role. iPierian has already developed a large cohort of iPS cells derived from SMA patient samples, and we have begun initial screening of molecules for their ability to correct an SMA disease phenotype in motor neuron cells.

Parkinson's disease is another intractable illness for which a fresh approach to drug discovery may yield maximum benefits. Parkinson's is far more common than SMA, affecting 1 person in 100 over the age of 60, and unlike SMA, it has no known biomarker and its cause, whether genetic or environmental, has not been determined.

What is known is that Parkinson's results from losses of cells in various parts of the brain, including a region called the substantia nigra (Nussbaum et al. 1997). The substantia nigra cells produce dopamine, a chemical messenger responsible for transmitting signals within the brain that allow for coordination of movement. Currently available drugs stimulate the production of dopamine, providing temporary relief of the trembling symptoms commonly identified with Parkinson's disease, but do nothing to halt the deaths of cells.

Like SMA, Parkinson's presents no clear target in the affected cells upon which to base a drug screen, so Parkinson's disease would be best studied phenotypically. However, with no way to harvest neurons from patients, screening drug candidates for Parkinson's has gone down the usual path of utilizing cell lines that are poor proxies for disease-relevant human cells and spending millions of dollars—with little progress to show for the effort and cost.

The iPS technology will enable sampling of skin cells from Parkinson's patients, reprogramming the skin cells into iPS cells, then differentiating the iPS cells into dopaminergic neurons that exist in the afflicted region of the brain, and using the neuronal cells to find compounds that change cellular behavior. The goal is a molecule that halts or reverses disease progression and can be demonstrated in a cohort of individual assays that closely parallels the actual patient profile to be treated clinically.

10.7 WALKING THE RISK/REWARD TIGHTROPE

One of the more pernicious myths of biotechnology was that drug development based on molecular biology would have inherently lower risks of failure than programs based on medicinal chemistry. This misconception was perpetuated by the industry's early concentration on replacement therapies; if recombinant insulin closely replicated the structure of human insulin, its activity in the body was easy to predict. However, as the industry exhausted the supply of replacement proteins and moved into the use of biologics as drugs, it became clear that the mathematics of the risk/reward ratio had not been repealed.

There are inevitable risks associated with the iPS technology platform, which is, after all, based on a science that did not exist a few years ago and a process that to date has mostly been performed in small quantities in very sophisticated laboratories. However, no potential for outsize returns would exist if the risks were not high as well. In the universe of known and unknown risk factors, the identification of the key obstacles that lie in the path will aid their resolution. One of the issues cited often in the scientific literature is that the cells differentiated from iPS cells may not truly reflect the actual somatic cells. While that may be true to a degree, it is known that the differences appear limited and inconsequential compared to the difference between a cell from a specific Parkinson's patient and one from a rodent or human neuroblastoma cell line. The predictive power of an assay based on patient-derived motor neurons generated from iPS cells will be greater than other tools available for drug screening or the elucidation of the disease pathway.

Another issue that is more apparent is the complexity of scaling the technology. It is one task to reprogram a few sample cells in small dishes in a laboratory; it is entirely another to create an automated system that can produce millions of cells to order, day in and day out, within specific tolerances and with a guaranteed level of purity. Yields have improved exponentially—early experiments reprogrammed fewer than 1 in 10,000 cells—but the field has a long way to go to achieve production volumes that would be needed for cell therapy. The biology needed to grow neurons is vastly more complex than growing generic tumor cells in a dish, so simple off-the-shelf solutions will not be adequate, and the sector is focused on developing differentiation protocols for a variety of cell types. As complex disease processes are addressed, a number of related cell types will need to be examined and incorporated into the relevant assays.

Each assay will face a specific scientific challenge for each disease, with a need to develop a suitably predictive phenotypic model. It is not enough to produce reprogrammed somatic cells for an assay if the particular cell type is not conclusively implicated in a disease. iPS cells reprogrammed from an SMA patient show clear differences from cells derived from the patient's healthy mother, but in ALS, symptoms take many years to develop so stressors and technologies to mimic the changes caused in cells by age and the environment will be needed.

10.8 A SOLID INTELLECTUAL PROPERTY PORTFOLIO

Beyond the science and engineering, biotechnology companies all typically face a level of legal risk. Most scientific discoveries have multiple innovators and determining ownership rights is often a long and litigious process. If the intellectual property underlying a new process is split among too many parties, any one of them may have a difficult time building a viable business.

Although Dr. Yamanaka's breakthrough work in mouse cells generated the headlines, we believe the first patent filings to demonstrate successful generation of human iPS cells took place in June 2007 by a team at Bayer Japan, whose work at a commercial company went largely unnoticed. Indeed, as late as October 2007, Dr. Yamanaka and others publicly stated that it was unclear when, if ever, human iPS cell research might succeed (Hyun et al. 2007). In a review of its operations following its merger with Schering, Bayer elected not to continue the iPS cell work in Japan, and iZumi, iPierian's predecessor, was able to acquire all the related intellectual property.

The patent estate forms the basis of iPierian's intellectual property portfolio and covers a broad range of subject matter including methods and applications of the technology. This has been further built through sponsored research at the Gladstone Institute and with a number of additional filings generated from internal research incorporating methods of differentiation and assays. The patent strategy, however, has been on the application of the technology, as we believe the ultimate intellectual property protection may result from patents derived from the iPS-based therapeutic discoveries and their method of use.

10.9 FINANCE BEYOND TERM SHEETS

The biotechnology industry's birth was initiated by venture capitalist Robert Swanson when he persuaded scientist Herbert Boyer to start a company that became Genentech. The same sequence of events occurred in the iPS field. Beth Seidenberg, a director at Kleiner Perkins Caufield & Byers, played a seminal role in the launch of iZumi and worked with Bob Higgins, Highland Capital's co-founder. The two firms provided the initial $20 million investment in the new company.

On the East Coast, Drs. George Daley, Douglas Melton, and Lee Rubin, faculty members at Harvard University, co-founded Pierian with Dr. Ashley Dombkowski, a managing director at MPM Capital, and Robert Millman, a well regarded patent attorney and partner at MPM. As a result of a merger of Pierian and iZumi, MPM, a Boston-based firm focused on biotech and medical device technology, invested $10 million in the merged company known as iPierian. FinTech Capital, a Japanese venture fund, invested an additional $1.5 million.

To achieve a goal as ambitious as reinventing the drug discovery paradigm requires significant capital. All these venture funds are devoted to building the company and supporting the level of innovation and opportunity represented by iPS cell technology. The venture capitalists continue to input their combined experience and counsel at the board level.

The iPS-platform is based on three principles:

- Game changing technology
- Great execution
- Outsize returns

The first point is clear: the shift from random screens, surrogate animal models, and serendipity to a patient-based, cell-specific approach will be the most compelling shift in drug discovery since the introduction of recombinant DNA. It is nearly impossible to imagine a scenario in which it would not be to medicine's advantage to have phenotypically relevant assays and population-based disease cohorts for drug screening and discovery.

At the end of the day, it is known that great execution is a matter of hiring great people and giving them the resources and a cultural environment to allow them to perform at the top of their potential. This includes the recruitment of leaders on stem cell science, whether as advisors or employees, to ensure that the work will always be at the forefront of the field. The maintenance of fruitful collaborative agreements with academic researchers is also of great value to a company.

Considering the billions of dollars now spent on drug discovery efforts that lead to suboptimal results, a new paradigm that reduces those costs and increases the number of leads that turn into drugs will substantially improve the returns on biopharmaceutical investment. Biotech companies have been built upon one or two successful molecules or upon production of essential tools for the work of others. At iPierian we have the chance to do both as we lead the application of cellular reprogramming in drug discovery and development in proprietary areas such as neurodegeneration and through broad partnerships in defined therapeutic areas.

REFERENCES

Dimos, J. T., K. T. Rodolfa, and K. K. Niakan, et al. 2008. Induced pluripotent stem cells generated from patients with ALS can be generated into motor neurons. *Science* 29: 1218–1221.

Geron Corporation. October 30, 2009. Press Release: Geron and FDA Reach Agreement on Clinical Hold 2009. Company and Regulatory Agency Define Path to Re-Initiate Human Trials for Spinal Cord Injury. Menlo Park, CA. http://www.geron.com/media/pressview.aspx?id=1195

Hyun, I., K. Hochedlinger, R. Jaenisch, and S. Yamanaka 2007. New advances in iPS cell research do not obviate the need for human embryonic stem cells. *Cell Stem Cell* 11: 367–368.

Kostova, F., V. Williams, J. Heemskerk et al. 2007. Spinal muscular atrophy: classifications, diagnosis, management, pathogenesis, and future research directions. *Journal of Child Neurology* 22; 926.

Nussbaum, R. L. and M. H. Polymeropoulos 1997. Genetics of Parkinson's disease. *Human Molecular Genetics* 6: 1687–1691.

Rosen, D. R., T. Siddique, D. Patterson et al. 1993. Mutations in Cu/Zn superoxide dismutase gene are associated with familial amyotrophic lateral sclerosis. *Nature* 362(6415): 59–62.

Takahashi, K. and S. Yamanaka 2006. Induction of pluripotent stem cells from mouse embryonic and adult fibroblast cultures by defined factors. *Cell* 126: 663–676.

Yamanaka, S. 2009. A fresh look at iPS cells. *Cell* 137: 13–16.

11 Discovery of Small Molecule Regenerative Drugs

Yen Choo

CONTENTS

11.1	Introduction	141
11.2	The Regenerating Body	142
11.3	Natural and Synthetic Regenerative Drugs	143
11.4	Regenerative Drug Screening	143
11.5	Using Stem Cells for Regenerative Drug Discovery	144
11.6	CombiScreen™	145
11.7	Use of Animal Models in Screening	146
11.8	Modes of Action of Regenerative Drugs	146
11.9	Disease Areas	147
11.10	Comparison of Drugs and Cell Therapy	147
11.11	Market Acceptance: Ethical Debate	148
11.12	Path to Market	148
11.13	Discovery and Development	148
11.14	Manufacture, Storage, and Distribution	149
11.15	Product Regulation, Marketing, and Reimbursement	150
11.16	Alignment with Pharma Model	151
References		151

11.1 INTRODUCTION

In the minds of today's scientific and lay majority, the basis of future therapies resulting from stem cell research is that master cells capable of giving rise to the various tissues of the body can be used to replace specific cell types lost as a result of aging or disease. However, lurking behind the basic premise of cell replacement therapy are a host of scientific and commercial difficulties that complicate the stem cell's path to market, transforming an apparently simple concept into a risky reality.

The notion that we can develop small molecule or biologic drugs that regenerate specific tissues has until recently seemed a fanciful proposition, but is a relatively straightforward undertaking given the industry's vast experience in drug development. Today there are at least 1,357 unique drugs approved for human use, of which 1,204 are small molecules and 166 are biologics (Overington et al. 2006).

Among these are familiar and successful examples of regenerative drugs such as erythropoietin (EPO) which today has a global market of approximately $12 billion. In contrast, only a handful of cell therapies have reached the market, none of which has produced revenues substantially over $30 million. Furthermore, as yet, there is no approved treatment based on embryonic stem cells, widely regarded as being the most promising type of stem cell.

This chapter explores the prospects of discovering and developing regenerative drugs based on small molecules and compares the discovery and development processes and business models to those of conventional cell-based therapies.

11.2 THE REGENERATING BODY

The adult human has a number of regenerating tissues, the best characterized of which is undoubtedly the blood. The hematopoietic system offers a paradigm of the role of stem cells in development (i.e., organogenesis), regeneration in response to injury, homeostasis, and disease (e.g., cancer). Since mature blood cells have relatively short life spans and may also be lost, for example, through bleeding, stem cells exist to replenish and maintain the cellular components of blood including red blood cells, megakaryocytes, macrophages, neutrophils, and lymphocytes.

The definitive hematopoietic system consists of adult stem cells (hematopoietic stem cells or HSCs) that are capable of reconstituting the entire blood system by giving rise to a hierarchy of specialized progenitor cells whose potency is progressively restricted. The HSC is capable of long-term self-renewal and differentiation toward all blood lineage phenotypes; multipotent progenitors such as the common myeloid progenitors (CMPs) and common lymphoid progenitors (CLPs) give rise to the two major blood lineages; while CMPs in turn give rise to more restricted progenitor types such as megakaryocyte/erythroid progenitors (MEPs) and granulocyte/macrophage progenitors (GMPs).

Many other stem cell systems exist to regenerate injured tissues or cells lost simply as a result of physiological turnover. In addition, we now know that an increasing number of cellular and structural microenvironments (niches) function in the maintenance and differentiation of resident stem cells. For example, HSCs reside within the bone marrow, close to endosteum and sinusoidal blood vessels; hair follicle stem cells are found in the bulge regions of follicles; gut epithelial stem cells inhabit the intestinal crypts; and satellite cells that regenerate skeletal muscle are dispersed beneath the basal laminae of myofibers.

While the above are notable examples of tissues capable of rapid regeneration, stem cell systems and niches have also been described in tissues that do not obviously turn over or regenerate. For example, the central nervous system generally struggles to recover from serious trauma such as stroke or spinal injury, nor does heart muscle heal readily after a myocardial infarction, yet both organs possess the machinery of cell regeneration. Different populations of multipotent cardiac progenitors have been described [c-Kit$^+$, Sca1$^+$, and side population (SP)] and can be purified from the adult heart, cultured in vitro and shown to differentiate into the three main cardiovascular cell types: cardiomyocytes, endothelial cells, and vascular smooth muscle cells. In the brain, neurogenesis has been observed in the subventricular zones (SVZs) of the lateral ventricles and the subgranular zone (SGZ) of the dentate gyrus in the hippocampus. Adult neural stem cells have been purified, shown to self-renew and to differentiate into all types of neural cells including neurons, astrocytes, and oligodendrocytes.

11.3 NATURAL AND SYNTHETIC REGENERATIVE DRUGS

Stem cells and/or progenitors receive instructive signals from their local microenvironment (niche) or even from a remote location (a different organ), leading to the selective regeneration of particular cell types. Some of these signals, including the familiar erythropoietin (EPO) and thrombopoietin (TPO), are soluble molecules whose cognate receptors are present on the surfaces of target cells. EPO is a glycoprotein hormone secreted by renal cells and bound by the erythropoietin receptor (EpoR) on committed erythroid progenitors (CFU-Es) in the bone marrow, promoting their survival and differentiation into red blood cells. EPO normally functions in red blood cell homeostasis and its absence, for example, in chronic kidney disease, leads to anemia. Importantly, when exogenous (recombinant) EPO is administered to humans, it effectively stimulates red blood cell production—indeed, EPO is the archetypal regenerative drug.

TPO is a hormone produced by the liver and kidneys that regulates the production of platelets by stimulating the production and differentiation of megakaryocytes in bone marrow. Excess TPO is sequestered by newly formed platelets, thus establishing a homeostatic mechanism that regulates platelet count. Recombinant TPO developed to treat thrombocytopenia (low platelet count) proved immunogenic in humans, providing the impetus to develop a non-peptidyl alternative. Eltrombopag (SB-497 115) is an oral, small-molecule agonist of the thrombopoietin receptor (TpoR) discovered by screening a chemical diversity library using a TPO-responsive cell line (Erickson-Miller et al. 2005). Marketed by GlaxoSmithKline (GSK) as Promacta, this is the first synthetic small molecule regenerative therapeutic to reach market.

Cellular regeneration in response to drug therapy has also been observed in the central nervous system. For example, chronic administration of antidepressants such as fluoxetine (Prozac) increases the proliferation of neural progenitors in the SGZs of mice (Encinas et al. 2006). The effects of antidepressants are mediated through changes in levels of serotonin and noradrenaline in the brain, which are in turn correlated to SGZ neurogenesis. While it is not known whether impaired neurogenesis is the cause of depression, stress-induced decrease in neurogenesis is known to be prevented by antidepressants. Similarly, there is now compelling evidence that the behavioral effects of antidepressants are at least partly dependent on the stimulation of neuron regeneration.

11.4 REGENERATIVE DRUG SCREENING

Past screens for regenerative drugs with in vivo efficacy (e.g., eltrombopag cited above) have been performed using immortalized cell lines expressing receptors for known regenerative factors and engineered with constructs that report specific activation of downstream signalling pathways. For example, Tian and co-workers used a luciferase reporter gene driven by a synthetic STAT-responsive promoter stably transfected into a G-CSF-responsive murine myeloid cell line to screen a library of chemical compounds (Tian et al. 1998). One compound, SB247464, was able to activate the reporter through the murine G-CSF receptor and also showed G-CSF-like activity in secondary assays such as primary bone marrow colony-forming unit granulocyte (CFU-G) assays and further in mouse in vivo experiments.

Such powerful screens are dependent on having a detailed understanding of the cytokines, receptors, and signalling pathways involved in tissue regeneration, which at present are obscure for almost all stem cell systems (with the notable exception of the blood).

11.5 USING STEM CELLS FOR REGENERATIVE DRUG DISCOVERY

While it has long been known that adult stem cell systems regenerate various tissues and are responsive to signalling molecules that can be used as drugs, the concept that stem cell technologies could be used to discover new regenerative therapeutics is relatively novel. Until recently, small molecule screens using stem cells (or differentiated stem cells) were performed with the aim of discovering more efficient differentiation protocols for the development of cell therapies (Zaret 2009) or in place of transformed cell lines or primary cells in the drug development process—target validation, efficacy, and toxicity assays—to characterize conventional (non-regenerative) drugs or drug targets (McNeish 2004).

Through such work, data began to accumulate showing that novel small molecules (c.f. growth factors) identified through screening could be used to promote pluripotency, differentiation, or de-differentiation of cells in vitro (Ding and Schultz 2004). Nevertheless, though it may have been possible to anticipate that such drugs might have future therapeutic applications, no practical drug discovery strategy was articulated by any researcher at this stage. On the contrary, a critical limitation of the early work is that cell-based screens employed ESC lines (Takahashi et al. 2003) or established/transformed multipotent cell lines (Wu et al. 2002 and 2004). Because the power of cell-based screening derives from the ability to interrogate the biology of cells, those screens would have been better carried out on physiologically relevant models.

While immortalized cell lines and ESCs have high proliferative capacities that greatly help establish high throughput screens, both cell types are very probably unlike any cells present in the adult body. Laboratory cell lines originally derived from adult tissues often display aberrant genetic and functional characteristics by the time continuous lines have been established and their value in screens is increasingly questioned. Likewise, pluripotent ESCs derived from preimplantation blastocysts are lost as an embryo develops and there is no evidence to suggest any trace of these remains to adulthood. Probably as a result, none of the differentiation-promoting compounds discovered in cell-based screens using ES cells has to my knowledge been reported to have a regenerative effect in animal models.

In view of the fact that the target cell of a regenerative drug must be present in the patient (e.g., an adult stem cell or, by analogy with the action of EPO on CFU-E, more likely a committed progenitor cell), cell-based screens aimed at discovering regenerative drugs should be built around physiologically relevant cell types. Unfortunately, primary adult stem cells and progenitors are rare: cardiac progenitor populations account for only 1 to 2% of cells in the heart, while HSCs represent only 0.003% of cells in bone marrow. Even in HSCs (live bone marrow is relatively easily sourced and an array of cell surface markers allow sorting of HSCs from other types by FACS), the purity of cell preparations is still only 50% (Kiel et al. 2005).

Isolating adult stem cells from the majority of other tissues is considerably more challenging. Furthermore, generating appreciable quantities of such cells for screening is likely to be challenging as it is difficult to expand adult stem cells in culture without loss of developmental potential.

11.6 COMBISCREEN™

Plasticell's solution to the problem of sourcing physiologically relevant progenitor cells for regenerative drug screening, termed CombiScreen (Choo 2005), is creating progenitor cells from in vitro from stem cells. Certain stem cells, for example iPS and ES cells, can be greatly amplified without loss of pluripotency and can give rise to all types of physiologically relevant stem cells, progenitors, and terminally differentiated cell types, as is apparent when ESCs are used to create transgenic animals.

If a stem cell can be differentiated into cell Z in vitro, then it is likely that its progenitor Y must also have appeared in vitro at an earlier stage of the differentiation process. Since highly effective differentiation protocols are performed stepwise, recapitulating natural developmental processes such as V → W → X → Y → Z (Salero and Hatten 2007; Mizuseki et al. 2003; Rathjen et al. 2002; Motohashi et al. 2007; Keirstead et al. 2005; Brustle et al. 1999; D'Amour et al. 2006), there are certain discrete stages in the process at which one could find physiologically relevant progenitors. Cells from that stage of differentiation, whether purified or in a background of other cell types, can be used in a drug screen to assay whether certain compounds enhance the appearance of terminally differentiated cells.

A complementary technology, called CombiCult™ (Choo 2002 and 2007) allows combinatorial screening of differentiation factors to discover highly effective sequential protocols for stem cell differentiation. The technology can be used to devise protocols yielding terminally differentiated cells or specific progenitor phenotypes.

A clear demonstration of the potential of these two stem cell technologies in regenerative drug discovery is the rediscovery of the G-CSF mimetic SB247464 in a screen. CombiCult was used to determine a highly effective, four-stage, serum-free differentiation protocol for the production of myeloid cells from mouse ESCs. When myeloid precursors produced in the third stage of differentiation were screened against a chemical library containing SB247464 in a 96-well plate format, the regenerative molecule and a G-CSF control were able to increase the number of phagocytic monocytes significantly over background levels. The screen relies on the presence of differentiated myeloid progenitors bearing the G-CSF receptor, since SB247464 does not exert appreciable differentiation activity on undifferentiated ESCs in this assay. As this molecule is known to be active in animals, the experiment clearly shows that in vitro-derived progenitor cells can be used in screening to identify molecules that act on physiologically relevant cells in vivo.

11.7 USE OF ANIMAL MODELS IN SCREENING

An interesting method of drug screening makes use of regenerative models studied in whole animals. While more dramatic feats of regeneration such as limb and tail regrowth in salamanders (Birnbaum and Sanchez Alvarado 2008) and heart regeneration in zebrafish (Poss et al. 2002) may be of questionable anatomical and/or physiological relevance to humans, different animals often share fundamental signalling and transcriptional mechanisms. The use of animal models also allows screening to be conducted in the context of the intact niche, which will likely be invaluable going forward.

Zon and colleagues used a live zebrafish embryo model of hematopoiesis to screen compounds for their effects on HSC induction in the aorta–gonad–mesonephros region (North et al. 2007). The screening readout was based on measuring the levels of *runx1* and *cmyb* transcripts required for mammalian hematopoiesis. The chemicals that enhance prostaglandin (PG) E2 synthesis increased HSC numbers, while those that inhibit PG synthesis had the opposite effect, suggesting a role for PGE2 in HSC formation. Remarkably, murine bone marrow exposed to PGE2 ex vivo prior to transplantation in irradiated mice was capable of producing three-fold more colony forming units.

11.8 MODES OF ACTION OF REGENERATIVE DRUGS

To date, the assumption is that regenerative drugs would most likely function by directly stimulating the appearance, replication, and differentiation of stem or progenitor cells. This is, of course, not the only anticipated mode of action of a regenerative drug. Advances in our understanding of stem cell biology have suggested further opportunities to intervene in regeneration.

One such prospect is drug-induced modulation of niche biology that can have a secondary effect on resident stem cells (Scadden 2006). The niche microenvironment comprises heterologous cell types and extracellular matrices, both of which are potential pharmacological targets. A case in point is parathyroid hormone (PTH) stimulation of osteoblasts in the bone marrow leading to an increase in the number of resident HSCs and increased survival rates following bone marrow transplantation in animals (Calvi et al. 2003). This finding raises the possibility that drugs developed to treat bone disease may find roles in hematopoietic stem cell modulation.

Instead of inducing the niche to be more hospitable to stem cells, it is also possible to better direct the stem cell toward the niche. Stem cell "homing" is the targeted migration of stem cells toward their maintenance niche or to a site of regeneration following injury. Homing of endogenous stem cells has not yet been modulated through pharmacological intervention, but directed trafficking of stem cells through biochemical engineering has been demonstrated. Ex vivo enzymatic treatment of mesenchymal stem cells (MSCs) to convert cell surface CD44 glycans into E-selectin/L-selectin ligands (normally found on hematopoietic cells) conferred enhanced osteotropism, leading to rapid recruitment of the modified cells in bone marrow (Sackstein et al. 2008). In the future, such control over stem cell trafficking to the niche may be effected by drugs, as is currently possible in the reverse process of niche-to-periphery migration (e.g., HSC mobilization).

11.9 DISEASE AREAS

The reader will have noticed that the above examples of regenerative drugs concern the blood. Briefly considering why this organ is the "low hanging fruit" in the field reveals certain principles of pharmacological action in regenerative medicine.

The first point is that the blood has evolved to turn over rapidly and regenerative drugs simply enhance an already efficient process, often through homeostatic mechanisms that function in routine regeneration. The situation in a quiescent organ such as the heart is more obscure. While the heart may possess stem cells, their potential to mediate regeneration is uncertain. If resident progenitors can be coaxed into regenerating cardiomyocytes, why does this organ not heal spontaneously?

Targeting a well understood regenerative system reduces the risk of failure, not simply due to the accommodating biology of the organ, but also because a large body of scientific knowledge facilitates many steps in drug discovery and development. In the case of the blood, understanding the various progenitor cells, their markers, regenerative factors, and extracellular and intracellular signalling components has clearly helped in designing primary and secondary screening assays and verifying the mechanisms of action of regenerative drugs both in vitro and in vivo.

A further important consideration is the nature of the pathology treated and, in particular, its effect on the stem cell and progenitor pools and their niches. A special feature of blood is that the stem cell niche is discrete from the location of the bulk organ and remains relatively intact in most blood diseases. Thus, for example, none of the approved indications for EPO (chronic kidney disease, AIDS, cancer chemotherapy, and perioperative use) destroys the bone marrow niche that can therefore effectively support red blood cell regeneration in response to drug therapy. However, chemotherapy damages the bone marrow and resident erythroid progenitors, and EPO is less effective in treating anemia caused by chemotherapy as opposed to anemia caused by renal disease. By extension it is not clear whether regenerative drugs would be effective in pathologies in which the niche is destroyed along with the organ, as for example in severe burn damage to the skin or a critical bone defect. In this case a tissue engineering approach by which a matrix that approximates the niche is transplanted together with stem cells might prove superior to drug therapy.

An additional consideration is whether the newly formed tissue is functionally autonomous or requires elaborate integration with existing cells. Circulating blood cells are fairly independent entities that do not form intricate connections with other cell types. Recapitulating brain circuitry through nerve regeneration may prove more difficult, but appears to occur in the case of antidepressant-induced regeneration (and in any case is likely a smaller challenge than integrating a mass of transplanted cells). Could the possibility of arrhythmias caused by the appearance of immature cardiomyocytes explain why the heart does not regenerate damaged areas?

11.10 COMPARISON OF DRUGS AND CELL THERAPY

What are the various considerations that determine whether regenerative drug therapy is preferable to conventional cell therapy? Clearly some questions are specific to the particular therapy, but a surprising variety of general issues discussed below weigh on the feasibility, risk, benefit, and commercial upside of regenerative therapies based on drugs or cells.

11.11 MARKET ACCEPTANCE: ETHICAL DEBATE

Between 2001 and the time U.S. President George W. Bush's departure from office in 2009, he declared human embryonic stem cell research a grave "moral hazard" and restricted federal funding of ESC research. This was a setback to the academic field and the perception that the world's largest market was apparently opposed to stem cells seriously damaged the nascent stem cell biotech industry.

While a number of good technical reasons (discussed later in this chapter) show why drug treatment is an attractive alternative to cell replacement therapy, an undeniable advantage of drugs is that they are universally acceptable forms of therapy. While it is tempting to think that the bad days of stem cell research are forever gone, the controversial policy shifts that took place in recent history illustrate that the pendulum could again swing back. The risk of this remains while the destruction of embryos is an emotional and divisive moral issue and political tool in the U.S. but should wane with the emergence of high profile stem cell treatments for serious illnesses.

11.12 PATH TO MARKET

Aside from ethical considerations, the respective merits of regenerative medicine business models based on drug or cell therapy are to be found in the apparent ease by which such therapies can be developed and brought to market. A general discussion of the key differences between the two therapeutic approaches is complicated by the fact that the commercialization routes of various cell therapies will vary, depending on the stem cell type and source. Significantly different technical, regulatory, and commercial pros and cons apply to pluripotent (embryonic or iPS) and adult stem cell therapies and to autologous and allogeneic adult stem cell treatments.

What is certainly true is that regulatory and commercial restrictions will almost always be more prescriptive for cell therapy as opposed to drug development, narrowing biotechs' ability to pursue certain technical and business options; for example, the vast majority of German and Japanese companies have been forced to focus on autologous therapies. In addition, the risk or uncertainty present in a given stage of cell therapy development is currently almost always greater than in the corresponding stage of the drug development process. The simple reason for this is that the latter is a fairly well trodden path with well known routes (complete with milestones indicating time lapse) between the various stages of development, while the path to market for a given cell therapy product is largely uncharted and arguably more obscure.

11.13 DISCOVERY AND DEVELOPMENT

The discovery stage is the most risky endeavour whether one is seeking to develop a cell or a drug therapueutic. Choosing to pursue a poor drug target or to develop the wrong cell type can be ruinous. Early critical decisions such as these are equally risky whether one is setting out to develop a drug or a cell therapy, however downstream there are technical differences in the efficiency and predictability of the respective discovery processes. Pharmaceutical drugs are most often discovered through screening chemical libraries to identify compounds with activity against a validated biochemical disease target. From hundreds of thousands (sometimes millions) of compounds used in a primary screen a few hundred "hits" are subjected to secondary assays, eventually whittling down the chemical diversity to a handful of promising leads. These are optimized and submitted to preclinical investigation; finally one of these compounds will progress to clinical trials. This drug discovery process is industrialized and the risk of failure at this stage is mitigated by industry experience in screening, as well as the sheer scale of experiments. In contrast, cell therapy development is still a cottage industry: here individual scientists work manually on small scale experiments aimed at e.g., isolating stem cells or efficiently directing their differentiation. It is difficult to forecast when the breakthrough will occur, or to foresee whether the cells produced will be suitable to progress to the next stage of development. The timeline uncertainty associated with the discovery of cell therapies translates to risk, and historically relatively few investors have had an appetite for this type of activity. Industrializing the process, for example by implementing high throughput screening technologies such as CombiCult (Choo 2008), will reduce the time and cost of discovery as well as the associated risk.

An important source of risk in discovery of drug therapeutics is toxicity in humans. A large number of drugs fail because of side effects on the heart and other organs, whether caused directly or through metabolites. Although the industry has procedures to predict drug toxicity and identify likely failures prior to commencing clinical trials, the threat of late-stage failure due to toxicity remains significant. In the case of cell therapy, the likelihood of toxicity per se is low, provided the transplanted material is free from adventitious agents introduced during production.

A specific concern for (mainly allogeneic) cell therapies is the risk of product failure arising from rejection by recipients' immune systems; this type of reaction is rare with most drugs. In addition, where components of the immune system are used in cell therapy, the opposite scenario (attack by the donor's immune system) can also take place, leading to graft-versus-host disease. A related issue (mainly concerned with pluripotent cell-derived therapies) is the possibility that any undifferentiated cells present in a transplant may give rise to tumors or other malignant tissue, or may wrongly take up residence in a stem cell niche that directs their differentiation to an unwanted cell type.

11.14 MANUFACTURE, STORAGE, AND DISTRIBUTION

Chemicals are usually much more straightforward to manufacture than cells, and after production are more stable and consequently easier and cheaper to store and distribute. These major differences between chemicals and cells confer a big advantage to developers of drug therapies in terms of the cost of goods and development risk.

The manufacturing of drugs is not completely without challenges but is by now standard industry practice; reproducible manufacture of pure product at high yield is routine. In contrast, producing large quantities of pharmaceutical grade human cells is a relatively new art: while eukaryotic cells have previously been used as factories to manufacture such products as vaccines, the cell has rarely been the end product. Cells are infinitely more complex than the sums of their chemical components, so achieving sufficient yield and purity of therapeutic cells in full-volume production is a highly complex operation. Attaining the level of manufacturing reproducibility and product stability required of therapeutics will be an even harder task, and presumably one for which the minimum standard will need to be defined on a case-by-case basis, leading to regulatory uncertainty.

Ironically the relative difficulty of manufacturing cell therapies may eventually prove an advantage over drugs, as it has for biologics. This is because easy-to-manufacture small molecule drugs lose significant market share to less costly generic products after the expiry of patent production. The technical complexity of producing cell therapies may prove a more effective barrier to entry for competitors than intellectual property. This is significant because comprehensive intellectual property protection for cell therapies may be difficult to secure given the difficulty of defining a product's essential characteristics.

A company wishing to manufacture cells is faced with the dilemma of whether to build plant facilities or outsource production. In-house production allows full control over the process but is prohibitively expensive. Outsourcing may be limited by a lack of know-how and/or manufacturing capacity. While the infrastructure for cell manufacturing is beginning to develop, large gaps currently exist in the value chain.

A further complication of cell therapies is quality control, a process that becomes more laborious and costly as production lot size reduces. In cell therapies, a lot is defined as a single dose (personalized therapy). Every sample must be individually quality controlled and a comparatively large share of the preparation must be sacrificed for this purpose.

The costs and complexities of manufacturing a therapy are also determined by the ability to distribute the therapy to patients. In contrast to drugs that are manufactured in bulk and kept in stock for long periods, cell therapies are likely to be made to order, partly because they are so expensive to manufacture and also because they have short shelf lives. Live cells must be transported more rapidly and stored with much greater care than drugs, requiring timely, "chilled chain" distribution. These unique factors require exquisite integration of demand with the supply chain (manufacturing, packaging, and distribution), which is surprisingly difficult to achieve.

11.15 PRODUCT REGULATION, MARKETING, AND REIMBURSEMENT

The approval of drug products by the FDA (and its equivalents outside the U.S.) is relatively straightforward compared to the regulation of cell therapy products. Their approval is a cumbersome process that may be difficult to navigate by a young biotech company. Particular attention must be paid to the classification of cell therapy products, especially where a delivery device or procedure is deemed an integral part of therapy. An additional serious complication is the lack of a harmonized framework across countries for regulating market approval for cell therapy products, making it more difficult for a company to access multiple markets simultaneously.

An important consequence of this, coupled with the higher cost of goods associated with cell therapy products compared to drugs, is that careful attention must be paid to the market. This is exceedingly complicated because cell therapies entering the clinic today will not reach the market for a number of years, by which time pricing pressure will have increased considerably as a result of healthcare reform. In addition, to ensure that insurers will pay for expensive cell therapies, they must be much more effective than drug alternatives available at the time of market entry. It is difficult for effective new drugs to compete in the marketplace, and this will be doubly true for all but the most innovative cell therapies.

11.16 ALIGNMENT WITH PHARMA MODEL

Despite the various advantages during the development process that drugs enjoy over cell therapies, perhaps the greatest asset of drugs is their familiarity to the pharmaceutical industry. Pharma will play an important role in the regenerative medicine value chain and its strengths lie in formulating, manufacturing, bottling, distributing, and selling medicines to patients. Crucially, pharma has no experience of selling cells to patients and historically has been wary of the organizational step changes required to accommodate new therapeutic modalities into its operations.

Undeniably, the success of biologics as therapeutics, coupled with the demise of the blockbuster drug model, forced big pharma to be more outward looking. As a result, we now see companies such as Pfizer and Roche considering the development of cell therapies. Nevertheless, accommodating the different marketing and sales models required for cell therapy will be a challenge. Mass market medicines are relatively simple products, sold by nonspecialist representatives and prescribed by general practitioners. Cells will be the basis of much more individualized and high-cost therapies and consequently will require a new, highly educated sales force to engage specialist physicians.

A final intriguing, if longer-term, opportunity for regenerative drugs is engaging Asia's emerging healthcare market. Fuelled by continued affluence, the diseases of the emerging economies will soon mirror those of the developed world. For instance, the incidence of diabetes in India and the Middle East is predicted to increase dramatically due to changes in lifestyles and life spans. Inevitably the largest market for regenerative therapies will soon be Asia, but the prospect of mass marketed cell therapies there seems remote.

REFERENCES

Birnbaum, K. D. and A. Sanchez Alvarado 2008. Slicing across kingdoms: regeneration in plants and animals. *Cell* 132: 697–710.
Brustle, O., K. N. Jones, R. D. Learish et al. 1999. Embryonic stem cell-derived glial precursors: a source of myelinating transplants. *Science* 285: 754–756.
Calvi, L. M., G. B. Adams, K. W. Weibrecht et al. 2003. Osteoblastic cells regulate the haematopoietic stem cell niche. *Nature* 425: 841–846.

Choo, Y. 2008. Use of combinatorial screening to discover protocols that effectively direct the differentiation of stem cells. In *Stem Cell Research and Therapeutics*, Shi, Y. and D. O. Clegg, Eds. Heidelberg: Springer.

Choo, Y. 2002. Cell Culture. EP1551954.

Choo, Y. 2005. Method for identifying a modulator of a cell signalling. PCT/GB2006/004483 WO: (Plasticell Limited, U.K.).

D'Amour, K. A., A. G. Bang, S. Eliazer et al. 2006. Production of pancreatic hormone-expressing endocrine cells from human embryonic stem cells. *Nat Biotechnol* 24: 1392–1401.

Ding, S. and P. G. Schultz 2004. A role for chemistry in stem cell biology. *Nat Biotechnol* 22: 833–840.

Encinas, J. M., A. Vaahtokari, and G. Enikolopov 2006. Fluoxetine targets early progenitor cells in the adult brain. *Proc Natl Acad Sci USA* 103: 8233–8238.

Erickson-Miller, C. L., E. DeLorme, S. S. Tian et al. 2005. Discovery and characterization of a selective, nonpeptidyl thrombopoietin receptor agonist. *Exp Hematol* 33: 85–93.

Keirstead, H. S., G. Nistor, G. Bernal et al. 2005. Human embryonic stem cell-derived oligodendrocyte progenitor cell transplants remyelinate and restore locomotion after spinal cord injury. *J Neurosci* 25: 4694–4705.

Kiel, M. J., O. H. Yilmaz, T. Iwashita et al. 2005. SLAM family receptors distinguish hematopoietic stem and progenitor cells and reveal endothelial niches for stem cells. *Cell* 121: 1109–1121.

McNeish, J. 2004. Embryonic stem cells in drug discovery. *Nat Rev Drug Discov* 3: 70–80.

Mizuseki, K., T. Sakamoto, K. Watanabe et al. 2003. Generation of neural crest-derived peripheral neurons and floor plate cells from mouse and primate embryonic stem cells. *Proc Natl Acad Sci USA* 100: 5828–5833.

Motohashi, T., H. Aoki, K. Chiba et al. 2007. Multipotent cell fate of neural crest-like cells derived from embryonic stem cells. *Stem Cells* 25: 402–410.

North, T. E., W. Goessling, C. R. Walkley et al. 2007. Prostaglandin E2 regulates vertebrate haematopoietic stem cell homeostasis. *Nature* 447: 1007–1011.

Overington, J. P., B. Al-Lazikani, and A. L. Hopkins 2006. How many drug targets are there? *Nat Rev Drug Discov* 5: 993–996.

Poss, K. D., L. G. Wilson, and M. T. Keating. 2002. Heart regeneration in zebrafish. *Science* 298 (5601): 2188–2190.

Rathjen, J., B. P. Haines, K. M. Hudson et al. 2002. Directed differentiation of pluripotent cells to neural lineages: homogeneous formation and differentiation of a neurectoderm population. *Development* 129: 2649–2661.

Sackstein, R., J. S. Merzaban, D. W. Cain et al. 2008. Ex vivo glycan engineering of CD44 programs human multipotent mesenchymal stromal cell trafficking to bone. *Nat Med* 14: 181–187.

Salero, E. and M. E. Hatten 2007. Differentiation of ES cells into cerebellar neurons. *Proc Natl Acad Sci USA* 104: 2997–3002.

Scadden, D. T. 2006. The stem-cell niche as an entity of action. *Nature* 441(7097): 1075–1079.

Takahashi, T., B. Lord, P. C. Schulze et al. 2003. Ascorbic acid enhances differentiation of embryonic stem cells into cardiac myocytes. *Circulation* 107: 1912–1916.

Tian, S. S., P. Lamb, A. G. King et al. 1998. A small, nonpeptidyl mimic of granulocyte-colony-stimulating factor [see comments]. *Science* 281: 257–259.

Wu, X., S. Ding, Q. Ding et al. 2002. A small molecule with osteogenesis-inducing activity in multipotent mesenchymal progenitor cells. *J Am Chem Soc* 124: 14520–14521.

Wu, X., S. Ding, Q. Ding et al. 2004. Small molecules that induce cardiomyogenesis in embryonic stem cells. *J Am Chem Soc* 126: 1590–1591.

Zaret, K. S. 2009. Using small molecules to great effect in stem cell differentiation. *Cell Stem Cell* 4: 373–374.

12 Adoption of Therapeutic Stem Cell Technologies by Large Pharmaceutical Companies

Alain A. Vertès

CONTENTS

12.1 Introduction 153
12.2 Constraints to the Pharma Business Model 154
12.3 Accessing External and Internal Innovation as Strategic Response 156
12.4 Stem Cell Technologies First Adopted by Big Pharma 160
12.5 Potential Big Pharma Business Models for Regenerative Therapies 161
12.6 Building New Franchises in the Terra Incognita of Therapeutic
    Stem Cells 164
    12.6.1 Patents 165
    12.6.2 Product Concept, Indications, and Markets 166
    12.6.3 Regulatory Requirements 166
    12.6.4 Manufacturing 167
    16.6.5 Logistics 168
    16.6.6 Delivery 168
    12.6.7 Pricing and Reimbursement 168
12.7 Perspective 169
Acknowledgment 169
References 169

12.1 INTRODUCTION

Pharmaceutical companies are implementing at an increasing pace stem cell technologies as discovery and diagnostic tools. In so doing, these companies pursue the objective of developing means to better understand the fundamental biology of human diseases and to optimize attrition rates as they bring to market new small molecules and biologics. For example, human embryonic stem cells (hESCs) and induced pluripotent stem (iPS) cells derived from hepatocytes or cardiomyocytes and cancer stem cells, or cells derived from diseased tissues enable practitioners to model diseases, screen compounds for efficacy, and assess compound safety using human-derived cellular screens that in principle more accurately model the molecular events that occur in patients.

The potential of stem cell technologies nevertheless also lies in therapeutic applications. The litmus test here is whether pharmaceutical companies are ready today to implement such emerging technologies despite numerous technical challenges remaining to be solved, and despite the road to the first blockbuster cell therapy being perhaps a decade or two away if one uses existing drug development statistics and comparables. What is more, the tension between immediate business threats such as managing the patent cliff that are faced by many a big Pharma and the long-term economic benefits that cell therapeutics represent, unless it is demonstrated cell therapeutics can indeed be developed faster and with a lower attrition rate than conventional drugs, tilts the balance in favor of partnering or developing novel drugs using well proven technologies to fill product portfolios, rather than venturing in the terra incognita of emerging therapeutic options such as iPS-derived stem cells or perhaps even also mesenchymal stem cell (MSC) therapeutics, and hematopoietic stem cell preparations. Entering the emerging field of cell therapeutics with "prudent optimism" constitutes one possible approach for Big Pharma that could help mitigate the risks involved in tackling the uncertainties that characterize emerging products and unproven markets.

Important considerations include the ability of Big Pharma to implement conventional partnering processes after consolidation has occurred in the therapeutic stem cell biotech microcosm or after cell therapeutic products with the appropriate science, business case, and strategic fit reach adequate clinical demonstration. Keys to corporate success in regenerative medicine may lie in two core competencies: (1) the ability to maintain momentum to deploy cell therapeutic technologies within a Big Pharma cultural environment, and (2) an internal nucleus of scientists and an external group of scientific advisors with the knowledge required to assess product opportunities arising from biotechnology companies. On the other hand, the transition from good to great for Big Pharma companies may very well reside in the ability to anticipate winning paradigm-changing technologies, and in turn place a discrete number of early bets, with a strong focus and top-down mandates to succeed in order to build a lasting competitive advantage.

History bears an essential lesson. In the field of monoclonal antibodies, the competitive forces at play in the Big Pharma world will ultimately bring to the market innovative drugs that meet heretofore unmet medical needs, making patients and payers the ultimate beneficiaries of the new developments, while providing the early Big Pharma adopters attractive economics that help strengthen shareholder value, fuel powerful innovation engines, and promise future product options and associated economic values.

12.2 CONSTRAINTS TO THE PHARMA BUSINESS MODEL

The historical economics generated by the pharmaceutical industry have been eroding slowly and consistently over the past decade (Scherer 1993; Foley 2004; Goodman 2009). The forces that induce this erosion include well identified parameters: (1) the complexity of innovation (high and exponentially increasing R&D

costs, diminishing rates of success); (2) long time scales for regulatory approvals; (3) competition with other industries for capital; (4) overall diminishing patent life, patent expiration, and competition from generics that decrease the price of a small molecule drug up to 80% in only a few months; and (5) healthcare budgets and price control policies (Vertès 2004).

Pharmaceutical companies have implemented a mix of strategies to cope with these constraints. Examples are operating a generic company branch (e.g., Novartis and Pfizer) and operating a pharmaceutical business unit within the framework of a conglomerate of loosely related industries (e.g., Johnson & Johnson). In addition, some companies have concentrated their efforts in the pharmaceutical core business to develop differentiated medicines such as disease-modifying, first-in-class, or best-in-class therapeutics, while pursuing complementary activities such as diagnostics and leveraging the synergies between the two domains to pioneer new areas such as personalized healthcare (e.g., Roche).

Other companies have created innovation centers to replicate within the realms of large multinational organizations the entrepreneurial cultures of biotechnology companies (e.g., GlaxoSmithKline, Merck, Pfizer). However, overall and with only a few exceptions, Big Pharma as an asset class no longer delivers double-digit growth (Goodman 2009). Among the key issues is that the internal rate of return (IRR) on small molecule R&D—a traditional strength of Big Pharma—is now 7.5%, that is, less than the industry cost of capital; it was approximately 12% between 1997 and 2001 (David et al. 2009).

This decrease can be ascribed, to some extent, to higher attrition rates and longer times required for R&D. Also, R&D costs have increased annually by 8% for the past few years while the pressure on drug pricing is stronger than ever (DiMasi et al. 2003; Collier 2009; David et al. 2009; Munos 2009). Despite biologics having delivered superior returns with IRRs of 13% on average, given higher average peak sales and a slower erosion of sales following patent expiration, it is unlikely that increased investment in these therapeutic modalities will suffice based on the still limited number of biologic drugs and the increasing competition of biosimilars (Aggarwal 2009; David et al. 2009).

Beyond managing these constraints by improving operations of the R&D value chain via decreased costs, increased speed, or streamlined decision-making to improve overall R&D productivity (Anonymous 2007; David et al. 2009), increased investment in research is frequently suggested as a necessary response to the business challenges faced by the pharmaceutical industry (Anonymous 2007; Garnier 2008; Jack 2009). This is important not only from an economic perspective, but also to ensure that a constant engine of innovation remains in place to bring to patients novel therapeutics with improved safety and efficacy and treat unmet medical needs with high associated morbidity or mortality rates. The factors of success for improving R&D productivity metrics include: (1) building a deep understanding of the pathophysiologies of diseases; and (2) developing biomarkers to stratify patients according to distinct biological conditions despite similar symptoms (Anonymous 2007). A typical example is chronic obstructive pulmonary disease (COPD), a little understood multifactorial disease involving inflammation and remodeling of the lungs. COPD is currently treated with drugs developed for asthma, although the two diseases exhibit fundamental differences (McEvoy and Niewoehner 1997; Barnes 2003; Agusti 2005). Another example is type 2 diabetes, a term that encompasses a variety of disease subtypes; it is important to understand whether each subtype is amenable to therapeutic intervention (McCarthy 2004; Anonymous 2007). Of course, the commercial viability of the resulting therapies must be congruent with the value creation expectations of Big Pharma and its core marketing competencies that still favor a blockbuster approach. However, Big Pharma is becoming increasingly aware that this particular business model no longer adapts to the technical and market realities of today (Gilbert et al. 2003; Campbell et al. 2007).

As an example, the commercial potential of a disease-modifying new drug for COPD is large. Despite low diagnosis rates, this market reached $7 billion in 2008 (U.S., France, Germany, Italy, Spain, United Kingdom, Japan) and is expected to reach $11 billion in 2018 (Jammen and Vasilakis-Scaramozza 2010). Likewise, the market for diabetes type 2 drugs is expected to reach $20 billion by 2016, driven by an increased incidence of the disease and the commercialization of novel therapies. The reliance on low cost generic drugs is expected to create barriers to the market penetration of novel products that will thus need to demonstrate significantly improved efficacy or safety attributes (Ahmad and Dreyfus 2007). Only prospects on such scales can justify the risk and investment necessary to tackle a totally different therapeutic modality at Big Pharma level.

However, Pfizer and Glaxo have both opened in 2010 new research units on rare diseases. Furthermore, novel treatments for orphan diseases or niche indications may constitute sufficient market pull for large biotechnology companies, as exemplified by the partnering deal between Genzyme and Osiris Therapeutics to develop a graft-versus-host disease (GvHD) treatment (Allison 2009; Mack 2009). Moreover, they could constitute appropriate technology development steps for developing other products such as for the diabetes type 1 market that is expected to reach $2 billion in 2017 (Gates and Dreyfus 2008); or for the Crohn's disease market that is expected to reach $4 billion in 2017 (for maintenance and acute treatments combined) (Anonymous 2009). A key to decreasing R&D time and costs is to test in humans the new hypotheses as soon as practical and as soon as appropriate proof of safety has been reached. The objective of this approach is that a comprehensive understanding of diseases can be generated early on in the development of novel therapeutic modalities and be put to use by improving efficacy attributes of the considered product candidates before carrying out costly development programs (Anonymous 2007).

12.3 ACCESSING EXTERNAL AND INTERNAL INNOVATION AS STRATEGIC RESPONSE

Unless compensated by the commercialization of novel products with appropriate market values, patent expiration of franchise blockbuster products that can generate revenues up to several billion U.S. dollars per year for Big Pharma results de facto in a sizable decrease in the operating incomes of these companies. This jeopardizes their attractiveness to investors and as a result their long-term survival along with effects on their capacities to innovate. On aggregate, from 2009 to 2012, the pharmaceutical industry as a whole will lose patent protection on drugs that generate $74 billion in sales (Pfrang et al. 2009).

Empirical evidence indicates that a patent cliff of this scale generates a market pressure that leads to increased consolidation in the industry (involving both Big Pharma and biotechnology companies alike), with later stage compounds and cost-cutting prospects as the primary rationales for mergers (Munos 2009; Pfrang et al. 2009). Whether consolidation is per se a positive driver for innovation is unclear (Munos 2009). Beyond mergers and acquisitions (M&As), partnering of late stage compounds is increasingly viewed as a strategic response implemented by virtually all pharmaceutical companies (Goodman 2009).

For example, the percentage of alliances covering mid- to late-stage compounds rose from 10 and 24% in 1999 to 24 and 35% in 2009, respectively (Pfrang et al. 2009). Competition increased among Big Pharma companies for compounds at the clinical development stage. In response, the industry moved up the innovation value chain to earlier phase opportunities including pre-proof-of-concept compounds targeted for strategic areas, with sound scientific fundamentals and sound intellectual property (IP) defined at minimum by both a robust portfolio of composition of matter patents and clear freedom to operate (Behnke et al. 2007). The underlying learning from these observations is that both strong M&A capabilities and strategic alliance capabilities are now mainstream and no longer constitute differentiating competition factors because all pharmaceutical companies are now equipped with appropriate business development departments to access external innovation (Quinn 2000; Vertès 2004; Whittaker and Bower 2007).

Two main categories of pharmaceuticals currently reach the market: (1) small molecules and (2) large molecules. The chemistry of small molecules constitutes the fundamental strength from which large pharmaceutical companies were built. Their growth stemmed from early successes in this arena and more often than not on chemical industry legacies. On the other hand, large molecules (biologics) came of age with the approval of Humulin, a blockbuster recombinant insulin commercialized in 1982, and Herceptin, a blockbuster product commercialized 25 years after the first monoclonal antibody was produced in 1975 (Köhler and Milstein 1975; Anonymous 2005; Shepard et al. 2008; Vertès 2009) (Figure 12.1).

In 2007, the total sale percentage share of biologics by Big Pharma and large biotechnology companies reached double digits only for Roche, Amgen, and Genentech (EvaluatePharma). This observation suggests that the market for biologics (several classes of compounds including hormones, monoclonal antibodies, growth factors, fusion proteins, cytokines, therapeutic enzymes, recombinant vaccines, blood factors, and anticoagulants) is still in a period of active growth (Aggarwal 2009). Both main categories of pharmaceuticals are based on well understood technologies and market precedents. For the sake of simplicity, progress in these two fields can be viewed as incremental innovation since the corresponding technologies are now core competencies of Big Pharma that can efficiently deploy their internal and external innovation engines. Incremental innovation is important to keep large companies competitive in the short- to medium-term and is best realized through excellence in implementation and well oiled processes.

Mabs

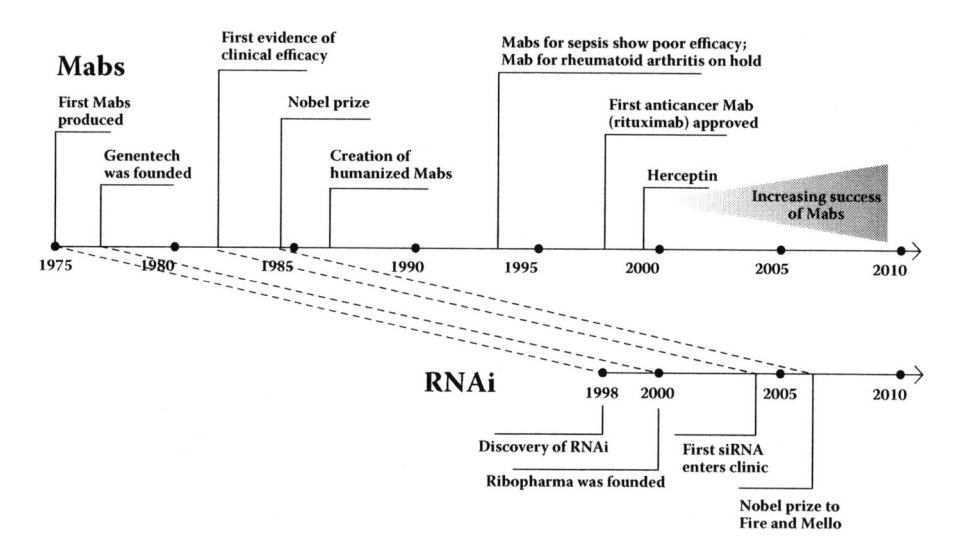

- First Mabs produced
- Genentech was founded
- First evidence of clinical efficacy
- Nobel prize
- Creation of humanized Mabs
- Mabs for sepsis show poor efficacy; Mab for rheumatoid arthritis on hold
- First anticancer Mab (rituximab) approved
- Herceptin
- Increasing success of Mabs

1975 ~1980 1985 1990 1995 2000 2005 2010

RNAi

1998 2000 2005 2010

- Discovery of RNAi
- Ribopharma was founded
- First siRNA enters clinic
- Nobel prize to Fire and Mello

FIGURE 12.1 Historic parallel between the development of monoclonal antibody and siRNA technologies. The commercialization of a radical innovation typically implies a large financial investment and a long timeline, as exemplified by the 25 years required for the first monoclonal antibody blockbuster to be commercialized.

Understandably, the pressure exerted by patent cliffs makes managers focus on short-term productivity goals. As a result, it is clear that the "race is on" for pharmaceutical companies to access late stage assets (focus on small molecules and biologics at the clinical stage of development) as a priority. This effort is exacerbated by the fact that many pharmaceutical companies face similar patent expiration hurdles and thus compete in the market for the same assets. Naturally, such focus on delivering short- to medium-term product options may make managers place on the back burner the parallel building of longer term product options.

This is clearly an issue because radical innovation is the main vector toward long-term growth that converts good companies to great ones. In adopting early (but not too early; see below) game-changing innovations, a company that has a strong position in one technology will maintain a competitive edge when the next technological paradigm comes of age (Galambos and Sturchio 1998; Leifer et al. 2000; Tushman and O'Reilly 2002). Clearly, the attention of managers to incremental innovation comes at the price of a diminishing ability to act on radical innovation (Christensen 1997). This is particularly true in large multinational corporations (Henderson 1993; Day et al. 2000; Colarelli et al. 2004) including Big Pharma.

Indeed, in large organizations, radical innovation may be inhibited not because managers do not understand the drivers of innovation (Markides 2009). More often than not they are keenly aware of them. In fact, the issue is acting upon new ideas (Markides 2009). Managers tend to favor what made them successful in the past, that is, what is familiar, what is mature, what is operationally simple, or what is near an existing solution (Shepard 1967; Ahuja and Lampert 2001; Tushman and O'Reilly 2002; Markides 2009). Moreover, radical innovation typically requires multiyear effort (a decade or more) and multimillion dollar cost. Radical innovation is also risky as exemplified by the electric automobile pioneering effort of General Motors (Welch 2008). Timing of entry into a new field is thus a critical parameter for large corporation managers to consider (Fenn and Raskino 2008). As was the case for General Motors, a premature entry may result in a loss of momentum and sunk costs that will act as deterrents to re-entry into the field after it has sufficiently matured (Welch 2008).

Notably, a typical hype cycle (Fenn and Raskino 2008) is observed during the early years when a paradigm-changing technology is implemented, as occurred with the commercialization of monoclonal antibody therapeutics. The initially high exuberance was quickly followed by a drop in interest and in the market value of the biotechnology companies active in the field. This resulted in a period of sell-offs until the emergence of a blockbuster product that rekindled the interests of scientists and investors to the new technology.

As a result, emerging technologies are best managed in large corporations such as Big Pharma through practices that differ from those typically applied to conventional development projects. For example, new entities such as idea incubators, innovation centers, and technology-focused centers of excellence alleviate the bureaucracy necessary to operate large firms and seem to constitute adequate proactive mechanisms to successfully conduct "internal entrepreneurship" (Ahuja and Lampert 2001; Colarelli et al. 2004; Roberts 2007). A highly diverse core group of internal "franchise entrepreneurs" and scientists who convey perspectives of various organizational groups and familiarize themselves with the fundamentals of the new technology and its markets are crucial (Colarelli et al. 2004; Roberts 2007). This is important for maintaining momentum, influencing change and advocating budget allocations to new projects, and appropriately assessing strategic alliance opportunities.

Building such internal capabilities is driven by a philosophy of "prudent optimism," whereby the new technology is probed, assessed, and de-risked by a strategy and implementation team. For example, the team may perform a landscape analysis of the start-ups active in the new area and conduct low-cost preliminary studies to enable the company to place a strategic bet at the appropriate time. Notably, in the cultural environments of large corporations, strong top-down strategic implementation mandates (as exemplified by the mechanisms underlying the creation at Pfizer of a regenerative medicine business unit under conditions of budget restrictions; Cookson and Jackson 2008) are crucial for translating ideas into commercial products. Cultural resistance to radically new concepts is a typical hurdle (Colarelli et al. 2004; Roberts 2007; Welch 2008). With visionary and proactive management, large corporations may successfully implement emerging technologies by leveraging their deep pockets, processes, infrastructures, and marketing capabilities. Moreover, alliances can be used to build competitive advantages in new technologies as both biotechnology companies specialized in the emerging technology of interest and Big Pharma present complementary capabilities (Hamilton and Singh 1992; Day et al. 2000).

The deployments at Roche of the monoclonal antibody and siRNA technologies are two interesting case studies. Their times of entry appear to coincide well with the appropriate scientific maturity of the technologies as estimated retrospectively by simple bibliometric methods and by the presence of biotechnology firms with the appropriate scientific competences (embodied in centers of excellence; Vertès 2009). The economic value that the monoclonal antibody technology brought to Genentech and Roche demonstrates that successful radical innovation brings long-term value to shareholders (Pollack 2009). It also poses the question whether the next critical success factor paradigm for Big Pharma after building strategic alliances and licensing departments may very well be in building specialized entrepreneurial departments to implement radical innovation.

Radical innovation is linked to high risk, high uncertainty, and high rewards. It is perhaps best defined as the creation of a new line of business for both the company and the market (Colarelli et al. 2004). Based on this definition, the technologies of therapeutic stem cells and regenerative medicine still constitute radical innovations because of the lack of commercial demonstration of the concepts (Isasi 2009; Pollack 2009) and the young age of the biotechnology companies active in the field (Lysaght and Hazlehurst 2003). As a result, their implementation in Big Pharma remains complex. Empirically, this translates in the market into deals between Big Pharma and biotechnology companies that involve relatively low payments, and particularly low upfront payments (Pollack 2009) or limited support from venture capital partnerships (Ledford 2008; see Chapter 5 by Prescott and Chapter 6 by Benjamin).

12.4 STEM CELL TECHNOLOGIES FIRST ADOPTED BY BIG PHARMA

The primary domains of research in which Big Pharma are active belong to the following therapeutic areas: (1) cardiovascular, (2) metabolic, (3) oncology, (4) inflammation, (5) CNS, and (6) infectious diseases. Investment by Big Pharma in the field of regenerative medicine (including therapeutic stem cells) has been modest to date (Pollack 2009; Smith 2009). In fact, Big Pharma embraced stem cell technologies first through the perspective of using stem cells as research tools to develop new small molecules (see Chapter 10 by Walker and Chapter 11 by Choo) that fit with the historical technical and implementation competences of these companies.

In addition to their usefulness in basic research to further fundamental knowledge of diseases in the areas listed above (Chien 2008), preparations of well differentiated cells such as hESCs isolated from blastocysts and iPS cells derived from the somatic cells of patients exhibiting certain diseases present the prospect of providing more accurate models of human diseases and thus establishing robust confidence in efficacies of candidate compounds (Stenger et al. 2001; McNeish 2004; Chien 2008; Hambor 2008; Lee et al. 2009; Chapter 10 by Walker). Similarly, stem cells are increasingly utilized in toxicology studies, particularly to assess cardiotoxicity, hepatotoxicity, genotoxicity, and epigenetic or reproductive toxicity (Davila et al. 2004; McNeish 2004; Webb 2009).

These tools have the potential to significantly reduce the attrition rates during conventional compound development and to enable practitioners to fail compounds faster and particularly before expensive clinical trials start, thus helping address a key issue of the pharmaceutical business (Wadman 2007). From 2002 through 2006, the industry success rates from lead initiations to the ends of Phase III clinical trials were 2% for small molecules and 12% for biologics (KMR 2007a and b); it will be informative to measure these metrics again after the use of stem cells for research has become routine.

This interest translates in the market to an increasing number of strategic alliances involving stem-cells-for-research technology platforms (Cookson and Jack 2008; Wallack 2008). Likewise, the cancer stem cells hypothesis, to which therapy-resistant cancer may perhaps be ascribed, constitutes an important development that can be leveraged immediately within the research organizations of Big Pharma (Reya et al. 2001; Dean et al. 2005; Clarke et al. 2006; Wicha et al. 2006),

Another close technological match with the traditional competences of Big Pharma is the development of compounds to modulate the activities of endogenous stem cells with the hope to promote in vivo their survival, proliferation, differentiation, reprogramming, and homing (or chemotaxis) characteristics (Ding and Schultz 2004; Chen et al. 2006; Fang et al. 2007; Daley and Scadden 2008; Xu and Ding 2008). While such an approach brings new challenges such as attaining appropriate confidence in safety, it fits well with the Big Pharma business model. Moreover, some compounds developed using more conventional approaches have already been observed to stimulate cell proliferation, e.g., Exendin 4, a glucagon-like peptide-1 analog that stimulates beta cell proliferation (Chen et al. 2006).

12.5 POTENTIAL BIG PHARMA BUSINESS MODELS FOR REGENERATIVE THERAPIES

The conventional Big Pharma business model is to develop off-the-shelf therapeutics (currently small molecules and biologics compounds) targeting mass markets. Based on this fundamental consideration, Big Pharma members do not exhibit the appropriate success factors to sell products that involve major service components. For example, autologous stem cell products require complex logistics and numerous, but very small scale, manufacturing operations (see Chapter 8 by Bravo and Blanco-Molina and Chapter 9 by Watt).

On the other hand, Big Pharma and autologous stem cell companies have areas of intersecting interest. For example, small molecules and biologics may enhance particular properties of stem cell preparations, as exemplified by the impact of dipeptidyl peptidase IV inhibitors on the engraftment properties of stem cells (Christopherson et al. 2004; Prabhash et al. 2009; Zaruba et al. 2009). These mutual interests may form the bases of strategic alliances such as out-licensing agreements from Big Pharma to autologous stem cell companies.

Based on their current stage of development, live cell therapeutics constitute terra incognita—products for the development of which Big Pharma presents few or none of the key success competencies required. Development of these products remains highly risky based on their lack of clinical demonstration (Isasi 2009) but admittedly they come with potentially high benefits (Figure 12.2).

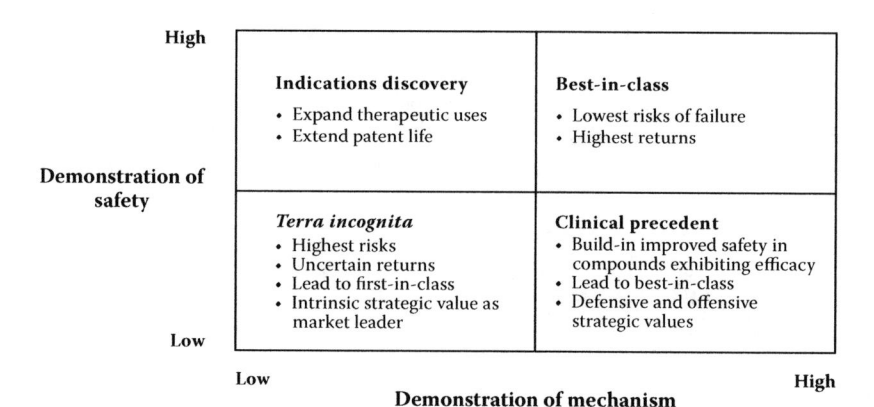

	Low — Demonstration of mechanism — High
High Demonstration of safety **Low**	**Indications discovery** • Expand therapeutic uses • Extend patent life	**Best-in-class** • Lowest risks of failure • Highest returns
	Terra incognita • Highest risks • Uncertain returns • Lead to first-in-class • Intrinsic strategic value as market leader	**Clinical precedent** • Build-in improved safety in compounds exhibiting efficacy • Lead to best-in-class • Defensive and offensive strategic values

FIGURE 12.2 Pipeline trade-off model. A balanced pipeline is essentially an expression of the optimization of operational risk according to two primary dimensions: confidence in safety and confidence in mechanism. Best-in-class products have typically yielded higher returns and benefitted from lower risks of failure and higher sales. Terra incognita compounds carry the greatest development risks and may not represent the highest returns (*Source:* Vertès, A. A. (2004). Towards a new valuation paradigm to create winning opportunities for venture capital companies. *Sloan Fellowship*, London Business School.)

Therapeutic stem cells fall into three main categories based on their possible clinical uses: (1) hESCs and iPS, (2) hematopoietic stem cells, and (3) mesenchymal stem cells (MSCs). Because of their pluripotencies, hESC and iPS can be expanded for many divisions and may be used to generate by differentiation virtually most or all types of specialized cells. Thus, in principle, they may have use in repairing damages to tissues or organs from disease or injury (Murry and Keller 2008; Passier et al. 2008; Lanza et al. 2009).

This prospect is of course extremely exciting (Bajada et al. 2008). However, the remaining hurdles to the successful commercialization of these cell preparations continue to hinder wide adoption of this technology for therapeutic purposes: (1) the ethical debate regarding the use of hESCs (Passier et al. 2008), (2) safety and particularly tumorigenicity risks (Hentze et al. 2006; Blum and Benvenisty 2008; Daley and Scadden 2008; Carpenter et al. 2009; Fink 2009; Yamanaka 2009), and (3) HLA barriers (Chidgey et al. 2008; Daley and Scadden 2008; Carpenter et al. 2009). Moreover, while preliminary signals of efficacy have already been reported in preclinical in vivo models (Passier et al. 2008), a clear-cut clinical demonstration is still lacking (Giordano et al. 2006).

On the other hand, adult stem cells (both hematopoietic and mesenchymal) offer the fundamental benefit that they can be readily isolated and expanded ex vivo. Unlike hESCs and iPS, differentiation is not a necessary step for generating a useful product from these cellular populations (Broxmeyer et al. 1989; Bobis et al. 2006; Kovacscovics-Bankowski et al. 2008; Nishino et al. 2009). Furthermore, beyond their natural ability to adhere to plastic, human MSCs in particular can be readily identified using a variety of positive (stro-1, CD106, CD73) and negative (CD11b, CD45, CD34, CD31, CD117) cell surface markers (Bensidhoum et al. 2004; Kolf et al. 2007). As a result, these cells can be manufactured via robust processes allowing well defined quality analysis and quality control procedures.

Manufacturability is a crucial consideration (Vertès 2009; Chapter 13 by Goswami and Chapter 14 by Preti); the technology of large molecules has been enabled primarily by the ability to manufacture large and complex molecules at industrial-scale volumes by fermentation. For example, insulin can be synthesized cost effectively using recombinant strains of *Escherichia coli* (Goeddel et al. 1979). It is also worth noting that MSCs can be isolated from a variety of widely available clinical samples including clinical waste: (1) bone marrow (Friedenstein et al. 1968; Kern et al. 2006), (2) umbilical cord blood (Kern et al. 2006; Avanzini et al. 2010), (3) placenta (Barlow et al. 2008; Battula et al. 2008), (4) Wharton's jelly (Seshareddy et al. 2008; Troyer and Weiss 2008), (5) dental pulp (Pierdomenico et al. 2005; Renard et al. 2007), and (6) adipose tissue (Zuk et al. 2004; Kern et al. 2006). This large availability of a variety of primary raw materials for manufacturing is an important parameter as it allows the development of an efficient manufacturing and logistics chain, thereby enabling the industrialization of live MSC therapeutics even in the absence of immortalized cell lines.

What is more, there are clinical precedents to the use of adult stem cell preparations derived from bone marrow for treating a variety of hematological and genetic diseases. The clinical practice of bone marrow transplantation is well established (Carpenter et al. 2009); notably, bone marrow comprises both hematopoietic and mesenchymal stem cells. It is also worth nothing that this technique dates back to 1951 and originates from the observation that mice survive lethal irradiation upon infusion of spleen or marrow cells (Lorenz et al. 1951; Thomas 2000; Buckley 2004). Hematopoietic stem cells can be used for autologous therapy or when a suitable histocompatibility complex match is possible (Broxmeyer et al. 1989).

While it is conceptually possible to generate and store appropriately a large number (up to a few thousand) of off-the-shelf hematopoietic stem cell preparations to match numerous patients, this intrinsic characteristic still represents an important initial complexity. On the other hand, MSCs appear to constitute the "low hanging fruit" of regenerative medicine: it is worth noting that these cells have been demonstrated clinically not to elicit proliferative responses from alloreactive lymphocytes and thus are "immunoprivileged" (Deans and Moseley 2000; Potian et al. 2003; Baksh et al. 2004; Barry and Murphy 2004; Le Blanc and Ringden 2005; Caplan and Dennis 2006; Keating 2006; Chamberlain et al. 2007; Kolf et al. 2007; Le Blanc and Ringden 2007; Nasef et al. 2007; Garcia-Castro et al. 2008; da Silva Meirelles et al. 2009; Djouad et al. 2009). Although definitive proof of clinical efficacy is still lacking (Javazon et al. 2004; Giordano et al. 2006), assuming appropriate efficacy, MSCs could be used allogeneically—an attribute that fits the business model of Big Pharma (Vertès 2009).

Nevertheless, MSCs are weakly HLA class I positive and thus questions remain regarding their safety when repeated infusions are used, for example, to treat chronic diseases (Dazzi et al. 2007; Tyndall et al. 2007). Moreover, it is possible that MSCs may play a role in the growth of epithelial solid tumors and this issue needs to be examined in more detail (Jorgensen 2009; Mishra et al. 2009).

In addition to their ability to evade the immune response and modulate it, MSCs exhibit a variety of attributes that can be leveraged to create novel therapeutic modalities: (1) secretion of a broad range of bioactive compounds including growth factors, cytokines, or chemokines (paracrine effects), (2) differentiation into a limited number of different lineages (chondrogenesis, osteogenesis, adipogenesis, myogenesis), (3) the presence of a molecular machinery to sense and respond to the local microenvironment (autocrine effects), and (4) accumulation at the site of molecular injury (pericyte nature, chemotaxis properties) (Deans and Moseley 2000; Potian et al. 2003; Baksh et al. 2004; Barry and Murphy 2004; Le Blanc and Ringden 2005; Caplan and Dennis 2006; Keating 2006; Chamberlain et al. 2007; Kolf et al. 2007; Le Blanc and Ringden 2007; Nasef et al. 2007; Crisan et al. 2008; Garcia-Castro et al. 2008; Bensidhoum et al. 2009; da Silva Meirelles et al. 2009; Djouad et al. 2009).

In addition, MSCs offer a platform to deliver drug-like effects because they do not appear to engraft, as observed, for example, in ischemic cardiac injury (Daley and Scadden 2008). Furthermore, MSCs could perhaps be used as delivery vehicles, for example for anti-cancer therapies (Jorgensen 2009). Based on these basic biological properties, MSCs appear to be potentially useful for developing novel therapies (Javazon et al. 2004; Uccelli et al. 2008) in numerous disease areas including (1) immunological diseases including autoimmune diseases (El-Badri et al. 2004; van Laar and Tyndall 2006; Le Blanc et al. 2008; Nöth et al. 2008; Dryden 2009), (2) cardiovascular diseases (Itescu et al. 2003; Brunnell et al. 2005), (3) CNS diseases (Jin et al. 2002; Cho et al. 2005), (4) pulmonary diseases (Lee et al. 2009), (5) diabetes (Timper et al. 2006; Liu and Han 2008; Sordi et al. 2009), (6) tissue engineering (Rosenbaum et al. 2008), (7) sepsis (Németh et al. 2009), and (8) wound healing (Badiavas et al. 2003).

MSCs therefore constitute an appropriate foundation for a technology platform that could be implemented to develop, manufacture, and commercialize products that strategically fit well with the business models of Big Pharma. Moreover, the competence factors that will be generated in so doing (in vivo and in vitro cell biology, cell characterization and analysis, small and large animal models, in vivo imaging techniques, manufacturing, logistics, formulation, and delivery) will most likely be translatable to the next wave of cell therapeutics that the iPS technology may enable in the future.

12.6 BUILDING NEW FRANCHISES IN THE TERRA INCOGNITA OF THERAPEUTIC STEM CELLS

The history of the technology of biologics and particularly that of humanized monoclonal antibodies (Köhler and Milstein 1975; Anonymous 2005; Shepard et al. 2008; Vertès 2009; Figure 12.1) suggests that the implementation of a new wave of therapeutics with totally novel attributes has the potential to revolutionize the biopharmaceutical industry and to provide great value to all stakeholders—patients, payers, and industries—despite initial resistance to investment and technical difficulties. "Adaptive medicine" represents a similar paradigm-changing concept that is as challenging as the monoclonal antibody concept was. The adaptive medicine concept could perhaps best be defined by answering the strategic breakthrough question whether novel medicines could: (1) optimize their responses to a disease environment, (2) present large safety margins, (3) exert similar effects at a large range of doses, (4) be activated only in diseased areas of the body, (5) be manufactured and distributed like other biologics, and (6) be leveraged to address previously unmet medical needs.

Allogeneic MSCs appear to have the potential to deliver most, if not all, of these characteristics. Indeed, stem cell therapeutics constitute a radical change from conventional approaches because live cell populations have been designed by millions of years of evolution to sense the conditions of their microenvironments and modulate their responses according to the local constraints they encounter. Interestingly, such sensing and responding properties of cells have long been harnessed by industrial microbiologists to design novel strains or processes (Palecek et al. 2002; Kobayashi et al. 2004). The reasons to believe that this vision of using MSCs to achieve some of the goals of adaptive medicine can become reality include (1) cell therapy is not a passing fad; it will transform medicine, (2) cells are not only transplants, they can act as drugs, (3) cell therapy's first paradigm-changing application is treating inflammation and autoimmune disease, (4) it is possible to protect the intellectual property of these new drugs, (5) they offer stunning efficacy with significantly reduced side effects, and (6) it is possible to consistently and cost efficiently manufacture these therapies and maintain their intrinsic attributes throughout the distribution chain.

The three fundamental pillars supporting these beliefs are: (1) strong science, (2) strong strategic fit, and (3) robust business case. The first two pillars have been discussed in the preceding paragraphs; the third pillar has seven essential components: (1) patents, (2) product concept, indications, and markets; (3) regulatory requirements, (4) manufacturing, (5) logistics, (6) delivery, and (7) pricing and reimbursement.

12.6.1 Patents

Patents have been granted for various stem cell-related inventions including composition-of-matter patents and patents covering the sources of human tissues from which stem cells are isolated (see Chapter 16 by Hitchcock and Crease and Chapter 17 by Resnick, Eisenstein, and McWilliams). The claiming strategies used by attorneys in the field include "(i) isolated stem cells; (ii) stem cell populations, differentiated and clonal, with and without expressing exogenous factors; (iii) methods to differentiate cells into various tissue types; (iv) differentiating stem cells in vitro and then implanting; (v) differentiating stem cells in vivo; (vi) use of cells to treat diseases or provide therapies" (Konski 2009).

Specifically, the presence and absence of a variety of cellular markers have been typically used by companies to define novel cellular populations and make composition-of-matter claims related to therapeutic adult allogeneic stem cells. Landscape analyses focused on iPS and hESCs demonstrate that the field is still emerging and that clusters of patents will continue to issue over the coming years (Bergman and Graff 2007; Konski and Spielthenner 2009; Webb 2009). Uncertainty about patents is almost always acute in an emerging field, as numerous patents are filed more or less simultaneously and few are granted, thus providing companies the freedom to operate until patent offices around the world make granting decisions. This constitutes an obvious business risk and intellectual property conflicts seem inevitable in the future. A sound intellectual property portfolio containing composition-of-matter patents, method-of-use patents, process patents, and freedom to operate is a prerequisite to a sound business. For this reason, detailed diligence is typically performed by Big Pharma when partnering with biotechnology companies.

12.6.2 PRODUCT CONCEPT, INDICATIONS, AND MARKETS

Differentiated products with superior efficacy and disease-modifying products will most likely be achieved when product attributes can be tailor-designed to treat specific indications. Since live cells used as therapeutics will adapt their biological responses to the specific molecular environments they encounter in each patient, it is more or less possible that various indications with fairly similar biological causes could be treated appropriately with the same cell preparations. Consequently, possible off-label use practice and its clinical and commercial impacts must also be assessed. Because MSCs have been subjected to numerous Phase I clinical trials without adverse events observed to date, it is possible to search earlier rather than later at the clinical stage for proof of efficacy and combine proof-of-concept and dose ranging trials (Orloff et al. 2009), especially for niche, orphan, or acute indications in which all other therapies failed (Giordano et al. 2006).

Success in such indications would pave the way for discovery efforts and further development leading to market expansion via indication discovery—a process whereby the field of use of a product is expanded in scope. Clearly, appropriate small and large animal model validation will still be necessary to accurately determine the right dose and more scientific depth must be covered and applied to models to understand sufficiently the biological responses of live cell preparations used as therapeutics and to optimize it (Vertès 2009). This is particularly true if one considers indications for which other treatments exist and particularly for treating chronic diseases.

Based on comparables from the industrial microbiology industry, numerous complementary techniques and countless experiments over two decades have been necessary to attain a sufficiently sophisticated understanding of the physiologies of bacterial populations in a reactor for performing specific biological tasks (Wendisch et al. 2006; Inui et al. 2007). Such intense scrutiny has not yet been applied to live cell preparations used as therapeutic agents.

Filters for indication selection are presented in Figure 12.3. Notably, it is unlikely that indications that require very large and expensive Phase III clinical trials will be researched or financed at first, given the risks and the competition for resources required for more conventional approaches.

12.6.3 REGULATORY REQUIREMENTS

While the regulatory framework is still evolving, a number of elements have been identified as important for designing preclinical animal studies, for example, by the

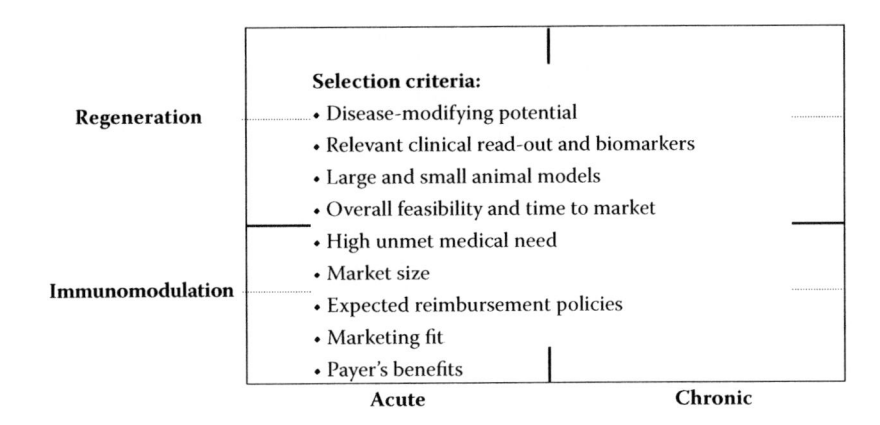

	Selection criteria:	
Regeneration	• Disease-modifying potential • Relevant clinical read-out and biomarkers • Large and small animal models • Overall feasibility and time to market	
Immunomodulation	• High unmet medical need • Market size • Expected reimbursement policies • Marketing fit • Payer's benefits	
	Acute	Chronic

FIGURE 12.3 Mesenchymal stem cell therapeutics and indication selection. Acute indications for which other available therapies have failed may constitute the most appropriate initial indications for live stem cell therapeutics. The new mechanisms leveraged may uniquely provide the complex molecular responses needed.

FDA as reported by Fink (2009): "(i) selection of relevant disease/injury models; (ii) testing of the product intended for clinical administration; (iii) using a route and method of delivery comparable to what is planned clinically; (iv) optimal timing of product administration with respect to disease/injury onset; and (v) an appropriate study duration that permits simultaneous assessment of potential adverse safety events and provides evidence of durable biological activity" (see Chapter 19 by Bravery and Chapter 20 Werner).

12.6.4 MANUFACTURING

As for any live cell preparation, manufacturing issues extend beyond compliance to good manufacturing practices, adequate production capacity to meet demand, the availability of raw materials, particularly the tissue samples from which stem cells are isolated (bone marrow, adipose tissue, placenta, umbilical cord blood), lot-to-lot reproducibility, and strong quality analysis and quality control (particularly determining the absence of contaminating agents). See Chapter 13 by Goswami and Chapter 14 by Preti. Of importance also is the possibility of using only animal-free ingredients, as a result of which developing serum-free media is an active area of research. Notably, lessons from the field of industrial microbiology can be applied here also, including the use of master and working cell banks and bioreactors.

Most likely, manufacturing know-how will constitute a strategic competence because the qualities and attributes (chemotaxis or "homing" properties) of live cell preparations (Bensidhoum et al. 2004; Karp and Teo 2009) will most likely depend on the medium used and the treatments applied, from tissue sample harvesting and distribution to the manufacturing plant to the fill–finish steps, and initial preservation of the individual stem cell doses. Last but not least, the cost of goods sold (COGS) is a crucial parameter that impacts pricing and profitability (Vanek 2008). While economies of learning are expected as cell therapeutic production increases, to become truly cost-effective, manufacturing must take place in a bioreactor. This points to the need to develop novel manufacturing processes and more accurately define the maximum number of passages (or number of cell divisions) to which cells can be practically subjected during ex vivo expansion.

16.6.5 Logistics

Storage at the points of bulk manufacturing, formulation, preservation, distribution, and use at the point of administration will probably exert a strong influence on the effectiveness of a product (see Chapter 14 by Preti). While learning from blood product and vaccine logistics issues can be applied here, it may be appropriate for a global company to implement a small number of manufacturing plants to serve major geographic markets in order to reduce the complexity of the logistics chain and mitigate production capacity limitation risks.

16.6.6 Delivery

The mode of administration of the stem cell preparation remains an important parameter that requires a convergence of practices involving preclinical research (large animal models) and the skills of clinicians (Scuderalli 2009). Systemic delivery via simple intravenous administration would be preferred but the observation of animal models treated with MSCs is that most of the cells injected become trapped in the lungs and do not reach the targeted organs (Fischer et al. 2009). Until the appropriate technology is developed to systemically deliver stem cells and appropriately reach target organs, local delivery is expected to remain a preferred route of administration via a catheter or simple surgery.

12.6.7 Pricing and Reimbursement

The reimbursement policy implemented in the major healthcare markets will have a significant impact, as the affordability of new drugs dictates the ability to develop a successful business, for example, to accompany the commercialization of live stem cell therapies (see Chapter 21 by Meurgey and Wille and Chapter 22 by Faulkner). Comparative effectiveness in particular will impact pricing and reimbursement policies. Evidence-based medicine is expected to become increasingly important as global health decision makers around the world request both economic and clinical evidence of new health technologies including cellular therapeutics (Faulkner 2009).

As reported by Faulkner (2009), significant reimbursement challenges are expected for "(i) costly therapies with only a marginal relative value proposition; (ii) competitive therapeutic areas; (iii) undefined patient populations when targeted populations are known; (iv) 'me-too' products or products with disadvantageous profiles; (v) products that are not cost-effective." As a result, ideal target product profiles should be defined early in the development process as a guide to conferring appropriate attributes to the new cell therapies so they can deliver superior efficacy and safety attributes to patients.

12.7 PERSPECTIVE

Despite the challenges that still lie ahead (e.g., tumorigenicity risks associated with the therapeutic use of hESC or iPS cell formulations, cost-effective manufacturing and delivery challenges even for MSC preparations), the fundamentals for commercializing live cell therapeutics by large pharmaceutical or biotechnology companies appear robust, with high likelihood of achieving efficient logistics, manufacturing methods, and supply chains. The result will be to provide healthcare professionals with off-the-shelf allogeneic cell therapies with superior efficacy and safety for a large array of indications for which conventional treatments have all but failed.

Nevertheless, even for the MSC therapeutics that are likely to become commercial realities in the not too distant future, pioneering efforts are still required by the pharmaceutical and biotechnology industries in partnership with academia to fully explore, on the one hand, the value of stem cell preparations as drugs and, on the other hand, techniques for leveraging their intrinsic biological (regenerative and immunomodulation) properties. Expanding from small molecules, biologics, and RNA therapeutics, genuine breakthroughs in the field of therapeutic stem cells would provide the medical professional with another class of pharmaceuticals to choose for treating patients with high and/or unmet medical needs. A clear and deeper understanding of the fundamental physiology of stem cell preparations, both as formulated drugs and after injections, is necessary to design products with superior therapeutic values that ideally complement the pharmacopeia achieved with other therapeutic modalities (Vertès 2009).

ACKNOWLEDGMENT

I thank Arnold Caplan for his critical review of the manuscript.

REFERENCES

Aggarwal, S. (2009). What's fueling the biotech engine? *Nat Rev Drug Discov* 27: 987–993.
Agusti, A. G. N. (2005). COPD, a multicomponent disease: implications for management. *Respir Med* 99: 670–682.
Ahmad, M. and J. Dreyfus (2007). *Type 2 Diabetes*. Waltham, MA: Decision Resources.
Ahuja, G. and C. M. Lampert (2001). Entrepreneurship in the large corporation: a longitudinal study of how established firms create breakthrough inventions. *Strat Mgt J* 22: 521–543.
Allison, M. (2009). Genzyme backs Osiris despite Prochymal flop. *Nat Biotechnol* 2: 966–967.
Anonymous (2007). Pharma 2020: the vision. Which path will you take? London: PriceWaterHouseCoopers.
Anonymous (2009). *Crohn's Disease*. Waltham, MA: Decision Resources.
Avanzini, M. A. and M. E. Bernardo et al. (2009). Generation of mesenchymal stromal cells in the presence of platelet lysate: a phenotypical and functional comparison of umbilical cord blood- and bone marrow-derived progenitors. *Haematologica* 94: 1649–1662.
Badiavas, E. V. and M. Abedi et al. (2003). Participation of bone marrow-derived cells in cutaneous wound healing. *J Cell Physiol* 196: 245–250.
Bajada, S. and I. Mazakova et al. (2008). Updates on stem cells and their applications in regenerative medicine. *J Tissue Eng Regen Med* 2: 169–183.

Baksh, D. and L. Song et al. (2004). Adult mesenchymal stem cells: characterization, differentiation, and application in cell and gene therapy. *J Cell Mol Med* 8: 301–316.

Barlow, S. and G. Brooke et al. (2008). Comparison of human placenta- and bone marrow-derived multipotent mesenchymal stem cells. *Stem Cells Develop* 17: 1095–1108.

Barnes, P. J. (2003). New concepts in chronic obstructive pulmonary disease. *Ann Rev Med* 54: 113–129.

Barry, F. P. and J. M. Murphy (2004). Mesenchymal stem cells: clinical applications and biological characterization. *Int J Biochem Cell Biol* 36: 568–584.

Battula, V. L. and S. Treml et al. (2008). Prospective isolation and characterization of mesenchymal stem cells from human placenta using a frizzled-9-specific monoclonal antibody. *Differentiation* 76: 326–336.

Behnke, N. and N. Hueltenschmidt et al. (2007). Partnering for proof-of-concept. *Bio-Pharma Partnering*. Winter.

Bensidhoum, M. and A. Chapel et al. (2004). Homing of in vitro expanded Stro-1$^-$ or Stro-1$^+$ human mesenchymal stem cells in the NOD/SCID mouse and their role in supporting human CD34 cell engraftment. *Blood* 103: 3313–3319.

Bergman, K. and G. D. Graff (2007). The global stem cell patent landscape: implications for efficient technology transfer and commercial development. *Nat Biotechnol* 25: 419–424.

Blum, B. and N. Benvenisty (2008). The tumorigenicity of human embryonic stem cells. *Adv Cancer Res* 133–158.

Bobis, S. and D. Jarocha et al. (2006). Mesenchymal stem cells: characteristics and clinical applications. *Folia Histochem Cytobiol* 44: 215–230.

Broxmeyer, H. E. and G. W. Douglas et al. (1989). Human umbilical cord blood as a potential source of transplantable hematopoietic stem/progenitor cell. *Proc Natl Acad Sci USA* 86: 3828–3832.

Brunnell, B. A. and W. Deng et al. (2005). Potential application for mesenchymal stem cells in the treatment of cardiovascular diseases. *Can J Physiol Pharmacol* 83: 529–539.

Buckley, R. H. (2004). Historical review of bone marrow transplantation for immunodeficiencies. *J Allergy Clin Immunol* 113: 793–800.

Campbell, D. and A. Chadwick-Jones et al. (2007). *Beyond the Blockbuster: Finding the Next Profit Zone in Pharmaceuticals through Business Design Thinking*. Boston: Oliver Wyman.

Caplan, A. I. and J. E. Dennis (2006). Mesenchymal stem cells as trophic mediators. *J Cell Biochem* 98: 1076–1084.

Carpenter, M. K. and J. Frey-Vasconcells et al. (2009). Developing safe therapies from human pluripotent stem cells. *Nat Biotechnol* 27: 606–613.

Chamberlain, G. and J. Fox et al. (2007). Concise review: mesenchymal stem cells: their phenotype, differentiation, capacity, immunological features, and potential for homing. *Stem Cells* 25: 2739–2749.

Chen, S. and S. Hilcove et al. (2006). Exploring stem cell biology with small molecules. *Mol Bio Syst* 2: 18–24.

Chidgey, A. P. and D. Layton et al. (2008). Tolerance strategies for stem-cell-based therapies. *Nature* 453: 330–337.

Chien, K. R. (2008). Regenerative medicine and human models of human disease. *Nature* 453: 302–305.

Cho, K. J. and K. A. Trzaska et al. (2005). Neurons derived from human mesenchymal stem cells show synaptic transmission and can be induced to produce the neurotransmitter substance P by interlukin-1 alpha. *Stem Cells* 23: 383–391.

Christensen, C. M. (1997). *The innovator's dilemma: when technologies cause great firms to fail*. Cambridge: Harvard Business Press.

Christopherson, K. W. and G. Hangoc et al. (2004). Modulation of hematopoietic stem cell homing and engraftment by CD26. *Science* 305: 1000–1003.

Clarke, M. F. and J. E. Dick et al. (2006). Cancer stem cells: perspectives on current status and future directions. *Cancer Res* 66: 9339–9344.

Colarelli O'Connors, G. and C. M. McDermott (2004). The human side of radical innovation. *J Eng Technol Mgt* 21: 11–30.

Collier, R. (2009). Drug development cost estimates hard to swallow. *Can Med Assn J* 180: 279–280.

Cookson, C. and A. Jack (2008). Pfizer to build 41 million pound UK stem cell centre. *Financial Times*, November 13.

Crisan, M. and S. Yap et al. (2008). A perivascular origin for mesenchymal stem cells in multiple organs. *Cell Stem Cell* 3: 301–313.

Dagani, R. (2005). The top pharmaceuticals that changed the world. *Chem Eng News.* 83(25).

Daley, G. Q. and D. T. Scadden (2008). Prospects for stem cell-base therapy. *Cell* 132: 544–548.

da Silva Meirelles, L. and A. M. Fontes et al. (2009). Mechanisms involved in the therapeutic properties of mesenchymal stem cells. *Cyotokines Growth Factors Rev* 20: 419–427.

David, E. and T. Tramontin et al. (2009). Pharmaceutical R&D: the road to positive returns. *Nat Rev Drug Discov* 8: 609–610.

Davila, J. C. and G. G. Cezar et al. (2004). Use and application of stem cells in toxicology. *Toxicol Sci* 79: 214–223.

Day, G. S. and P. J. H. Schoemaker et al. (2000). *Wharton on Managing Emerging Technologies.* Hoboken, NJ: John Wiley & Sons.

Dazzi, F. and J. M. van Laar et al. (2007). Cell therapy for autoimmune diseases. *Arthritis Res Ther* 9: 206–215.

Dean, M. and T. Fojo et al. (2005). Tumour stem cells and drug resistance. *Nat Rev Cancer* 5: 275–284.

Deans, R. J. and A. B. Moseley (2000). Mesenchymal stem cells: biology and potential clinical uses. *Exp Hematol* 28: 875–884.

DiMasi, J. A. and R. W. Hansen et al. (2003). The price of innovation: new estimates of drug development costs. *J Health Econ* 22: 151–185.

Ding, S. and P. G. Schultz (2004). Role for chemistry in stem cell biology. *Nat Biotechnol* 22: 833–840.

Djouad, F. and C. Bouffi et al. (2009). Mesenchymal stem cells: innovative tools for rheumatic diseases. *Nat Rev Rheumatol* 5: 392–399.

Dryden, G. W. (2009). Overview of stem cell therapy for Crohn's disease. *Expert Opin Biol Ther.* 9: 841–847.

El-Badri, N. S. and A. Maheswari et al. (2004). Mesenchymal stem cells in autoimmune disease. *Stem Cells Develop* 13: 463–472.

EvaluatePharma. http://www.evaluatepharma.com

Fang, Y. Q. and W. Q. Wong et al. (2007). Stem cells and combinatorial science. *Comb Chem High Throughput Screening* 10: 635–651.

Faulkner, E. C. (2009). Navigating the storm: considerations for health benefits providers in the evolving managed care business environment. http://www.hcwbenefits.com/documents/Faulkner.pdf.

Fenn, J. and M. Raskino (2008). *Mastering the Hype Cycle.* Cambridge: Harvard Business Press.

Fink, D. W. J. (2009). FDA regulation of stem cell-based products. *Science* 324: 1662–1663.

Fischer, U. M. and M. T. Harting et al. (2009). Pulmonary passage is a major obstacle for intravenous stem cell delivery: the pulmonary first–pass effect. *Stem Cells Develop* 18: 1–9.

Foley, S. (2004). Drug companies seek new cures as threats to their profits multiply. *Independent*, July 21.

Friedenstein, A. J. and K. V. Petrakova et al. (1968). Heterotrophic bone marrow: analysis of precursor cells for osteogenic and hematopoietic tissues. *Transplantation* 6: 230–247.

Galambos, L. and J. L. Sturchio (1998). Pharmaceutical firms and the transition to biotechnology: a study in strategic innovation. *Bus History Rev* 72: 250–278.

Garcia–Castro, J. and C. Trigueros et al. (2008). Mesenchymal stem cells and their use as cell replacement therapy and disease modeling tools. *J Cell Mol Med* 12: 2552–2565.

Garnier, J. P. (2008). Rebuilding the R&D engine in Big Pharma. *Harvard Bus Rev* May: 69–77.

Gates, C. and J. Dreyfus (2008). *Type 1 Diabetes*. Waltham, MA: Decision Resources.

Gilbert, J. and P. Henske et al. (2003). Rebuilding Big Pharma's business model. *In Vivo* 21(10): 1–10.

Giordano, A. and U. Galderisi et al. (2006). From the laboratory bench to the patient's bedside: an update on clinical trials with mesenchymal stem cells. *J Cell Physiol* 211: 27–35.

Goeddel, D. V. and D. G. Kleid et al. (1979). Expression in *Escherichia coli* of chemically synthesized genes for human insulin. *Proc Natl Acad Sci USA* 76: 106–110.

Goodman, M. (2009). Pharmaceutical industry financial performance. *Nat Rev Drug Discov* 8: 927–928.

Hambor, J. E., Ed. (2008). Breaking the stem cell technology barrier: designing renewable, physiologically functional cells for modern drug discovery. In *World Stem Cell Report 2008*. Siegel, B. et al., Eds. Washington: Genetics Policy Institute.

Hamilton, W. F. and H. Singh (1992). The evolution of corporate capabilities in emerging technologies. *Interfaces* 22: 13–21.

Henderson, R. (1993). Underinvestment and incompetence as responses to radical innovation: evidence from the photolithographic alignment equipment industry. *RAND J Econ* 24: 248–271.

Hentze, H. and R. Graichen et al. (2006). Cell therapy and the safety of embryonic stem cell-derived grafts. *Trends Biotechnol* 25: 24–32.

Inui, M. and M. Suda et al. (2007). Transcriptional profiling of *Corynebacterium glutamicum*: metabolism during organic acid production under oxygen deprivation conditions. *Microbiology* 153: 2491–2504.

Isasi, R. M. (2009). Registration of stem cell–based clinical trials: a scientific and ethical imperative. In *World Stem Cell Report 2009*. Siegel, B. et al., Eds. Washington: Genetics Policy Institute.

Itescu, S. and M. D. Schuster et al. (2003). New directions in strategies using cell therapy for heart disease. *J Mol Med* 81: 288–296.

Jack, A. (2009). Pharmas urged to focus spending on R&D. *Financial Times*, June 13.

Jammen, R. E. and C. Vasilakis-Scaramozza (2010). *Chronic Obstructive Pulmonary Disease*. Waltham, MA: Decision Resources.

Javazon, E. H. and K. J. Beggs et al. (2004). Mesenchymal stem cells: paradoxes of passaging. *Exp Hematol* 32: 414–425.

Jin, H. K. and J. E. Carter et al. (2002). Intracerabral transplantation of mesenchymal stem cells into acid sphingomyelinase-deficient mice delays onset of neurological abnormalities and extends their life span. *J Clin Invest* 109: 1183–1191.

Jorgensen, C. (2009). Link between cancer stem cells and adult mesenchymal stromal cells: implications for cancer therapy. *Regen Med* 4: 149–152.

Karp, J. M. and G. S. L. Teo (2009). Mesenchymal stem cell homing: the devil is in the details. *Cell Stem Cell* 4: 206–216.

Keating, A. (2006). Mesenchymal stromal cells. *Curr Opin Hematol* 13: 419–425.

Kern, S. and H. Eichler et al. (2006). Comparative analysis of mesenchymal stem cells from bone marrow, umbilical cord, or adipose tissue. *Stem Cells* 24: 1294–1301.

KMR Biologics Study (2007a). http://www.kmrgroup.com

KMR General Metrics Report (2007b). http://www.kmrgroup.com

Kobayashi, H. and M. Kaern et al. (2004). Programmable cells: interfacing natural and engineered gene networks. *Proc Natl Acad Sci USA* 101: 8414–8419.

Köhler, G. and C. Milstein (1975). Continuous cultures of fused cells secreting antibody of predefined specificity. *Nature* 256: 495–497.

Kolf, C. M. and E. Cho et al. (2007). Biology of adult mesenchymal stem cells: regulation of niche, self–renewal and differentiation. *Arthritis Res Ther* 9: 204–234.

Konski, A. F. (2009). *Patenting Mesenchymal Stem Cell Technology in the United States*. Chicago: Foley & Lardner.

Konski, A. F. and D. J. F. Spielthenner (2009). Stem cell patents: a landscape analysis. *Nature Biotechnol* 27: 722–726.

Kovacscovics-Bankowski, M. and P. R. Streeter et al. (2008). Clinical scale expanded adult mesenchymal stem cells prevent graft-versus-host disease. *Cell Immunol* 255: 55–60.

Lanza, R. and J. Gearhart et al. (2009). *Essentials of Stem Cell Biology*. San Diego: Academic Press.

Le Blanc, K. and F. Frassoni et al. (2008). Mesenchymal stem cells for *the treatment of steroid-resistant, severe, acute graft-versus-host disease: Phase II study*. Lancet 371: 1579–1586.

Le Blanc, K. and O. Ringden (2005). Immunobiology of human mesenchymal stem cells and future use in hematopoietic stem cell transplantation. *Biol Blood Marrow Transpl* 11: 321–334.

Le Blanc, K. and O. Ringden (2007). Immunomodulation by mesenchymal stem cells and clinical experience. *J Intern Med* 262: 509–525.

Ledford, H. (2008). In search of a viable business model. *Nat Rep Stem Cells*, October 30. http://www.nature.com/stemcells/2008/0810/081030/full/stemcells.2008.138.html.

Lee, G. and E. P. Papapetrou et al. (2009). Modelling pathogenesis and treatment of familial dysautonomia using patient-specific iPSCs. *Nature* 461: 402–406.

Lee, J. W. and X. Fang et al. (2009). Allogeneic human mesenchymal stem cells for treatment of *Escherichia coli* endotoxin-induced acute lung injury in the ex vivo perfused human lung. *Proc Natl Acad Sci USA* 106: 16357–16362.

Leifer, R. and C. M. McDermott et al. (2000). *Radical Innovation: How Mature Companies Can Outsmart Upstarts*. Cambridge: Harvard Business Press.

Liu, M. and Z. C. Han (2008). Mesenchymal stem cells: biology and clinical potential in type 1 diabetes therapy. *J Cell Mol Med* 12: 1155–1168.

Lorenz, E. and D. Uphoff et al. (1951). Modification of irradiation injury in mice and guinea pigs by bone marrow injections. *J Nat Cancer Inst* 12: 197–201.

Lysaght, M. J. and A. L. Hazlehurst (2003). Private sector development of stem cell technology and therapeutic cloning. *Tissue Eng* 9: 555–561.

Mack, G. S. (2009). Osiris seals billion-dollar deal with Genzyme for cell therapy. *Nat Biotechnol* 27: 106–107.

Markides, C. (2009). *The Innovation Solution*. London Business School. http://www.london.edu/newstandevents/news/2009/11/The_Innovation_Solution_1045.html

McCarthy, M. I. (2004). Progress in defining the molecular basis of type 2 diabetes mellitus through susceptibility gene identification. *Human Mol Genet* 13: R33–R41.

McEvoy, C. E. and D. E. Niewoehner (1997). Adverse effects of corticosteroid therapy for COPD. *Chest* 111: 732–743.

McNeish, J. (2004). Embryonic stem cells in drug discovery. *Nat Rev Drug Discov* 3: 70–80.

Mishra, P. J. and P. J. Mishra et al. (2009). Mesenchymal stem cells: flip side of the coin. *Cancer Res* 69: 1255–1258.

Munos, B. (2009). Lessons from 60 years of pharmaceutical innovation. *Nat Rev Drug Discov* 8: 959–968.

Murry, C. E. and G. Keller (2008). Differentiation of embryonic stem cells to clinically relevant populations: lessons from embryonic development. *Cell* 132: 661–680.

Nasef, A. and Y. Z. Zhang et al. (2007). Selected stro-1-enriched bone marrow stromal cells display a major suppressive effect on lymphocyte proliferation. *Int J Lab Hem* 31: 9–19.

Németh, K. and A. Leelahawanickul et al. (2009). Bone marrow stromal cells attenuate sepsis via prostaglandin E$_2$-dependent reprogramming of host macrophages to increase their interleukin-10 production. *Nat Med* 15: 42–49, 462.

Nishino, T. and K. Miyagi et al. (2009). Ex vivo expansion of human hematopoietic stem cells by a small–molecule agonist of c–MPL. *Exp Hematol* 37: 1364–1377.

Nissim, A. and Y. Chernajovsky (2008). Historical development of monoclonal antibody therapeutics. *Hbk Exp Pharmacol* 181: 3–18.

Nöth, U. and A. F. Steinert et al. (2008). Technology insight: adult mesenchymal stem cells for osteoarthritis therapy. *Nat Clin Pract Rheumatol* 4: 371–380.

Orloff, J. and F. Douglas et al. (2009). The future of drug development: advancing clinical trial design. *Nat Rev Drug Discov* 8: 949–957.

Palecek, S. P. and A. S. Parikh et al. (2002). Sensing, signalling and integrating physical processes during *Saccharomyces cerevisiae* invasive and filamentous growth. *Microbiology* 148: 893–907.

Passier, R. and L. W. van Laake et al. (2008). Stem cell-based therapy and lessons from the heart. *Nature* 453: 322–329.

Pfrang, P. and S. Elinson et al. (2009). *Acquisitions versus Product Development: An Emerging Trend in Life Sciences*. New York: Deloitte.

Pierdomenico, L. and L. Bonsi et al. (2005). Multipotent mesenchymal stem cells with immunosuppressive activity can be easily isolated from dental pulp. *Transplantation* 80: 836–842.

Pollack, A. (2009). Pfizer acquires a stem cell therapy. *New York Times*, December 21.

Pollack, A. (2009). Roche agrees to buy Genentech for $46.8 billion. *New York Times*, March 12.

Potian, J. A. and H. Aviv et al. (2003). Veto-like activity of mesenchymal stem cells: functional discrimination of responses to alloantigens and recall antigens. *J Immunol* 171: 3426–3434.

Prabhash, K. and N. Khattry et al. (2009). CD26 expression in donor stem cell harvest and its correlation with engraftment in human haematopoietic stem cell transplantation: potential predictor of early engraftment. *Ann Oncol*, September 16.

Quinn, J. B. (2000). Outsourcing innovation: the new engine of growth. *Sloan Mgt Rev* 41: 13–28.

Renard, E. and S. Lopez-Cazaux et al. (2007). Les cellules souches de la pulpe dentaire. *Compt Rend Biols* 330: 635–643.

Reya, T. and S. J. Morrison et al. (2001). Stem cells, cancer, and cancer stem cells. *Nature* 414: 105–111.

Roberts, E. B. (2007). Managing invention and innovation. *Res Technol Mgt*, January–February: 35–54.

Rosenbaum, A. J. and D. A. Grande et al. (2008). The use of mesenchymal stem cells in tissue engineering. *Organogenesis* 4: 23–27.

Scherer, F. M. (1993). Pricing, profits, and technological progress in the pharmaceutical industry. *J Econ Persp* 7: 97–115.

Scuderalli, M. (2009). The delivery dilemma. http://www.nature.com/stemcells/2009/0908/090806/full/stemcells.2009.104.html.

Seshareddy, K. and D. Troyer et al. (2008). Method to isolate mesenchymal-like cells from Wharton's jelly of umbilical cord. *Methods Cell Biol* 86: 101–119.

Shepard, H. A. (1967). Innovation-resisting and innovation-producing organizations. *J Bus* 40: 470–477.

Shepard, H. M. and P. Jin et al. (2008). Herceptin. *Hbk Exp Pharmacol* 181: 183–219.

Smith, D. (2009). Creating partnership with large pharma? Getting ready for the big day. In *World Stem Cell Report 2009*, Siegel, B. et al., Eds. Washington: Genetics Policy Institute.

Sordi, V. and M. L. Malosio et al. (2009). Bone marrow mesenchymal stem cells express a restricted set of functionally active chemokine receptors capable of promoting migration to pancreatic islets. *Blood* 106.

Stenger, D. A. and G. W. Gross et al. (2001). Detection of physiologically active compounds using cell-based biosensors. *Trends Biotechnol* 19: 304–309.
Thomas, E. D. (2000). Bone marrow transplantation: a historical perspective. *Med Ribeirão Preto* 33: 209–218.
Timper, K. and D. Seboek et al. (2006). Human adipose tissue-derived mesenchymal stem cells differentiate into insulin, somatostatin, and glucagon expressing cells. *Biom Biophys Res Commun* 341: 1135–1140.
Troyer, D. and M. L. Weiss (2008). Concise review: Wharton's jelly-derived cells are a primitive stromal cell population. *Stem Cells* 26: 591–599.
Tushman, M. L. and C. A. I. O'Reilly (2002). *Leading through Innovation*. Cambridge: Harvard Business Press.
Tyndall, A. and U. A. Walker et al. (2007). Immunomodulatory properties of mesenchymal stem cells: a review based on an interdisciplinary meeting, etc. *Arthritis Res Ther* 9: 301–331.
Uccelli, A. and A. Moretta et al. (2008). Mesenchymal stem cells in health and disease. *Nat Rev Immunol* 8: 726–736.
van Laar, J. M. and A. Tyndall (2006). Adult stem cells in the treatment of autoimmune diseases. *Rheumatology* 45: 1187–1193.
Vanek, P. G. (2008). Lonza cell therapy. *Regen Med* 3: 237–241.
Vertès, A. A. (2004). Towards a new valuation paradigm to create winning opportunities for venture capital companies. *Sloan Fellowship*, London Business School, London.
Vertès, A. A. (2009). Creating an effective clinical delivery plan for cell therapies. In *World Stem Cell Report 2009*, Siegel, B. et al., Eds. Washington: Genetics Policy Institute.
Wadman, M. (2007). New tools for drug screening. *Nat Rep Stem Cells*. http:www.nature.com/stemcells/2007/0712/071220/full/stemcells.2007.130.html.
Wallack, T. (2008). Harvard stem cell research gets boost. *Boston Globe*, July 25.
Webb, S. (2009). The gold rush for induced pluripotent stem cells. *Nat Biotechnol* 27: 977–979.
Welch, D. (2008). GM: live green or die. *Business Week*, May.
Wendisch, V. F. and M. Bott et al. (2006). Metabolic engineering of *Escherichia coli* and *Corynebacterium glutamicum* for biotechnological production of organic acids and amino acids.
Whittaker, E. and D. J. Bower (2007). A shift to external alliances for product development in the pharmaceutical industry. *R&D Mgt* 24: 249–260.
Wicha, M. S. and S. Liu et al. (2006). Cancer stem cells: an old idea: a paradigm shift. *Cancer Res* 66: 1883–1890.
Xu, Y. and S. Ding (2008). A chemical approach to stem cell biology and regenerative medicine. *Nature* 453: 338–344.
Yamanaka, S. (2009). A fresh look at iPS cells. *Cell* 137: 13–17.
Zaruba, M. M. and H. D. Theiss et al. (2009). Synergy between CD26/DPP-IV inhibition and G-CSF improves cardiac function after acute myocardial infarction. *Cell Stem Cell* 4: 313–323.
Zuk, P. A. and P. Benhaim et al. (2004). Stem cells from adipose tissue. *Hbk Stem Cells* 2: 425–447.

13 Role of Tool and Technology Companies in Successful Commercialization of Regenerative Medicine

Joydeep Goswami and Paul Pickering

CONTENTS

13.1 Introduction 177
13.2 Business Models in Regenerative Medicine 178
13.3 Cells for Research and Screening.. 178
13.4 Cells for Therapy 180
    13.4.1 Panel 1: Cell Culture Automation.. 184
        13.4.1.1 Direct Labor Substitution... 184
        13.4.1.2 Workflow Streamlining... 185
    13.4.2 Panel 2: Product Risk and Benefit Profile.. 185
        13.4.2.1 Exogenous Risks 185
        13.4.2.2 Intrinsic Risks 186
13.5 Conclusions... 187
References... 187

13.1 INTRODUCTION

Regenerative medicine has the power to radically change the healthcare system from management and mitigation of serious illnesses to interventions that can cure at a single stroke. While the promise of the field is accordingly very high, it is perhaps only at the beginnings of its development, as measured by the number of commercial regenerative medicine approaches on the market today and the appetites of traditionally risk-tolerant investors such as venture capitalists and large pharmaceutical companies. In this chapter, we will seek to parse out the key elements of commercialization in the context of regenerative medicine with a specific focus on cell-based therapies and outline with case studies the roles tools and technology providers can play in expediting the commercialization process.

13.2 BUSINESS MODELS IN REGENERATIVE MEDICINE

The emerging field of regenerative medicine provides several unique opportunities and challenges for companies that supply tools, reagents, and services to the industry. These opportunities and challenges can be classified into three broad categories:

- Scale—providing enough cells necessary for treatment or research, often from limited donor tissue
- Economics—providing cells at a price that makes therapy affordable or screening for drugs worth the extra cost
- Safety—cells free of undesired contaminating cell types and adventitious agents

From the perspective of the tools industry, the two broad types of uses for cells in regenerative medicine are (1) cells for research and screening and (2) cells for therapy.

13.3 CELLS FOR RESEARCH AND SCREENING

Research and screening for regenerative medicine requires cells in a variety of stages to study disease progressions and the effects of various molecules in treating these disease states. Until recently, the study of many of these diseases was limited to using animal cells or immortalized human cells. Unfortunately both of types have their drawbacks. Disease progression and regeneration processes in animal cells are often different from those dynamics in human cells.

Immortalized cells often lose their ability to function as normal cells in terms of physiological function, response to external stimuli, and other factors. Improvements in technologies to isolate and grow primary and stem cells from various human tissues and obtaining cells after differentiation of embryonic stem cells exhibit the potential to generate useful tools for drug discovery. For example, it is now possible to obtain dopaminergic neurons in various stages of differentiation from human embryonic stem cells that can serve as potentially invaluable tools for studying Parkinson's disease.

In general, cells for research and screening have lower bars for safety than cells used in therapy and can be produced in non-good manufacturing practice (GMP) or non-good laboratory practice (GLP) environments. However, they are often required in large numbers (more than a billion), particularly for secondary or tertiary screens for new molecules. Cost remains one barrier to wider adoption of these cells. In general, human primary and stem cells are five to ten times more expensive than their animal or immortalized counterparts. This sets the bar fairly high to prove that the screening results from these cells are better than the current ones used. To date, this has been demonstrated to be the case with human hepatocytes and other cells, but cell choice must be evaluated on a disease-by-disease basis. The cost of cells currently limits their use to secondary and tertiary screening.

Efforts are in progress to improve the usefulness of these cells by providing more defined culture conditions and niches. For several cell types, serum-free culture media have provided better characteristics, consistency, and often higher yields from a starting population. Systems such as co-culturing (e.g., hepatocytes with Kuppfer cells), emulating blood flow over cells, or providing a three-dimensional (3D) environment using materials such as Algimatrix™ (Gerecht-Nir et al. 2004; Dvir-Ginzberg et al. 2004) may provide more physiologically relevant models in an in vitro context.

Use of cells in research or screening often requires the study of gene and protein expression in response to disease progression or treatment with molecules. The traditional methods of post facto analysis (PCR, Western blot, ELISA, and FACS) provide answers but mostly require removal of cells from the environment and sacrificing them as part of the analysis. This precludes real-time and temporal analyses because the cells may differentiate into other types.

Engineering cells with promoter–reporter systems provides a good way to "follow" cells as they react and morph in response to external stimuli. In general, however, transfection of stem and primary cells faces two hurdles. First, these cells are often harder to transfect than "workhorse" lines such as Chinese hamster ovary cells and others. Second, the behavior of the transfected genes is often affected by the site of integration of the construct. Transfection using traditional lipids or electroporation has been difficult for many primary and stem cell lines (Masuda et al. 2009).

Virus-based transfection using lentiviral, retroviral, or adenoviral systems and more recently baculovirus-based systems such as BacMam (Kost et al. 2005) offer higher transfection efficiencies in a broad variety of cells, but creating viral particles requires specific laboratory clearances and expertise and longer lead times. Recent advances such as nucleofection (Lonza's Nucleofector®) and microporation (Invitrogen's Neon™) provide flexibility and higher transfection efficiencies, while overcoming some of the hurdles of cell death observed with traditional approaches (Masuda et al. 2009).

One of the drawbacks of traditional methods of gene transfection into cells has been that the genes integrate at random into the genome. This often leads to off-target effects that randomly and unpredictably perturb the system that is being studied. Technologies such as Cre and Flp have provided some control using directed integration into a defined site, but random excision and silencing of inserted genes during differentiation have been reported (Sauer 1994). More recent technologies such as the Rosa 26 locus (Irion et al. 2007) and technologies such as Invitrogen's Target™ and Jump-in™ (Thorpe et al. 2000) provide a much improved solution to unidirectional site-specific integration.

The problem of gene silencing can be handled by using insulators upstream of the genes. This requires technologies such as IRES or Invitrogen's Multisite Gateway that can package multiple genes in the same entry vector, and thus deliver them in one shot to a specified locus. These technologies also allow researchers to study the expression of more than one gene at a time, e.g., which genes in a pathway are turned on and off and at what times during differentiation.

A completely different approach to avoid random integration is to use episomal constructs that do not integrate into the genome. Systems like adenovirus and BacMam provide this advantage. Episomal constructs are, however, diluted on replication of the cells—a particular problem in stem cells. Most cell types can be retransduced multiple times with BacMam without apparent toxicity to overcome this problem of dilution. In addition, technologies such as Epstein-Barr virus nuclear antigen (EBNA) that allow replication of episomal constructs can be used in conjunction with BacMam.

There are three basic models for commercialization of screening tools. Companies can:

- Provide products such as cells and reagents or services that enable others to create assays to screen molecules for regenerative medicine. This model fits well with the traditional tools provider model and utilizes the infrastructure already in place at these companies. The risk assumed by the company is minimal and the benefit to the customer is somewhat limited, especially if the company is unfamiliar with handling cells. Most tool and reagent companies operate on this model.
- Contract to develop models for specific diseases, screen a library of compounds using the model, and deliver the results to the customer. This model offers customers a good solution to use the power of stem and primary cells while they are still building up their internal capabilities to work with these cells. The biggest hurdle is often interpretation of results obtained from a third party and the ability to compare the data with results obtained from traditional high throughput screening. This model is often adopted by start-ups that specialize in particular cell types, but is also used by larger tool companies that have service arms. The collaboration model is often for multiyear deals by which a customer pays for some dedicated resources and a substantial fraction of the payment is reserved for successful model development. Contract research organizations (CROs) have capabilities in adsorption, distribution, metabolism, excretion, and toxicology studies, but have generally not adopted this model for the more cutting edge cell types such as embryonic cells and cardiomyocytes. Several large pharmaceutical companies are also setting up deals with academic institutions such as Harvard University, Scripps Research Institute, and the Salk Institute. This is especially true in the burgeoning area of induced pluripotent stem cells (iPSCs) and disease-specific iPSCs.
- Develop their own models and compounds and plan to partner with a more traditional pharma or biotech to commercialize the compounds. This is so far a rarer model given the amount of investment required and lack of interest from venture capitalists to fund this type of model. However, companies such as iPerian are starting to adopt this model based on being able to obtain disease-specific iPSCs and create models that are specific to neurodegenerative diseases (see Chapter 10 by Walker).

13.4 CELLS FOR THERAPY

Successful commercialization of regenerative medicine depends on a strong business plan that can be divided into three key elements:

Successful Commercialization of Regenerative Medicine

- Proposed intervention and its ability to meet a clearly defined, reimbursable, unmet medical need at a scale that offers an attractive revenue stream
- Ability to control manufacturing costs to generate gross margins to at least be comparable to margins of industries with analogous regulatory clearance processes e.g., biologics license application (BLA), new drug application (NDA), and 510K or premarket approval application (PMA) in the U.S. for the biotech, pharma, and medical device industries, respectively
- A credible and ideally accelerated path through the regulatory hurdles including plans to manage gaps in regulatory coverage created by the evolution of regenerative medicine

Investors in regenerative medicine-based approaches must be assured that management teams have clearly thought through all the key elements and have well developed plans to address them. Selection of the intervention and concomitant indication strategies are covered elsewhere in this book and are also well summarized by McAllister et al. (2008). The ability to manufacture cost effectively and expeditiously and navigate regulations can be greatly influenced by the choice of appropriate technology partners. From the perspective of a tool or reagent company, a few distinct cell therapy models must be served. These models are based on the extent of centralization of processing of cells and the degree of manipulation of cells (Figure 13.1).

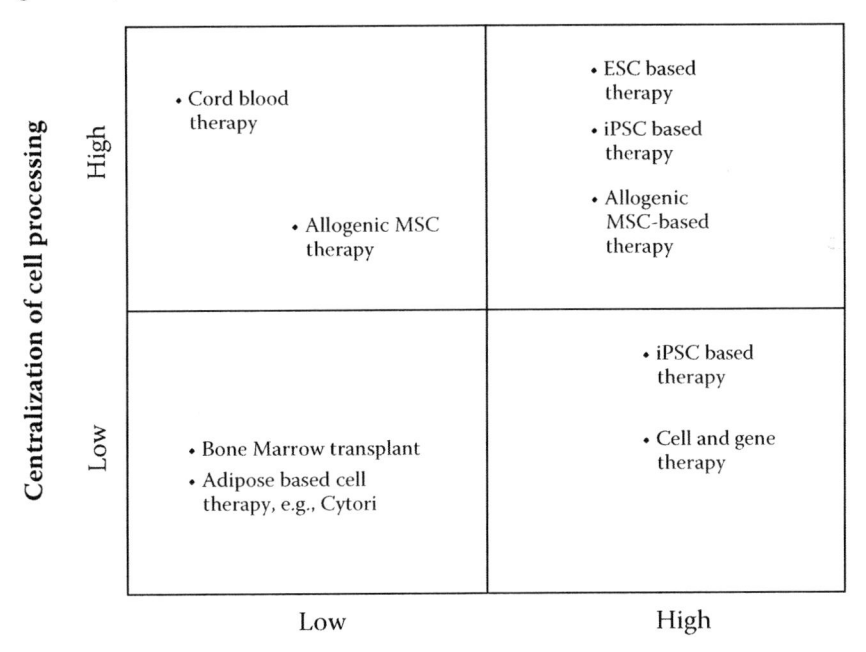

	Low	High
High	• Cord blood therapy • Allogenic MSC therapy	• ESC based therapy • iPSC based therapy • Allogenic MSC-based therapy
Low	• Bone Marrow transplant • Adipose based cell therapy, e.g., Cytori	• iPSC based therapy • Cell and gene therapy

Centralization of cell processing

Degree of cell manipulation

FIGURE 13.1 Classifying therapies based on extent of manipulation and centralization of processing.

Bone marrow transplants and therapies involving minimally processed autologous cells, e.g., Cytori's fat cell-based therapies (Alt et al. 2009), are usually performed at the point of care and require very little manipulation of the cells (usually a few enzymes and centrifugation). On the other hand, therapies such as cord blood also require very little processing of cells but the cells are often processed centrally and cryopreserved. Generally, tool-based companies focusing on minimally processed cell therapies tend to focus on instruments (e.g., specialized centrifuges) and disposables (e.g., cartridges, cryopreservation pouches). The primary objectives are ensuring a sterile, closed environment and enabling simplified (often automated) functionality with minimal intervention. There are also service models, in which the service provider isolates the right types of cells, cryopreserves, and banks them. An example of such a service is that provided by cord blood banks (see Chapter 9 by Watt).

Therapies with more extensive processing, for example, those which would not comply with the current FDA guidance (2009) for cells or nonstructural tissues (processing that does not alter the relevant biological characteristics of cells or tissues), present a whole different and often complex set of challenges around scale, economics, and safety. The way these challenges are addressed are often quite different based on the degree of centralization of cell processing. Therapies such as cell and gene therapy, where cells are infused with a gene of interest and then delivered to the patient (e.g., T cell therapy overexpressing specific cytokines) fall into the category of highly manipulated and decentralized processes.

Autologous iPSC-based therapy—using primary cells from a specific patient, reprogramming the cells, and differentiating them back into a desired cell type for treatment of the same patient—also falls into this category. Cells can be obtained from donors, undergo processing (expansion, differentiation, purification, transfection), and be reintroduced into the patient at the point of care. Generally these therapies are autologous. The scale for each treatment (single dose) tends to be fairly small (for a single patient). In this respect, such therapies are closest to being considered personalized regenerative medicines. Since the scale of the manufacturing operation (footprint) is small, the focus shifts toward economics and safety. Safety can be addressed by reagents produced under GMP conditions and compatible with FDA 510K classification.

Such media and culture reagents are already available for several cell types including iPSCs, MSCs, and keratinocytes among others. Increasingly such reagents are also free of xeno (non-human) components. Reactor systems that are automated and are closed during process operations such as media changes and passaging can dramatically increase the safety of cells for treatment. In addition, the ability to grow a large number of cells in a small volume reduces the need for handling multiple flasks that can increase the risk of error. Further, to minimize the risk of contamination, systems, especially those that come in contact with the cells, should be modular and disposable.

Another aspect of safety is having the tools to assay the cells for purity, sterility, and potency prior to delivering the dose to the patient and then remove any undesired cell types. The assessment today is primarily carried out with cell specific antibodies and quantitative PCR.

The economics of the single dose process are mostly driven by the man hours of processing and paperwork involved to ensure that the process, reagents, and instruments are compatible with clinical use. Interestingly, the types of qualified reagents and reactors described above also radically reduce the labor involved in ensuring the safety and compliance with Institutional Review Board (IRB) and Food & Drug Administration (FDA) standards. Another aspect of reducing effort at the clinician's end requires robust reagents and protocols to isolate, grow, and differentiate cells. This ensures that minimum modification, monitoring, and intervention are required.

Several business models have been devised to serve this market. Companies can supply novel reactor configurations and associated disposables as a consumable stream. For economies of scale, these reactor configurations must be compatible with multiple cell types. Another opportunity is supplying GMP grade, 510K-compatible culture and isolation reagents. After-sales support is a critical element of providing these reagents. Custom-produced reagents represent another option although the small volumes required often make the cost of the reagents too high. New service models may also emerge for commonly used but complex unit operations such as transfection. The biggest limitation here is the lack of consistent demand and the difficulties in shipping and recovering cryopreserved cells. Multisite handling of cells raises concerns of sterility and contamination.

Therapies in which the cells are processed in a central facility and distributed to multiple points of care—the "cells in a bottle" approach—are closest to the traditional model followed by pharma and biotech. These therapies generally are for cells that can be expanded severalfold and can be used for allogeneic treatment. The scale of manufacture can run into hundreds of billions or trillions of cells per batch. Providers of these therapies generally make substantial investment in GMP facilities to process cells in a sterile environment. The economics are primarily driven by productivity of the media and reduction of steps of processing (automation, less cell passaging, and less processing equipment). In addition to the common safety concerns of prevention of contamination with adventitious agents and maintenance of sterility, given the extent of expansion, karyotypic stability of the cells is often a concern that needs to be measured and addressed.

To address this problem, instrument providers supporting these customers are focused on providing reactors that can be semi-automated and offer a large processing capability with a small footprint. Equipment is often customized for the needs of specific customers. The focus of reagent providers is to have culture reagents that can deliver sufficient expansion and targeted differentiation of cells in minimum time while maintaining cell potency and minimizing chances of karyotypic instability. The focus is on GMP-grade and 510K-approved (or at least compatible) reagents that are also xeno-free or chemically defined to minimize safety concerns.

In this context there is a strong push to serum-free formulations, since studies in several cell types have shown that cells grown without serum are not only safer but more robust and potent. Given the scale of operations, media formulations can be customized to the needs of a particular process. There is a move on several cell types towards suspension culture which radically changes the footprint of equipment and effort required. With several stem cell types, the development of reagents to purify specific populations of cells (e.g., remove embryonic stem cells from a population of differentiated cells) to ensure consistency between batches is gaining in importance.

13.4.1 Panel 1: Cell Culture Automation

The processing complexity of cells cultured for therapeutic applications combined with strict adherence to highly technical protocols in demanding operating environments often requires intensive use of expensive specialist skill sets. This often leads to labor and associated facility overhead that represent the greatest costs cell therapy developers must manage. In any cell therapy workflow, multiple unit operations may include tissue digestion, cell isolation, expansion, differentiation, purification, washing, final formulation, fill, and finish. The length and complexity of unit operations associated with cell culture often extend from weeks to months, making cell processing the focus of much labor productivity analysis.

Automation offers the opportunity to minimize or even eliminate the need for costly and highly skilled labor associated with cell therapy processes. Aside from minimizing skilled labor costs, related benefits include enhanced process reproducibility and compliance along with greatly reduced adventitious agent risk. Further savings in facility capital and operating expenses are also possible due to the ability to reduce considerably the operational footprint that must be subject to strict levels of environmental control and monitoring (Fitzpatrick 2008). Finally, employing an automation strategy greatly facilitates operational scalability, especially in relation to new locations that may be required from a logistical view due to shelf life issues.

Several cell therapy companies have successfully implemented automation, most notably Organogenesis which, after bankruptcy in 2001, was able to bring its cost of goods target low enough to achieve sustainable profitability (Lysaght et al. 2008). Implementation of automation has been a key part of Organogenesis's success. In an industry more commonly associated with cost of goods reaching tens of thousands of dollars levels, Organogenesis achieved per-dose costs less than a tenth of common levels (McAllister et al. 2008; McKay 2009).

Cell culture automation needs significant design, implementation, and capital investments. Often implementations need to be customized to the specific needs of a particular culture process, reducing the ability to enjoy economies of scale for the automation platform. Significant service and support are likely to be required post-implementation, reducing potential labor savings. Finally intrinsic variability in the cell biology may put automation out of the realm of consideration. This is an especially important issue in autologous-based approaches in which patient-to-patient variability in the starting material is an intrinsic factor of the process. Two main approaches can be employed in relation to cell culture automation: direct labor substitution and workflow streamlining.

13.4.1.1 Direct Labor Substitution

In this instance the existing culture vessel, often in cell therapy applications a T-flask or some two dimensional single or multilayer variant thereof, is kept, but external manipulations of that vessel such as media changes and cell passaging are

Successful Commercialization of Regenerative Medicine

handled through automation. Here the labor of a highly skilled operator is directly substituted by automation. In addition to the general cost savings outlined above, this approach has the benefit of maintaining a high level of process consistency. In principle this transitional consistency mitigates regulatory concerns which arise whenever a new processing approach is sought to be implemented in a clinical environment. An example of this type of approach in an "off the shelf" format is the CompaT SelecT™ plaform developed by The Automation Partnership (Thomas et al 2008).

In spite of their flexibility and relatively ready implementation into a preclinical or clinical stage process, the sophisticated robotics required to perform such extensive manipulation of existing culture ware result in substantial cost, mitigating against their use at low throughputs.

13.4.1.2 Workflow Streamlining

The second approach to culture optimization is a more semi-batch approach using proprietary bioreactor designs with predefined disposable fluid paths. By letting the fluids do the work, a significant level of sophistication can be removed from the automation employed, resulting in greatly reduced capital costs. An example of this approach is Aastrom's CPS system (Koller et al. 1998). Aside from the lower capital costs, additional advantages may be realized from the proprietary nature of the bioreactor such as a small culture footprint due to enhanced surface area-to-volume ratio and more efficient mass transfer. The lower capital costs also improve the feasibility of the so called point-of-care-based cell therapies.

In contrast to the direct labor substitution approach, the ability to "plug in" this type of automation becomes more difficult from a validation standpoint, depending on the configuration and design of the bioreactor.

13.4.2 Panel 2: Product Risk and Benefit Profile

In determining whether to allow a commercial license and marketing of a new drug or biologic entity, regulatory bodies seek to balance patient benefit against unmet medical need based on the safety profile of the proposed intervention. Although safety and efficacy vary by product, this section seeks to highlight common risks associated with cell-based therapies and categorized as exogenous (initially external to the cell) and intrinsic (part of the underlying cell biology).

13.4.2.1 Exogenous Risks

One of the greatest exogenous risks to the safety of a cell therapy arises from the introduction of adventitious agents. The two most common potential sources are operators during cell manipulation and raw materials derived from animal or human sources. Unlike small molecules and to a significant extent more conventional protein-based biologics, after a lot of material is contaminated with an adventitious agent, very few options allow its removal through further processing. Consequently the patient risks posed by adventitious agents are serious and must be addressed proactively in the design of a manufacturing process for clinical use.

13.4.2.1.1 Cell Manipulation Risks

Humans are often exposed to and carry pathogenic organisms. The agents of greatest concern are viral and mycoplasma contamination. Aside from well controlled and monitored operating environments and trained and appropriately gowned operators, one of the best ways to minimize exogenous risks is to implement functionally closed processing systems, often characterized as disposables. The use of automated systems to remove operators from proximity to a product offers an additional level of risk mitigation against exogenous adventitious agents.

13.4.2.1.2 Raw Material Risks

Raw materials by definition come into direct contact with the cell product. Unlike the cell manipulation case, there is a 100% probability that any adventitious agent present in a raw material will be exposed to cells used for therapeutic intervention. In addition to viral and mycoplasma risks, animal origin components such as fetal bovine serum present further risks of transmissible spongiform encephalopathy (TSE) introduction; cross-species immunologic reactions are possible.

Although steps such as irradiation can be taken to mitigate adventitious agent risks for such animal origin components, due to the consequences of a failure in this approach, the preference now is to remove such components entirely from the culture process. To this end, Life Technologies developed a suite of complementary xeno-free systems whose only intrinsic adventitious agent risk comes from components derived from screened human blood and used in the national blood product supply chain (Lindroos et al. 2009; Rodriguez-Piza et al. 2009).

13.4.2.2 Intrinsic Risks

Most cell therapies are dependent on a particular genetically stable phenotype and their effectiveness often depends on maximizing the desired population for delivery to the patient. Conversely, it is often necessary to remove certain undifferentiated cell populations from a therapeutic dose because they could potentially harm the patient.

- **Undesirable cell populations**—With several stem cell types, the development of reagents to purify specific populations (e.g., removal of ESCs from a population of differentiated cells) to ensure consistency of batches is gaining in importance. The most common path today is to use antibodies on conjugated magnetic beads that recognize surface markers on cells, for example, Invitrogen's SSEA-4 Dynabeads®. It is more common to remove (deplete) the undesirable cell types rather than isolate the desired cell type because there is no need to remove the antibody or the bead from the desired type. Bead technology can bring about several log reductions in numbers of undesirable cells and the beads can be manufactured under GMP conditions. The cost of GMP antibodies in early clinical trials remains an issue; however, new technologies and appropriate risk mitigation for recombinant antibodies could help address this issue. There is also a potential to use small molecules that selectively kill the undesired populations of cells while leaving the target population unharmed. Another idea is the use of suicide

genes that are turned on only by promoters that are active in the undesirable cell types. However, other issues involved in transfecting foreign DNA into a clinical cell population make this an option of last resort.
- **Genetic stability**—To ensure lower costs and sufficient doses for patients, it is critical to be able to rapidly expand cells in many therapies. Unfortunately, the manyfold expansion of cells in an ex vivo environment can also trigger karyotypical abnormalities. Karyotypic abnormalities and mutations are generally stochastic events, but the wrong culture conditions can accelerate the pace at which the mutations accumulate. Culture reagents and conditions to grow and differentiate cells for therapy must be screened for maintenance of karyotypic stability over multiple passages. Also, current methodologies such as G banding to assess karyotype are often subjective and insufficiently accurate. More sensitive and less subjective methods to assess karyotype that are also cost effective need to be developed.

13.5 CONCLUSIONS

The successful commercialization of regenerative medicine platforms requires detailed analysis concerning issues of scale, cost, and safety. Tool providers offer significant expertise in addressing many of these challenges through all states of a regenerative medicine workflow (isolation, expansion, differentiation, and storage). In this chapter we have sought to outline the depth of thought required to resolve these issues and discuss options available from tool providers to address them. Developers of commercial regenerative medicine platforms should be encouraged to engage early with tools-based partners to leverage their experiences in order to minimize the resources and time required to get a product to market.

REFERENCES

Alt, E., Pinkernell, K., Scharlau, M. et al. (2010). Effect of freshly isolated autologous tissue resident stromal cells on cardiac function and perfusion following acute myocardial infarction. *Int J Cardiol* (in press; corrected proof available online), May 13.

Dvir-Ginzberg, M., Elkayam, T., Aflalo, E. et al. (2004). Ultrastructural and functional investigations of adult hepatocyte spheroids during in vitro cultivation. *Tissue Eng* Nov–Dec, 10(11–12): 1806–1817.

Gerecht-Nir, S., Cohen, S., Anna-Ziskind, A. et al. (2004). Three-dimensional porous alginate scaffolds provide a conducive environment for generation of well vascularized embryoid bodies from human embryonic stem cells. *Biotechnol Bioeng* 88: November 5, 2004.

Fitzpatrick, I. (2008). Cellular therapy success through integrated automation: anticipating scale-up of vaccines and other live cellular products. *Bioprocess Int* 6(9): 32–37.

Irion, S., Luche, H., Gadue, P. et al. (2007). Identification and targeting of the ROSA26 locus in human embryonic stem cells. *Nat Biotechnol* Dec, 25(12): 1477–1482.

Koller, M. R., Manchel, I., Maher, R. J. et al. (1998). Clinical-scale human umbilical cord blood cell expansion in a novel automated perfusion culture system. *Bone Marrow Transpl* 21: 653–663.

Kost, T.A., Condreay, J. P., and Jarvis, D. L. (2005). Baculovirus as versatile vectors for protein expression in insect and mammalian cells. *Nat Biotechnol* 23(5): 567–575.

Lindroos, B., Boucher, S., and Chase, L. et al. (2009). Serum-free, xeno-free culture media maintain the proliferation rate and multipotentiality of adipose stem cells in vitro. *Cytotherapy* 11: 958–972.

Lysaght, M. J., Jaklenec, A., and Deweerd, E. (2008). Great expectations: private sector activity in tissue engineering, regenerative medicine, and stem cell therapies. *Tissue Eng A* 14: 305–315.

Masuda, K., Ishikawa, Y., Onoyama, I. et al. (2010). Complex regulation of cell-cycle inhibitors by Fbxw7 in mouse embryonic fibroblasts. *Oncogene* 9.

McAllister, T. N., Dusserre, N., and Maruszewski, M. et al. (2008). Cell-based therapeutics from an economic perspective: primed for a commercial success or a research sinkhole? *Regen Med* 3: 925–937.

McKay, G. (2009). Leading in regenerative medicine. Paper presented at Cell Therapy Industry Summit, Carlsbad, CA.

Rodriguez-Piza, I., Richaud-Patin, Y., Vassena, R. et al. (2009). Reprogramming of human fibroblasts to induced pluripotent stem cells under xeno-free conditions *Stem Cells* 28: 36–44.

Sauer, B. (1994). Site-specific recombination: developments and applications. *Curr Opin Biotechnol* 5: 521–527.

Thomas, R. J., Chandra, A., Hourd, P. C. et al. (2008). Cell culture automation and quality engineering: a necessary partnership to develop optimized manufacturing processes for cell-based therapies. *J Assn Lab Automation*, June 2008, 3: 152–158.

Thorpe, H. M., Wilson, S. E., and Smith, M. C. (2009). Control of directionality in the site-specific recombination system of Streptomyces phage phi C31. *Mol Microbiol* 38: 2000.

U.S. Food & Drug Administration, Center for Biologics Evaluation and Research (2009). *Guidance for Industry: Minimally Manipulated, Unrelated Allogeneic Placental/Umbilical Cord Blood Intended for Hematopoietic Reconstitution for Specified Indications.*

14 Key Considerations in Manufacturing of Cellular Therapies

Robert A. Preti

CONTENTS

14.1 Introduction 190
14.2 Lessons Learned 191
14.3 Quality Systems Considerations and Manufacturing Infrastructure 193
14.4 Cell Therapy Manufacturing Program 194
    14.4.1 General Methods for Product Isolation and
              Prevention of Contamination and Cross Contamination 194
    14.4.2 Process Descriptions and Process Flow 196
          14.4.2.1 Cord Blood Manufacturing Process 196
          14.4.2.2 Dendritic Cell Manufacturing Process 197
    14.4.3 Facilities 197
    14.4.4 Cell Therapy Patient Components and Product Flow 198
14.5 Quality Systems 199
    14.5.1 Elements of Quality Systems 199
          14.5.1.1 Document Control 199
          14.5.1.2 Change Control 199
          14.5.1.3 Internal Auditing 200
          14.5.1.4 Supplier Qualification and Auditing 200
          14.5.1.5 Quality Agreements 200
          14.5.1.6 Personnel Training and Qualification 200
          14.5.1.7 Job-Specific Skills 201
          14.5.1.8 Deviation and Corrective and Preventive Action
                   (CAPA) Programs 201
          14.5.1.9 Non-Conforming Materials and Products Systems 201
          14.5.1.10 Laboratory Control Systems 202
          14.5.1.11 Outcome Analysis: Models for Manufacturing 203
14.6 Make versus Buy: The Decision to Outsource 204
    14.6.1 Cost Sensitivity 204
    14.6.2 Autologous versus Allogeneic Production 207
    14.6.3 Transportation 208
    14.6.4 Delivery at Bedside 209

        14.6.5 Fresh versus Frozen Products 209
        14.6.6 Automation 210
14.7 Conclusions 210
References 211

14.1 INTRODUCTION

Much like a developing child, the evolving and ever maturing cell therapy and regenerative medicine industry has faced, continues to face, and will continue to face its share of challenges. As if not treacherous enough, the many moving parts associated with developing products in a developing industry pile risk on top of risk, leaving more uncertainty than comfort as we progress an idea toward a therapeutic. In many respects, as a therapy's "parents," we have not surprisingly tended to focus on those risks and on their uniqueness. We do this for a simple reason: we are often stumped for solutions and while we look to previous industries (drugs, devices, blood, biologics) for answers, indeed clues, to solve some of them, we are often left with approaches that approximate what we need but fall short of filling our special niche in just the right way.

And at the risk of running this parenthood metaphor just a bit too far, what inventor among us doesn't have just the most special, unique, and precious concept to nurture from its infancy through commercial reality? Over the past 20–25 years through which this latest surge of the potential of the cell has continually gained momentum and promise, we have learned that while the challenges do certainly abound, as with previous industries, the solutions will form around and from those challenges. And they are directly responding to the logistic, financial, regulatory, clinical, and scientific obstacles in a manner proportional to their cause, for it should not surprise us that we encounter such difficulties on the path towards a goal so lofty and unforgiving as nothing less than "cure."

Perhaps this is especially true as we attempt to devise robust and affordable manufacturing platforms for these most innovative and unique therapies. And while the precise solutions for each and every specific obstacle have yet to be devised, we are sure that it is not a path to go alone. There are lessons from the industries that have come before us, there are regulatory authorities eager to participate in the process with the common goal of providing safe and effective medicines, and there are a host of complementary technology/solutions providers, individuals and organizations, out there, each with its own "value-add," and together this patchwork of multi-faceted expertise is every day pushing this cottage industry of today into the mainstream medicine of tomorrow. So we have learned quite simply that a focused, collaborative, and responsive "team" approach, employing an integrated strategy, weaving different solutions, and approaching all gaps with integrated perspective, is required to deliver a solution that addresses the multiple impacts.

The basis for understanding what constitutes a sound manufacturing paradigm is the recognition that as products develop from concepts to commercial realities, they also change. Among other considerations, whether such changes are planned or unexpected, negative or positive, they have scientific and regulatory implications that may result in modulation of the safety, identity, and/or potency of a cell product.

Such changes can result in poor outcomes, a requirement for repeat clinical trials, costly delays, and regulatory concerns, each of which may impact the commercial success of the product.

A rational product characterization plan, including a multiparametric product assessment developed sequentially in coordination with a clinical development program and representing a direct correlation to (or reasonable approximation of) the putative mechanism of action, forms the platform upon which release for patient use will be based. Such a plan serves as the spine around which all change is controlled during product development. The application of a sound comparability strategy based on solid product characterization activities is the sole manner through which one can ensure that process, reagent, equipment, facility, and other changes do not negatively impact the cell product they are intended to improve. Further, this element of change control not only has impact during product development per se, but allows for comparability assessment during technology transfer, platform leverage through clinical product pipeline expansion, multisite manufacturing, process qualification, and automation, and thus has far reaching implications on cost of goods.

As direct measures of mechanism of action are functional by nature, a sound product characterization also includes assessment of real-time, readily readable, and robust assays to serve as functional surrogates, following proper association and validation of the relevant functional characteristics of the cell product that may be employed then as practical release assays. Given the central importance of a well characterized product, it is recommended that efforts to identify and correlate direct and surrogate indicators that represent at a minimum a putative mechanism of action are made at as early a stage of clinical development as possible.

14.2 LESSONS LEARNED

It is instructive to examine the successes and mistakes of the past to inform our path to the future. In this regard, it is critical to consider during clinical development all the pieces in an integrated and coordinated fashion to get the job done right—the first time. Integral to this philosophy and well understood is that one must develop a product from the early stages with commercialization in mind. But what does this truly mean?

Does it mean that at Phase I, one should build commercial manufacturing capacity, automate the process, and attempt to validate assays including potency as would be expected by the end of Phase III clinical testing? Should the cost of goods be anticipated to be high, as with most early stage products in this industry? When should one automate the manufacturing processes in an attempt to mitigate these costs? Is Phase I too early? Worse, is Phase III too late? And is automation simply a matter of finance or is there a more elemental rule that can assist a developer in moving a product through the continuum of development?

Over the course of development, some obstacles present themselves and some are overlooked more commonly than others. The following list describes these two levels. While the two lists overlap to a degree and one could argue placement of some items from one list to another, Level I obstacles generally involve overarching issues with respect to clinical development. Level II is more specifically related to manufacturing:

Level I Obstacles
Patient accrual rate for clinical trials
Market acceptance and penetration
Reimbursement

Level II Obstacles
Product characterization (critical)
Cellular source material
Staffing and staff training
Product stability
Distribution and delivery to patient
Automation
Scale up and scale out
Product data collection and systems to analyze it
Raw material comparability
Robustness of supply chain
Cost of goods

The remainder of this chapter will focus on the Level II obstacles, as they more directly relate to the manufacturing of the cell product. To this end, the first consideration is that, as a general rule, the product of early clinical development is most certainly not the product that will ultimately be delivered to the patient in commercial distribution. As previously noted, through process development, starting before clinical trials and continuing through approval, perhaps even beyond, products by definition "develop"; they change. And changes have regulatory considerations, perhaps even forcing repeat clinical trials should the character of product change beyond its ability to, or more to the point beyond its ability to be demonstrated to bridge to the product of previous preclinical and clinical data generated around it. It is within this context that the concepts of Quality Systems and Comparability establish themselves as critical to the successful manufacturing and development program. The overall project philosophy during product development, therefore, involves a critical knowledge of the cell product and process, employing a blended risk-based regulatory strategy that leverages scientific understanding relative to process control and exploits a "continuum approach" to compliance.

For many years, developers relied on a "process is product" philosophy that belied surrender to the difficulties of identifying meaningful product characteristics that tied directly to ill-defined mechanisms of action for cell products. It is not a difficult leap to determine why so many products cost so much over time to develop, why it is difficult to convince regulatory agencies of their potential in a data-driven and robust manner, and why the pace of successful commercialization of regenerative medicine products has been disappointing to date.

Where does this leave us? A successful manufacturing strategy, built and executed in consultation with the United States Food & Drug Administration (FDA), European Medicine Agency (EMA), and other applicable regulatory authorities, based on multiparametric characterization, all with an eye to comparability, must

Key Considerations in Manufacturing of Cellular Therapies

build in potency and biological characterization approaches as early as possible. Thus re-engineered, Level II obstacles can be addressed in a stepwise fashion, and using the rearrangement with comparability through product characterization can sustain the changes inherent in successful drug development. Multiparametric characterization will generate data throughout development to control manufacturing process changes, define stability and shelf-life logistics, sustain manufacturing facility changes, and support a deep pipeline of disease indications and clinical trials.

Based on this reasoning, it is clear that process is in fact not product and that a focus on understanding the product is the key to successful commercial development of cell therapies and drugs that preceded them. The rearranged Level II obstacles solved through multiproduct characterization then look like this:

- Product characterization (critical)
- Cellular source material (donor and/or patient)
- Product stability
- Distribution
- Automation
- Scale up
- Scale out
- Raw material
- Comparability
- Robustness of supply chain
- Reduced overall cost of goods

14.3 QUALITY SYSTEMS CONSIDERATIONS AND MANUFACTURING INFRASTRUCTURE

If the finer points of comparability, robustness, and deliverability revolve around the product characterization issues discussed above, the foundation resides in the soundness of the quality systems that define the manufacturing program. In order for a manufacturer to fully develop its infrastructure, it is important to understand multiproduct manufacturing and facility design within FDA's current regulatory strategy for cellular therapy products.

While many manufacturers intend to process only a single product, in reality most off-the-shelf cell products involve multiple donors and different cell lines, with the extremes tested for autologous or other patient-specific therapies. Most manufacturers therefore must contend with the challenges associated with the processing of various cells types including those regulated through the Current Good Tissues Practices (cGTPs) cited in 21 Code of Federal Regulations (CFR) 1271 and Current Good Manufacturing Practices (cGMPs) cited in 21 CFR 210 and 211 as promulgated by the FDA. The topics include, but are not limited to cord blood, bone marrow, apheresis, and the manufacture of a series of cell therapy products for clinical trials: dendritic cells, T cells, expanded cord blood, mesenchymal stem cells, skin, embryonic cells, and other types that may require testing along the approved cell product pathway.

A cell product may involve genetic manipulation, combinations with a variety of biomaterials, and extensive culture in a diverse array of platforms that vary widely in their degree of sophistication. These processes require a fair amount of expertise, specific trade secrets, knowledge, and intellectual property, all of which must be taught within the cGMP culture to manufacturing operators, quality control (QC) analysts, and quality assurance (QA) associates.

At the GMP crossroad we find the "culture of GMP," one of the most underestimated definitions in the industry. While we cannot ignore the importance of a well constructed cGMP-grade facility, the elements of cGMP are most critically evidenced in the culture of the employees, the structure of the institution, and the operational components of day-to-day life in a manufacturing environment—not simply in the physical construction of the facility. While our discussion of infrastructure will start at the level of the facility, it is important to keep in mind that the facility must be designed to help meet cGMP requirements and consequently provide an environment suitable for the safe and repeatable processing of cellular therapeutics. Facilities, however, represent only part of the compliance with cGMP requirements. It is helpful to bear this in mind related to the following discussion of the design, chemistry, manufacturing, and controls (CMC), and infrastructure of a facility for producing cellular therapies.

In addition to environmentally controlled suites (controlled environment rooms or CERs) for cell processing, each facility has cryopreservation and quality control capabilities including product testing and characterization. A typical cell therapy manufacturing facility contains ISO 5 biosafety cabinets (BSCs), and ISO 6 and ISO 7 CERs for aseptic processing of cellular therapies. These facilities often provide cryopreservation capabilities and quality control testing laboratories capable of performing routine and perhaps not-so-routine QC testing: flow cytometry, polymerase chain reaction (PCR), chromatography measurements (ELISA), automated cell counting (CBC), microbiology, and cellular morphology analyses. Key facility design considerations include:

- Process segregation and controls adequate for processing of multiple patients' cells and cellular products simultaneously in separate CERS
- Process segregation and controls adequate for processing of cells from multiple patients simultaneously in the same CERs
- Control procedures adequate to prevent contamination, cross-contamination, and mix-ups

14.4 CELL THERAPY MANUFACTURING PROGRAM

14.4.1 General Methods for Product Isolation and Prevention of Contamination and Cross Contamination

Patient-specific cellular therapy production processes present common challenges that must be overcome by a combination of facility attributes, process design, and procedural controls:

- Starting patient components are cell-containing biological materials that cannot be effectively sterilized and may be contaminated with exogenous or endogenous human pathogens.

- Patient (donor) screening and testing prior to collection of biological materials may not be completely effective in identifying all the potential adventitious agents present.
- Some patients who are seropositive for certain pathogens (e.g., cytomegalovirus or CMV) may be accepted for treatment under clinical exclusion criteria described in the company's investigative new drug (IND) submission. These patients' components or products must be handled by methods that will ensure that other patients' materials do not become contaminated with the known pathogen.
- Finished cellular therapy products cannot be effectively sterilized; aseptic manufacturing procedures are required to prevent contamination during in-process manipulations.
- The maximum time between donation of cellular components by a patient and return of the processed, patient-specific product for re-implantation may be limited and cryopreservation of patient products may not be possible. In some cases a product must be shipped for re-implantation before sterility assays can be completed.
- It may difficult or impossible to obtain replacement materials from a patient if problems occur during the manufacturing process.
- A manufacturing facility for patient-specific, cellular therapy applications will contain a large number of individual patients' products in process simultaneously; procedures must be in place to prevent mix-ups and cross-contamination.

To meet these general concerns, a facility must be designed for aseptic processing. The proposed manufacturing procedures must be performed in functionally closed systems, including double containment, blood bags, special cell separation devices, and other presterilized, disposable systems. All sterile connections must be performed within a Class 100 BSC. All media, salt solutions, and other similar materials must be purchased as presterilized solutions. If water is required for any step in a process, only sterile water for injection (WFI) may be used.

Individual patient components should be de-packaged. Primary containers of an individual patient's materials must be placed into a disposable, covered, plastic "tote" labeled with the patient-specific production batch number. Additional labels for all containers and/or disposable vessels required for this manufacturing procedure may be included within the tote and a checklist for all of the contents may be attached or contained in the batch record. To prevent product mix-ups, it is acceptable to employ manual and/or electronic systems to assist in securing chain of custody and proper identification for every patient cellular material and process component, keeping in mind that enhancement through the use of electronic systems is subject to the data security and electronic signature requirements of 21 CFR 11 rules for electronic records promulgated by the FDA.

The covered totes may be transferred to the manufacturing area. All in-process components and final product must be kept closed within patient-specific containers or segregated by other equivalent means. The only exceptions to this containment are when the materials are oversized, actively processed, or stored on a shelf dedicated to that patient's product within a temperature-controlled incubator chamber. Each manufacturing room must be used for only one product type at a time. Each such room may have as many as two (or more) separate work stations with all of the required processing equipment for the production process. However, only one patient's materials may be in process at a time in a BSC. To ensure adequate prevention of cross-contamination, validated line clearance and cleaning procedures including wipe-down of all work surfaces with a disinfectant solution between operations performed on different patients' materials must be followed.

While the standard procedures must be sufficient to prevent contamination, cross-contamination, and mix-up (there is never a guarantee that a specific patient's components are free of adventitious agents), special procedures should be followed for materials derived from known, intentionally selected seropositive patients. Seropositive materials should be placed into dedicated incubators that contain only other seropositive materials. All disposable plastic totes and other supplies for seropositive patient materials should be clearly labeled to indicate this status. They must not be transported or handled at the same time as seronegative patient materials. After completing the line clearance following manufacturing steps for seropositive patient materials, the operators exposed during these activities must exit the classified area. Personnel accidently exposed to a hazardous spill during seropositive activities must shower and change into new scrubs before gowning if it is necessary for them to re-enter the classified production areas.

Materials and supplies for each series of process steps for a single patient product are best assembled within a separate area or "kit room" and loaded into clean, sealed containers for transfer to the appropriate work station. As a method of ensuring proper processing, a bill of materials should be included in the batch production record (BPR) for the patient's product lot. All materials not used during the defined production steps on an individual patient's materials should be removed as waste during the line clearance procedure. All labels issued with the production batch record should be reconciled at the conclusion of the manufacturing procedures.

14.4.2 PROCESS DESCRIPTIONS AND PROCESS FLOW

The following are process flow summary process descriptions for two typical planned products to be manufactured within a facility. As generic processes presented for illustration purposes, they should also be suitable for other similar cellular therapy products and/or processes. The description of cord blood intended for autologous or first and second degree relative administration is included for reference. A typical dendritic cell manufacturing process is also described along with recommendations for process control within a facility.

14.4.2.1 Cord Blood Manufacturing Process

Cord blood processing and storage may fit into the category of "361" products described in the proposed 21 CFR 1271 rule* as minimally manipulated. Assuming

---

* The "361" categorization is cited in 21 CFR 1271 and based on Section 361 of the U.S. Public Health and Safety Act found in Volume 42 of the United States Code.

Key Considerations in Manufacturing of Cellular Therapies

processing follows the 21 CFR 1271 regulations, the cord blood units received should be depackaged within the accession room and materials kits obtained from the materials kit assembly room or equivalent area as described above. While only a single product unit may be processed in "open" steps (sterile connection and cell suspension transfer) at a single time within a work station, other closed processing steps such as centrifugation, cell settling, precooling, and controlled rate freezing may be performed on more than a single cord blood unit at a time. After filling the final product container, the cord blood unit is transferred to the product freezing room for controlled rate freezing and long-term storage.

14.4.2.2 Dendritic Cell Manufacturing Process

Patient leukapheresis components are received and transferred to the accession room for depackaging and then into a specified CER. If required, the covered tote may be stored in a refrigerator before the start of the production process. A materials kit for the initial steps of the process is transferred from the released materials area to the kit assembly room. Labels with production batch numbers from the covered, disposable plastic tote are placed on all vessels to be used in processing. Only one patient's materials may be open at the work station during these process steps.

Tumor antigen materials, if required, are stored in an appropriate, properly qualified storage vessel (e.g., in a −80°C freezer), perhaps within the production CER. At the conclusion of initial activities for this patient, the culture bags containing the in-process product may be placed on a single incubator shelf, labeled for that production lot. Samples taken during the process are transferred to the QC laboratory over a controlled and defined route. Cell counts and other test data required during processing are communicated to the operators.

14.4.3 FACILITIES

Certain general attributes and finishes should be employed within the classified areas of a clinical manufacturing facility. The level of control and stringency of design are inversely proportional to the degree of proven, data-demonstrated, inherent aseptic characteristics of the specific process. Typically, early in development, processes that use open systems and multiple manual manipulations, systems, and facilities must participate in preventing contamination, cross-contamination, and the introduction of communicable diseases during manufacture. Conversely, although it is difficult to imagine a scenario in which a controlled environment is not required, conditions may be somewhat relaxed through the use of a closed system designed to prevent breach of aseptic conditions during manufacture. Critical to this determination is the data-driven definition of "closed" typically obtained through challenge studies of the closed system using, for example, *Brevundimonas diminuta* or other suitable microbe to enable challenge studies to demonstrate the aseptic robustness of the system.

Although environmental controls may differ in degree, certain general principles apply and must be considered in the design and operation of a suitable facility for cell manufacturing. The heating, ventilation, and air conditioning (HVAC) system should be monitored and alarmed for a number of important parameters including temperature, humidity, air flow, and pressure. The pressure levels within each room may be controlled by a static, fixed damper system or by a dynamic computer-controlled, building management system. In both cases, the HVAC system must be fully validated before initiation of production for clinical products. In addition to mechanical and operational parameters for the HVAC equipment, important room parameters and ranges include air flow (the rate at which fixed and prescribed air changes per unit of time occur) through terminal (ceiling use) HEPA (high efficiency particulate arresting) filters and air is returned through low exhaust returns.

Control of room pressure (pressure relationships between adjacent areas when all doors are closed) allows for directional control over air flow, protecting aseptic spaces from incoming flow of outside contaminants, and also, if properly and strategically balanced, protecting the external environment from efflux of hazardous, airborne contaminants used in or inherent to the manufacturing process or materials.

Control of room temperature and humidity is designed to discourage microbial growth. Ongoing, routine monitoring of both non-viable particles per unit air volume and viable particles is standard for evaluating the cleanliness of aseptic processing areas.

Ideally, minimum room finishes for classified areas include floors of seamless welded vinyl construction with integral coving at walls and corners, and walls and ceilings. Suspended ceiling tiles should ideally not be used in classified areas, although there are designs that are suitable. Penetrations for HEPA filters and lights should be sealed and the edges smoothed with silicone caulk or the equivalent. All CER surfaces should be designed and installed to facilitate frequent cleaning with harsh agents and disinfectants. Surfaces must be smooth with minimum projections to reduce dust collection and air flow disruption.

After design and construction of a facility, its operation is defined by structured monitoring, rigid flow of air, product, samples, personnel, materials, and wastes. All these factors are codified in a series of detailed standard operating procedures (SOPs) to which staff must comply. Logs, worksheets, and other documentation of actual conditions must be maintained.

14.4.4 Cell Therapy Patient Components and Product Flow

The movement of patient components and the resulting patient-specific products is carefully choreographed to provide protection from mix-ups and cross-contamination involving products of different origins. The biological materials are received and immediately transferred to a designated room or area for depackaging, inspection, and initiation of the production batch record for the specific patient's product. The unpackaged blood bags or other incoming primary containers are placed into a covered plastic tote used for isolating each patient's materials and in-process product throughout the manufacturing process. The materials are then transferred to the appropriate production room.

Production activities based on the manufacturing process for each product type are performed in a designated production room. Only one product at a time is open or processed at a single work station. All process steps are performed within disposable closed bag (or disposable separation unit) systems. Connections are made in a Class 100 BSC. Completed patient-specific products are placed in a pass-through and subsequently transferred to a final packaging and/or cryopreservation area.

14.5 QUALITY SYSTEMS

Quality systems ensure quality of systems and procedures and should achieve continuous improvement through a corrective and preventive action (CAPA) program. The foundations a of a quality systems are:

- 21 CFR 210–211: Current Good Manufacturing Practices in Manufacturing, Processing, Packaging, or Holding of Drugs
- 21 CFR 610: Biological Products, General; Subpart B, Establishment Standards
- 21 CFR 1271: Current Good Tissue Practice for Manufacturers of Human Cellular and Tissue-Based Products
- 21 CFR Part 11: Electronic Records; Electronic Signatures
- Product consistency testing (PCT) requirements as defined in SOPs, processes, methods, and specifications
- Client-specific requirements as defined in client SOPs, processes, methods, and specifications
- Good documentation practices

14.5.1 ELEMENTS OF QUALITY SYSTEMS

Quality systems contain many components: document control, change control, internal auditing, supplier qualification and/or auditing, quality agreements, personnel training and qualification, job-specific skills, deviation and CAPA programs, non-conforming materials and products systems, laboratory control systems, and outcome analysis. All these systems are described below.

14.5.1.1 Document Control

Document control is administered by QA staff. Procedures cover numbering, formats, change control, and distribution of compliance-critical documents (SOPs, forms, batch records, methods, specifications, validation protocols, and reports). Implementation and deletion/absolution of documents are controlled by QA. SOPs should be subjected to periodic reviews to ensure they are current and accurate.

14.5.1.2 Change Control

Document change control procedures apply to all compliance-critical documents. They are used to add new documents to the system and change existing documents. The document change control form details the change, reason for the change, justification for the change, impact of the change, and identifies all other documents and procedures affected and lists requirements for implementation such as training, qualification, process validation, and so forth.

Equipment change control (ECC) is a two-step process. Initially an equipment change is identified, documented, reviewed, justified, approved or rejected, and implemented if approved. This process determines what levels of testing, qualification, validation, and documentation are needed to make changes to the qualified or validated system. After approval of the change control, the system may be changed. After successful execution and approval of follow-up activities, the system is available for use. The ECC is considered closed after QA documents the changes and attaches any follow-up requirements to the file. ECC is also used to add new equipment to a system or make changes to existing equipment and/or systems. ECCs for new equipment and/or systems are used to determine and document the level of qualification or validation versus calibration. The ECC procedure captures changes to equipment, instrumentation, facilities, utilities, and computer systems.

14.5.1.3 Internal Auditing

Internal audits of the GMP environment are performed regularly by in-house QA. In addition, outside consultants may be engaged. Audit findings are reported to the functional area department head, senior QA and QC management, and executive management. Observations are reviewed and agreed-upon corrective action plans including responsibility and estimated completion date are determined in conjunction with QA, and QA monitors the execution of the plan to completion.

14.5.1.4 Supplier Qualification and Auditing

Qualification or approval of vendors of key processing components is dependent on the services or materials provided and, if required, is based on a sliding scale that considers the type of material and its use in the process and the clinical phase of development. The procedure can range from evaluation of a vendor-completed questionnaire to a vendor visit to a due diligence audit or to a full GMP on-site audit. In early stage trials, pretested or prereleased materials and reagents for use in manufacture of a product are received with appropriate certificates of analysis (CofAs) or certificates of conformance (CofCs).

14.5.1.5 Quality Agreements

Quality agreements between suppliers and developers identify and define their respective roles and responsibilities. Quality agreements are designed to assure that the manufacture and control of a product for clinical or commercial purposes complies with cGMPs, GTPs, and/or other applicable requirements. Both parties agree to and execute a formal quality agreement that identifies primary and back-up quality department contacts responsible for adherence to the agreement. A quality agreement may be incorporated as part of an overall service contract between the parties or may be a stand-alone document.

14.5.1.6 Personnel Training and Qualification

QA conducts basic cGMP and cGTP training for new employees involved in good practice quality guidelines (GXP) activities. Minimum annual retraining typically provides a brief refresher of existing requirements and an in-depth review of any changes in the requirements since the last training. QA also conducts periodic training as the quality system is revised and upgraded to ensure that employees involved in GXP activities are aware of the changes. In addition, training is conducted on all new and/or revised SOPs. A training assessment is performed by area management for hands-on and/or document qualifications training as needed. Completed training documentation is maintained in each employee's training records file.

14.5.1.7 Job-Specific Skills

Each functional group leader is responsible for assessment of the training requirements for each new employee and for determining an appropriate job skills training plan based on position requirements. Appropriate documentation is completed upon successful execution of the training. Completed documentation is maintained by the applicable department in each employee's training records file.

14.5.1.8 Deviation and Corrective and Preventive Action (CAPA) Programs

A deviation from or a discrepancy in executing an approved procedure or from a GMP requirement is handled according to an approved deviation SOP. Such deviations and discrepancies may involve equipment, materials, systems, manufacturing methods, or test procedures; they may be planned or unplanned. When a deviation occurs or is discovered, the immediate supervisor is notified and a deviation report is issued to ensure that management is made aware of the situation. An investigation into the root cause of the deviation is conducted and extends to other lots of the same product and/or use of common materials. A determination is made as to the impact of the deviation on product quality.

Any immediate step taken as a remedial corrective action is documented. Preventive actions are based on root cause analysis and intended to prevent initial occurrences and recurrences. If applicable, preventive actions are identified and tracked to completion by QA. CAPA systems are designed to drive continuous improvement by preventing a deviation or discrepancy from recurring. This includes trending to ensure that repetitive issues can be identified and addressed. QA management reviews and approves the investigation and CAPA and based on the results determines whether the final product is approved and released or rejected and subjected to final disposition.

14.5.1.9 Non-Conforming Materials and Products Systems

QA maintains a system to ensure that materials that fail to conform to specified requirements cannot be inadvertently used or released. Non-conforming materials are handled through the deviation system and an investigation is conducted to document the non-conformance. For out-of-specification results, unless the test allows repeat of analysis based on results of controls or standards, a determination is made whether the out-of-specification test result is due to an assignable laboratory error. If that is the case, the incident is documented into the deviation system and the test repeated. If no assignable laboratory error is found, the investigation will extend to other aspects including the manufacture of the out-of-specification product. Final disposition of the affected final product is determined by QA.

Cellular products with sterility test requirements that reveal positive microbial culture results are handled as non-conforming materials under the deviation system to determine the source of contamination, impact, and disposition of the product and/or according to the SOP that outlines the notification and reporting requirements for human cell and tissue products (HCT/Ps) and conditions under which a culture-positive HCT/P may or may not be distributed for infusion.

14.5.1.10 Laboratory Control Systems

Specific requirements must be established for materials and final products. Incoming materials are received, inspected, and released as per approved procedures. In all cases these actions are based on written procedures, specifications, and requirements. In-process and final product test requirements including sterility assays are detailed in approved specifications and/or SOPs. Product is tested to ensure conformance to specifications or requirements. Test results are recorded and documented, included in batch records, and reviewed in the course of batch record approval and release.

14.5.1.10.1 Equipment

All GMP equipment should be assigned a unique master equipment file (MEF) number and the number must be affixed to the equipment. MEFs containing all information and records relating to each individual piece of equipment (calibration, preventive maintenance, and repairs) are maintained.

A validation policy and a master validation program must be established to define the equipment to be validated, level of validation, installation qualification (IQ), operation qualification (OQ), performance qualification (PQ), and/or calibration requirements typical for the different types of equipment. Validation protocols and reports are generated for each piece of equipment. An equipment change control system should be implemented to capture and approve all changes. The change control will also capture the addition of new equipment along with validation, additional validation, and revalidation requirements.

GMP equipment must be on a calibration and preventive maintenance schedule. SOPs should exist for each equipment type and model and describe, use, cleaning and sanitization, calibration, and preventive maintenance procedures.

14.5.1.10.2 Materials Control System

GMP materials are logged in upon receipt, assigned an internal number for tracking purposes, and quarantined until released for use. A green release status label is applied at time of release and shows part number; shelf life; expiration date, retest or reevaluation date; and release status. An accountability log records the part number, amount issued, use location, and the balance remaining. This allows forward and backward traceability. The part number is also recorded in the batch record. Rejected or quarantined materials are labeled as such and kept in a clearly identified area.

14.5.1.10.3 Production Control System

All GMP batch records and labels are issued by QA from an approved master batch record that is assigned a unique lot number. Critical steps are documented and verified. Within these records completed GMP batch data are reviewed and approved by QA. For GMP-manufactured products, QC performs in-process and final product control testing according to approved procedures as required (in-process testing may be performed by a qualified operations technician in early stage clinical phase). Results are recorded and compared to established specifications to determine whether the material meets requirements and the results are

Key Considerations in Manufacturing of Cellular Therapies

included in the batch record. Product is shipped to the user and/or stored under appropriate conditions (for example in liquid nitrogen) as per procedure.

Quality Assurance must authorize the final product release of GMP batches after review and approval of the batch record and test results to ensure the product and record are complete and conform to current specifications and any discrepancies or omissions have been resolved.

14.5.1.10.4 Product Tracking and Labeling System

All manufacturing procedures are carried out according to approved SOPs, master production records (MPRs), and batch production records (BPRs) for each patient lot or batch. To track patient cells, the patient collection should be assigned a dedicated number—the unique lot or batch number assigned to the BPR for the manufacture of the product. The unique number is contained on all labels on containers used for the processing, storage, and sampling of product during the manufacturing process.

14.5.1.11 Outcome Analysis: Models for Manufacturing

Successful cell therapy development and commercialization require a sophisticated, multidimensional, cooperative approach in specialized facilities and the processes are both product- and clinical phase-dependent. Separating product from process and having an inadequately coordinated approach will most certainly leave the process unfinished and the job undone. Therefore the natural question is: How does one most effectively navigate through this maze to enhance chances of achieving commercial success?

It is clear that relying on hope as a strategy is inadequate—too many things can go wrong. In addition to Murphy's law, which most certainly should not be ignored, the strategy must be designed to withstand interruptions in funding, clinical hold, loss of key personnel, changes in plans, and a host of other unforeseen delays in the process. A solid business case must be made, therefore, to determine the right balance between in-house resources and contract services to create the right team to tackle the issues and devise the most robust clinical development program that will optimize focus, leverage, value creation, and risk mitigation.

Examination of each of these important considerations will allow a developer to construct the right business model to weather the storms that may arise. First and foremost, the developer must consider value creation and in this context design a business model that reflects that of a service provider or product company and takes into account risk versus reward. The developer must be sure the whole entity is worth more than the sum of the parts. To best determine this, the organization must determine its focus, i.e., determine what it is good at. Have a laser beam focus on clinical development and drop the other issues, utilizing the right balance of internal and external resources to meet the needs. Focus must also be determined within the context of risk mitigation and prepare for changes of plans through fleet-of-foot flexibility, maintenance of reserves, and freedom from yesterday's direction, in part or whole, in the face of new evidence and circumstances. Finally, the developer must consider what it can use to best leverage its own and outside resources. In addition to cost mitigation provided by leveraging capacity and infrastructure as needed, leverage provides targeted access to specific and well informed expertise on an issue-related basis.

14.6 MAKE VERSUS BUY: THE DECISION TO OUTSOURCE

To this end, a developer must consider a balance between outsourcing and the use of in-house resources. In-house execution affords a level of control over production schedules, staff, and general operations that contractors may not be able to provide. Targeted access to expertise and episodic use of infrastructure and expertise as needed supports the case for establishing a partnership with a reputable service provider, especially during early development when the periodic requirements for staff and physical assets do not justify the expense to provide them. Further, establishing a GMP culture, the importance of which we earlier established, may be more challenging in an underutilized environment than in one that regularly reaches capacity. However, it is essential that a developer selecting a contract manufacturing organization as a partner ensures through reputation, audit, and reference checks that, in addition to being compliant and having controlled systems, the contractor has the capacity and competence to perform the desired functions within the timelines required for the clinical program.

For situations in which the developer believes that the technical competency required for a manufacturing process may not reside with a contractor, many contract manufacturing organizations (CMOs) will work with a developer to establish creative arrangements through which the developer's staff may oversee or even carry out manufacturing and/or testing while the CMO remains responsible for the balance of activities required for the program. In any case, the developer should consider employing a person to regularly oversee the contractor in the plant to ensure the success of the manufacturing program. This person's mission is critical to the success of the entire clinical development program. At later stages of clinical development, particularly during commercial distribution, the economics and throughput may better lend themselves to in-house manufacturing, although as discussed elsewhere in this chapter, overflow and surge capacity may best be managed through a relationship with a reputable and capable service provider.

14.6.1 COST SENSITIVITY

Cost of goods is a key consideration of the make versus buy decision and is also instructive about the economics of cellular medicines. This section breaks down the costs of manufacture in an attempt to assist developers make decisions regarding outsourcing. Table 14.1 shows a capital expense (CapEx) breakdown for establishment of a typical manufacturing facility. Typical industry rates for labor are included for reference in Table 14.2.

While the cost varies depending upon the product and the process, over a range of product it is possible to examine a typical ratio between these costs as shown in Figure 14.1. First, it is worth noting that variations in work flow affect cost of manufacture and ultimately therefore cost of goods during commercial manufacture. These variations include facility and staff utilization, the balance of QC testing

TABLE 14.1
Capital Expenditure Requirements for Typical Manufacturing Facility

Description	Unit Cost ($)	Number Required	Total Cost ($)
Freezers for Storage			
Controlled rate freezer[a]	9,495	2	18,900
Freezer for primary processing	20,078	1	20,078
Mechanical freezer to store serum samples	8,000	1	8,000
Work Stations and Related Equipment			
Work station unit (including tube sealer, centrifuge, miscellaneous equipment)	14,074	7	98,515
Facility-Related Equipment			
Facility unit (including walk-in refrigerator [CER], −20°C freezer [23 cu. ft. capacity, manual retrieval device, and 2 advanced overwrap systems[a]	100,240	1	100,240
Other Major Processing Equipment			
CBC/cell counters	18,000	2	36,000
Flow cytometer	120,000	1	120,000
Red top tube centrifuge	9,700	2	19,400
Environmental, Access Control, and Alarm Systems			
Environmental monitoring system	39,966	1	39,966
Access control system	13,936	1	13,936
Build-Up of Original Infrastructure			
Clean room	499,500	1	499,500
Technician lab space, freezer room, etc.	156,000	1	156,000
Liquid nitrogen bulk tank installation	55,650	1	55,650
Technician space, furniture, fixtures	118,000	1	118,000
Architect fees[b]	75,000	1	75,000
Data processing equipment	68,000	1	68,000
Telephone and communications equipment	75,000	1	75,000
Office furniture and fixtures	150,000	1	150,000
Other (building permits, licenses, CofA, accreditation, etc.)	100,000	1	100,000
Building contingency 10% of CapEx 16, 17, 99	34,865	1	34,865

[a] Two units include one for back-up.
[b] One-time payment for design.

against manufacturing requirements, lot size (one-off batches versus cell banks), requirements for different containment levels, high- and low-volume products, and the need for dedicated space or shared space for a campaign. While these factors also vary based on whether the production is of allogeneic or autologous nature, the length and complexity of the manufacturing process, and the intensity of the

TABLE 14.2
Typical Industry Salary Ranges

Position	Salary Range (Annual)
Manufacturing Ops and QC Analyst (Hourly)	
Level I	$32,500 to $42,000
Level II	$40,000 to $52,000
Level III	$50,000 to $65,000
Quality Assurance (Salaried and Hourly)	
Associate I	$32,500 to $42,000
Associate II	$40,000 to $55,000
Associate III	$52,000 to $75,000
Management (Salaried)	
Supervisor	$55,000 to $75,000
Manager/Senior Manager	$70,000 to $125,000
Director/Senior Director	$80,000 to $175,000

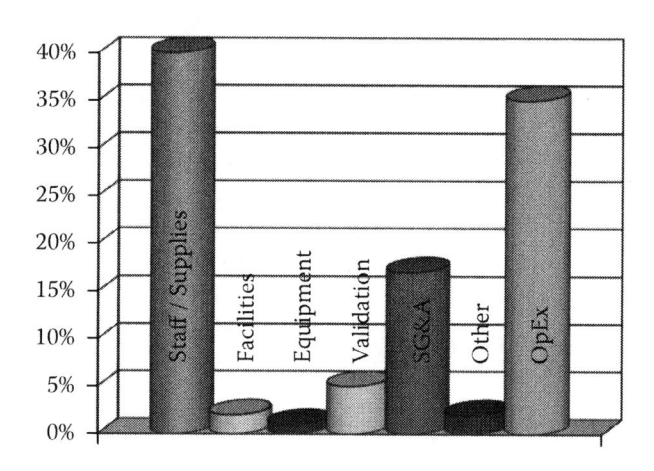

FIGURE 14.1 Cost sensitivity analysis: "Typical" cell therapy manufacturing cost ratios.

supplies, the general message holds true that 70 to 80% of costs are attributable to facility operating expense and direct labor.

It follows that during periods of peak activity, when staff is utilized at capacity, the cost of goods is as efficient as it can be. Whether staff members are employed by a contract manufacturer or are full-time employees of the developer, there is little difference in cost. In fact, in-house manufacturing may provide a greater degree of control and save money in this scenario due to margins applied by contracting services. This scenario assumes that staff can be trained and will be sufficient to cover vacations, sickness, and other non-productive events. At commercial production level, it may be attractive to consider bringing manufacturing in house so that it becomes routine and predictable and staffing may be efficiently absorbed. However, during clinical development, accrual is difficult to predict, volumes are low, and the frequency of utilization of manufacturing is sporadic.

With multiple project ongoing concurrently, a CMO can more readily absorb utilization inefficiency through application of staff, equipment, and facilities to other projects. However, even during commercial production, the initial surge in market use pursuant to a backlog of patients eligible for the treatment overinflates the expected annual market acceptance and causes a surge that may be difficult to staff. Furthermore, fluctuations in demand and increasing market acceptance will require additional capacity that will be utilized inefficiently until capacity is reached, only to be followed by another cycle of inefficiency as demand outstrips efficiently utilized capacity. In such a scenario, it is not difficult to predict a role for a contract manufacturer that can smooth out fluctuations in efficient use of staff for a developer. Also, in the interest of risk mitigation, multiple facilities belonging to one contractor or use of a contractor to provide redundancy protects against catastrophic shutdown (or interruption) of operation at a facility and allows uninterrupted production and satisfaction of market demand.

14.6.2 Autologous versus Allogeneic Production

Cell-based medicines will most certainly require greater reimbursement due to relatively high costs of goods and their greater potential to result in cures over palliative medicine. While the debate about costs of goods for regenerative medicine and ways to mitigate them rages on, it often is waged in the discussion of autologous and allogeneic medicines. Cell-based products can be roughly divided into three categories, defined by cell source in relation to intended recipient:

- Autologous
- Allogeneic, patient-specific
- Allogeneic, off the shelf

From the standpoint of cost of goods and manufacturing infrastructure, the autologous and patient-specific allogeneic products are essentially identical. Without automation, scale-up activities equate to scale-out and involve materials, labor, equipment, and infrastructure to grow at a rate proportionate to the commercial demand for the product. Clearly, at steady-state distribution, one can leverage overhead and physical infrastructure. Cost-of-goods analyses show that a reasonable business model within many of the existing paradigms is limited in nature and reduced in potential to generate the traditional and attractive margins expected at the end of clinical development and beyond the commercial launch of a therapeutic.

In contrast, off-the-shelf allogeneic products from which many patient doses can be derived from a single donor's cell material clearly present a cost-of-goods advantage over patient-specific products. This model approaches the leverage of classic biotechnology products—lots of increasing size that require marginally additional material, labor, and facilities. However, there are two important considerations that should not be overlooked in the analysis. First, patient-specific medicines avoid or reduce the impact of product rejection due to mismatch of tissue type between donor or source material and recipient of the therapeutic. While certain cell types including mesenchymal stem cells (MSCs) appear to be less immunogenic, at least during the initial post-treatment phase, not all therapeutics have similar properties or mechanisms of action that will lend themselves to the uses of cells with these properties. While there is no argument that large batches of cells stored in dose-sized aliquots attractively resemble bottles of traditional therapeutics, the issues of cold-chain distribution, shipping, preparation, bedside preparation, and administration must be taken into account in the cost-of-goods equation.

Further, the more complex a cell product, especially combinations of cells with or without biomaterial scaffolds and those that rely on maintenance of their organizational structures (bladders, skin, blood vessels) for activity, it becomes increasingly more difficult to envision their production, cryopreservation, and storage on a large, off-the-shelf scale. Clearly, both patient-specific and off-the-shelf therapeutics have their places in regenerative medicine. Close attention must be paid to business models, transportation systems, final packaging, reimbursement, and how easily a product lends itself to automation and robustness before choosing the superiority or commercial feasibility of one over the other.

14.6.3 Transportation

Regenerative medicine requires a delivery system designed to maintain adequate product control from cellular acquisition to reinfusion. A key component of such a system is an adequate logistics and transportation network capable of accomplishing the coordinated movement of cellular material to accommodate patient schedules. The system must be capable of handling cellular materials from collection through manufacturing and delivery to the patient and now involves access to and partnerships with specialized package delivery airlines and networks of ground couriers.

Requirements for commercial distribution, likely involving a network of approaches, will most certainly be forged into a system dedicated to cellular therapy and will comply with relevant control elements of cGMP to result in a very efficient procedure for shipping human cells to and from physicians' offices, medical centers, and laboratories for treatment or long-term preservation. Using SOPs and qualified approaches, the system will provide a secure chain of control for clinical materials.

The necessary attributes of a transportation system include documented verification of shipment status and integrity at every step in the chain of possession, from the sender to the shipper through ground couriers and airport transfer points to the final recipient. Via radio frequency identification and delivery (RFID) technology, door-to-door cell transportation services for environment-sensitive and time-critical packages can provide (1) integrated, secure package handling and monitoring of package integrity at critical checkpoints, (2) established recovery procedures to be activated en route in the event of deviations from specified environmental parameters, and (3) en route verification to ensure timely delivery of viable packages to final destinations within specified shelf-life parameters, by alternate means if necessary.

14.6.4 DELIVERY AT BEDSIDE

Traditional medicines enjoy the benefit of ease of administration at bedside. Most of them are delivered intravenously, by injection, and as oral medications and practitioners have become accustomed to these routes of administration. In contrast, regenerative medicines will by and large require sophisticated understanding of therapy-specific preparation and administration procedures. While one may envision that, at least from a regulatory perspective, the finished formulation or end of manufacturing is defined as the packaged product at the manufacturing plant, the reality is that should a practitioner fail to store, prepare, or administer a regenerative product in an appropriate manner, use of the product in a very small percentage of cases may be unsafe, but more likely will have negative impact on the efficacy of the therapeutic.

Such an outcome would have devastating impact on the medical and patient acceptance of these novel therapeutics. Therefore, despite the regulatory and financial implications, "manufacturing" should be considered to end at the bedside and the manufacturer needs to pay continued attention to the process and control mechanisms at the clinic. Two critical features of administration of cell-based products are the delivery apparatus and the instructions for use. It is wise to think about these factors early in development and, based on a strong product profile, work on solutions for ease-of-use deliverability early in the development process.

14.6.5 FRESH VERSUS FROZEN PRODUCTS

As previously discussed in earlier sections, a final formulation involves biologic, operational, financial, and regulatory considerations. Adding a cryopreservation component early in the development process is intended to extend shelf life, enable off-the-shelf inventory, and thus enhance the deliverability of a therapeutic. A developer should also be aware of consequent challenges related to cold-chain storage and distribution, controlled and documented storage at the clinic, and delivery and bedside administration of the cryopreserved therapeutic.

Product development efforts aimed at final packaging must be coordinated with clinical trial design, training, and commercial distribution logistics, again supported by a strong product profile, to determine the best path forward for each therapeutic. Clearly a one-size-fits-all philosophy will not apply to all therapeutics, nor will a blanket bias toward cryopreserved products. While it is attractive to envision the large scale manufacture and distribution of blockbuster cell therapies from a single facility, certain considerations surround the decision to cryopreserve (or not) if production involves multisite manufacturing, storage depots, and exclusive centers of excellence at which only certain therapies may be received. In any case, careful attention must be paid very early in the development cycle to the shelf lives of both incoming cellular material and the outgoing final product.

Clinical development may be severely hampered or even terminated if expiry times are short. Some products may be simply unable to go through the entire range of manufacturing processes within their given expiry times. Staffing, physical infrastructure, and scheduling factors are directly linked to the stabilities of raw materials and final products. Short shelf lives put pressure on manufacturing to provide adequate, multishift staffing and facilities to accommodate limited expiries on raw materials. They also exert pressures on clinics that may be forced to operate off schedule to accommodate limited final product expiry. Careful consideration of shelf life for both frozen and fresh products enables market acceptance and deliverability in ways often underestimated, especially early in clinical development.

14.6.6 AUTOMATION

Automation clearly plays a significant role in realizing the promise of cellular therapy and regenerative medicine. As a rule in this developing industry, we have access to tools and solutions that were originally designed and/or adapted for other industries and medicinal products. As a result we have become masters of adaptation, using retreads including regulations, tools, delivery paradigms, experience, and even reimbursement models. This is a reasonable strategy for development due to the availability of these solutions and our familiarity with them.

However, properly engineered and applied automated solutions have the potential to enhance process robustness and reduce both labor costs and costs of goods for cell-based medicines, but two related considerations, timing and money, currently limit the implementation of automation in cell therapy applications. The design of tailor-made automation systems implies that the processes are mature enough and sufficiently well understood to justify the expense of tailored manufacturing solutions that are likely to be sufficiently permanent to enable a return on their investment. This assumes that the money is available at early development, which in most cases, it is not. Therefore, it is clear that the time for automation has yet to arrive for most cellular therapies. However, to establish a sound product development strategy, it is critical to include automation specialists as part of the multidisciplinary product clinical development team to consider current and future automation of processes early in development, for small changes and decisions made early in development may lend themselves more readily and efficiently to automation later in development.

14.7 CONCLUSIONS

Much progress has been made over the past 20 years as the commercial realities of cell therapies have been realized and the age of wide-scale application of blockbuster products is soon to be realized. As with other industries, the unique challenges and solutions to overcoming the challenges to realizing this goal are becoming increasingly clear and will change the way medicine is delivered.

In recognition of the intimate interrelations of the various gaps identified during the development process, teams of people representing multifaceted expertise (facilities design, quality, manufacturing, engineering, finance, automation, reimbursement, science, clinical medicine, regulatory, and other key skills), along with choreographed cooperation between contract manufacturing organizations and in-house resources will demonstrate that a collaborative, sound clinical development plan built atop a multiparametric product characterization foundation within the development continuum, along with a needle-to-needle assessment of the product, are steadily moving more innovative cell-based products to market. The single great thought that is key to continuing this progress in this field is that *product is both product and process*. Manufacturing not only supports clinical development; it exerts wide-ranging impacts on every facet of it.

REFERENCES

Current good manufacturing, processing, packing, or holding of drugs. 21 CFR 210. http://www.access.gpo.gov/nara/cfr/waisidx_09/21cfr210_09.html

Current good manufacturing practice for finished pharmaceuticals. 21 CFR 211. http://www.access.gpo.gov/nara/cfr/waisidx_09/21cfr211_09.html

Electronic records; electronic signatures. 21 CFR 11. http://www.access.gpo.gov/nara/cfr/waisidx_09/21cfr11_09.html

Human cells, tissues and cellular and tissue-based products. 21 CFR 1271. http://www.access.gpo.gov/nara/cfr/waisidx09/21cfr1271_09.html

15 State of the Global Regenerative Medicine Industry

R. Lee Buckler, Robert Margolin, and Sarah A. Haecker

CONTENTS

15.1 Introduction	214
15.2 Definitions	214
15.3 The Business of Regenerative Medicine	215
15.3.1 Success? Record for Cell Therapy Companies	217
15.3.2 Non-Cell-Based Segment	218
15.4 Stakeholders: Synopsis of Regenerative Medicine Companies and Products	218
15.4.1 Cell Therapy Companies and Products	218
15.4.1.1 Phase III Pivotal Stage Products	219
15.4.1.2 Commercial Products	220
15.4.1.3 Supporting Technologies for Regenerative Medicine Products and Companies	221
15.4.2 Non-Cell-Based Regenerative Medicine Companies and Products	222
15.5 Core Components of Regenerative Medicine	222
15.5.1 Matrices and Scaffolds	223
15.5.1.1 Collagen	223
15.5.1.2 Gelatin	223
15.5.1.3 Silk	223
15.5.1.4 Fibrin	224
15.5.1.5 Laminin	224
15.5.1.6 Hyaluronic Acid	224
15.5.1.7 Biological Acellular Matrices	224
15.5.2 Growth Factors	225
15.5.3 Cells	226
15.5.3.1 Cell Type	228
15.5.3.2 Cell Procurement	231
15.5.3.3 Expansion	231
15.6 Conclusion and Prognostication	232
References	233

15.1 INTRODUCTION

This chapter is designed to provide a summary of the regenerative medicine industry with a core focus on products, product development, and commercialization (rather than basic science or academic discovery). We hope to shed light on the characteristics or features of regenerative medicine technologies, the technologies likely to be critical components of regenerative medicines, and both current and projected trends for the industry.

15.2 DEFINITIONS

In Chapter 1, Professor Dame Julia Polak defines regenerative medicine as "aiming to replace, repair, or restore normal function to a given organ/tissue by delivering safe, effective and consistent living cells" (Mason and Dunnhill 2008). What Mason and Dunnhill suggest is that regenerative medicine's central focus is human cells—a different and broader focus than the delivery of cells. We submit that while cell therapy is concerned wholly with the *delivery* of cells, regenerative medicine is about *affecting* cells—delivering or applying cells, molecules, or other biologic or non-biologic agents to affect an in vivo therapeutic response.

The Alliance for Regenerative Medicine's current working definition is:

> Regenerative Medicine is a rapidly evolving interdisciplinary field in health care that translates fundamental knowledge in biology, chemistry and physics into materials, devices, systems and therapeutic strategies, including cell-based therapies, which augment, repair, replace or regenerate organs and tissues.

We consider *regenerative medicine* an umbrella term that includes both cellular and non-cellular products used to augment, replace, repair, or regenerate human cells, tissue, or organs, to restore or establish normal function (Mason and Dunnhill 2008). It is, indeed, a multidisciplinary field representing discrete therapeutic modalities including cell therapy, gene therapy, tissue engineering, bioengineering, and molecular therapies.

Regenerative medicine may or may not utilize cells ex vivo as or as part of an administered therapeutic. Regenerative medicine may activate cells therapeutically in vivo through the use of large and small molecule biologics (drugs, antibodies, etc.), non-cell-based gene therapy, non-cellular biomaterials (e.g., de-cellularized matrices derived from but no longer containing live cellular products), and/or other acellular components (e.g., synthetic scaffolds).

The terms *tissue engineering*, *regenerative medicine*, and *cell therapy* are often employed confusingly and interchangeably. Because the genesis of tissue engineering was largely in biomaterials, the term focuses predominantly on the replacement or regeneration of human tissues. Historically this was done with synthetic or acellular materials but more recently tissue engineering has included combinations of such materials with live cells. Thus, tissue engineering is a subset of regenerative medicine and certainly overlaps with cell therapy where it involves the ex vivo use of cellular products.

Cell therapy is the prevention or treatment of diseases, defects, or other disorders by the clinical administration (transplantation) of cells that have been selected, treated, and/or otherwise engineered ex vivo. Such therapeutic cells can be immune, progenitor, primary, or stem cells, originating from autologous, allogeneic (somatic or embryonic), or xenogeneic sources (e.g., encapsulated pig islets (Living Cell Technologies)), transplanted to replace diseased or damaged tissue, secrete factors that stimulate tissue regeneration, or regulate (up or down) a desired immune response.

Most cell therapy is a subset of regenerative medicine that concentrates on the "replacement or regeneration of human cells, tissue, or organs, to restore or establish normal function." Cell-based immunotherapy is the exception and involves the use of cells to trigger an innate immune response in vivo to fight disease. The modality among these cell-based vaccines and immunotherapies is not replacement or regeneration, but rather invoking an immune response. In this sense, such a therapy is not regenerative.

For this chapter, a cell therapy is defined as a cell-based product manufactured ex vivo using autologous, allogeneic (somatic or embryonic), or xenogeneic cells as part of the product administered to a recipient for therapeutic purposes. Included are cell-delivered gene therapies. Also included—albeit stretching the use of the word *therapy* to some extent—are bio-aesthetic regenerative treatments such as Fibrocell's Laviv™ (azfibrocel-T).

Some limit the definition of cell therapies to only those regulated or approved by the applicable regulatory agency, for example, those regulated under Section 351 of the Public Health and Safety Act in the U.S. (commonly referred to as "351 products"). Using the American regulatory system as the example, we prefer a more expansive definition of cell therapies to include products that are regulated but do not require IND/BLA approval (commonly called "361 products" referring to the governing Section 361 of the Public Health and Safety Act) (Carpenter 2009).

However, we exclude simple, unexpanded stem cell transplantation (SCT) for any and all indications, thus excluding the uses of SCT in regulated markets as well as unregulated SCT services. We do not limit products to only those that have received formal regulatory approval in the jurisdictions in which they are commercially available. Most jurisdictions provide formal or informal provision for "simple" cell therapy products (e.g., a cell population selected from a patient's bone marrow draw and then reinfused) to be made commercially available without formal approval per se. While some suggest such treatments are more like services than products, there are some cases where the cells or processes may have sufficient proprietary protection so as to be considered products in the classic sense rather than simply services in which a non-proprietary commodity is offered for sale.

15.3 THE BUSINESS OF REGENERATIVE MEDICINE

As in other multidisciplinary fields, regenerative medicine companies do not present a homogenous type of company, business model, product, or technology. For example, very different types of commercial infrastructures are required to produce and market autologous, bio-engineered bladders versus a non-cellular matrix that triggers skin regeneration: a small molecular that activates cellular response in vivo versus an autologous immunotherapy with a 72-hour shelf life; or an allogeneic cell therapy that can be mass-produced and inventoried versus a personalized autologous product created from a patient biopsy and suitable only for that patient.

This is true even among companies within the same category, such as autologous cell therapy. Cytori's point-of-care device isolates and purifies a cell population collected by a physician and reinfused by the physician shortly after harvest. This is a very different business model (driven by equipment placements and sales of disposables) than a service-based model like Genzyme's Carticel, which is similar in that it is also an autologous cellular product but involves surgical cell procurement, a cellular manufacturing facility, and a surgical implantation in addition to the logistic of shipping a fresh product with limited stability.

In addition to the sector presenting multiple business models, there are as yet few commercial successes in the sector to mimic. Even adventurous venture capital has largely shied away from the sector because early investments in tissue engineering and cell therapy performed poorly and an investment community with a long memory is reluctant to repeat such history until a successful commercial precedent appears (Parson 2008; see Chapter 5 by Prescott and Chapter 6 by Benjamin). Furthermore, regenerative medicine is, for the most part, a disruptive technology (Mason and Dunnhill 2008). Such technologies inherently lack proven commercial precedents but have certainly rewarded early investors. However, the hope inherent in the promise of this new genre of therapies has generated enormous consumer demand in millions of patients worldwide. Even though the investment community has remained cautious to invest in many regenerative medicines, private philanthropy, charity organizations, and the public have supported translational research and clinical trials within the stem cell/regenerative medicine sector. The most notable non-traditional funding entity is the California Institute for Regenerative Medicine (CIRM), a California stem cell agency created through the passage of Proposition 71, a statewide ballot measure which provided $3 billion in funding for California stem cell research projects. Since CIRM's inception in 2005, over $1 billion has been allocated to universities, research institutes, companies, and other translational stem cell and regenerative medicine projects.

Another example is JDRF, which partnered with Maryland-based Osiris Therapeutics in 2007 to accelerate the discovery, development, and commercialization of the stem cell therapeutic Prochymal for the treatment of newly diagnosed type 1 diabetics. Under the partnership agreement, JDRF provided $4 million in funding to support the Phase II clinical trial. In May 2010, Osiris received FDA Orphan Drug Designation for the use of Prochymal as a treatment for type 1 diabetes mellitus.

In January 2010, the UK Stem Cell Foundation and the MS Society jointly donated £1 million toward MS-related stem cell research projects. A few months later, in the spring of 2010, the Motor Neurone Disease (MND) Association announced funding for a stem cell research program between scientists at the University of Edinburgh, Kings College in London, and Columbia University of New York to study the biochemical pathways altered in MND patients. Lastly, in June of 2010 The New York Stem Cell Foundation announced an Innovator Award Program that will allocate $24.5 million to fund several translational research projects over the next 5 years.

At the risk of being trite, the key to commercial success for regenerative medicine will be therapeutic efficacy. While the "better" product does not always prove to be a commercial success over a better managed mediocre product, it is difficult to imagine that a therapy that proves remarkably more effective than the existing standard of care will fail to succeed commercially. Similarly it is difficult to imagine that such a product will not be reimbursed by insurers.

15.3.1 SUCCESS? RECORD FOR CELL THERAPY COMPANIES

In what Chris Mason, the director of the Stem Cell and Regenerative Medicine Bioprocess Group at University College London, calls *Regenerative Medicine 1.0*, he estimates that 70 tissue engineering companies failed between 1985 and 2002 or emerged mere shadows of their former high-flying valuations. In the current era of *Regenerative Medicine 2.0*, he cites several factors suggesting that companies are now much more focused on the business and commercialization prospects of their technologies than were their technology-driven predecessors (Mason 2007).

Cell-based therapeutics have been available for over 40 years in the form of stem cell transplants for immune system reconstitution after radioactive therapy. Numerous cell-based therapies are now available to patients around the world in the areas of oncology, wound and cartilage repair, and retinal repair. However, for a number of reasons this sector has yet to realize material commercial success.

The non-cellular side of the industry represents by far the largest portion of revenue earned from regenerative medicines to date. This segment crossed the $1 billion mark in annual revenue in 2006—still relatively small compared to other more mature biotechnology and pharmaceutical sectors (Smith 2008). No cell product has yet to generate $500 million in revenue (Parson 2008). There are a number of reasons cell therapy products have not produced the kind of commercial success investors aim to achieve (McAllister 2008; Lysaght 2008). Some of these reasons are discussed briefly below.

Some suggest the situation is not a failure but rather it represents the natural maturation of a nascent technology sector in what Mason and Dunnhill call typical of the "pattern for pioneering start-up" (2008).

The products brought to market to date are first-generation technologies that typically would not be expected to be produce blockbuster commercial success.

These products were brought to market ahead of a clear regulatory framework that covered them. The lack of guidance allowed products to hit the market prematurely, before having proven the type of efficacy that would now be expected of an approved product. Although it is arguable, some suggest that such vetting rigor simply transfers the failure rate to pre-market, thus translating into a higher likelihood of commercial success for those products which make it to market.

At the time these products were brought to market, insurers were not yet ready to reimburse them. Consequently, in the United States, there was often a time lag between product approval and the procurement of commercially viable levels of reimbursement. This clearly impacted both profitability and clinical adoption.

Many of the companies that brought the existing cell therapy products to market suffered considerably in the financial downturn around the turn of the century. This left them financially crippled (if not insolvent) and certainly poorly capitalized to finance the kind of strategic market launch that might have otherwise triggered broader market adoption. Furthermore, the disruptive nature of these cell therapy products engendered significant resistance from clinicians and companies invested in the success of traditional therapeutics.

Many of the existing cell therapy products brought to market to date have not demonstrated sufficiently better clinical efficacy compared with existing standards of care and thus have not been able to overcome resistance to clinical adoption.

Some argue that while cell therapy products (if this refers to products requiring formal regulatory approval) have yet to demonstrate the kind of profitability to be considered commercial successes, stem cell transplantation as a clinical service has become a commercially successful industry that generates significant revenue for hospitals, clinicians, and suppliers of ancillary products and services. This has led some to postulate that cell therapy is destined to be a hospital-based service rather than a collection of proprietary products developed and marketed by private industry. While this will certainly be true of some types of cell therapy, no serious analyst of the industry believes this will be universally true.

15.3.2 Non-Cell-Based Segment

Regenerative medicine products surpassed an estimated worldwide revenue of $1 billion in 2006 and the top ten regenerative medicine products are estimated to have generated over $1.1 billion in 2007 (Smith 2008). The top four products generated well over 90% of this revenue. Not unlike other biotech sectors, well over half of that revenue came from one product (Medtronic's InFUSE)—a growth factor with matrix product for spinal fractures, facial fractures, and open tibial fractures. In a 2008 article, Devyn M. Smith summarized the regenerative medicine business as follows:

> The top 10 Regenerative Medicine products on the market generated ~$1.2B in aggregate revenue last year [2007]. Interestingly, all of these products target "simple" tissue such as skin, bone and cartilage. Only two regenerative medicine products generated more than $100M in revenue (InFUSE by Medtronic and Alloderm by Lifecell) and neither have a cellular component. In fact, only one of the top five products in the regenerative medicine sector is composed of cells (Carticel, produced by Genzyme). While five of the top ten Regenerative Medicine products are cell based, they only account for approximately $165M in annual revenue in aggregate—which in aggregate would not be qualify as one of the top 200 drugs on the market in 2007 (Smith 2008).

The business models for non-cellular regenerative medicine products mimic traditional pharmaceutical, biotechnology, and device business models.

15.4 STAKEHOLDERS: SYNOPSIS OF REGENERATIVE MEDICINE COMPANIES AND PRODUCTS

15.4.1 Cell Therapy Companies and Products

An estimated 275 companies globally have at least one cell therapy product on the market or in development (clinical and preclinical). The Cell Therapy Group (www.

State of the Global Regenerative Medicine Industry

celltherapygroup.com) is currently tracking more than 700 companies defined as stakeholders in the cell therapy sector of regenerative medicine. This includes the 275 therapeutic companies noted above with approximately 320 cell-based therapeutic products on the market or in clinical or preclinical development. These products can be approximately broken down into the following stages:*

Phase	Number of Products
Preclinical (estimated)	78
Phase I	77
Phase II	89
Phase III	28
Commercial (marketed in at least one country)	44

A recent article presented a different dataset based on a search of the ADIS Insight database using "cell replacement therapy" as the search parameter (McKernan 2010). The result was 68 cell-based therapies in clinical development, over 90% of which were company-sponsored.

15.4.1.1 Phase III Pivotal Stage Products

As of November 2009, approximately 28 Phase III or pivotal cell therapy industry-sponsored trials were in progress. They are broken down approximately as follows:

Product Type	Number
Autologous	14
Allogeneic	5
Allogeneic–autologous combination	3
Allogeneic–autologous combination with drug	1
Autologous with gene modification	1
Allogeneic with gene modification	1
Allogeneic with device	3

Half of the products are fresh (predominantly the autologous products). The breakdown of indications is as follows:

Indication	Number
Cardiac repair	8
Wound or skin repair	6
Oncology	6
Eye diseases and conditions	2
Unique conditions	6

---

* Note that these numbers are limited to industry-sponsored trials. The preclinical number is likely much smaller than the true number because this data is limited to products in only the latest stages of preclinical studies.

In a breakdown of cell types, 10 involve differentiated cell types; 8 are described as types of MSCs (marrow stromal cells, mesenchymal stem cells); 4 are based on immune system cells; 2 are described as cord blood-derived stem cells; and the remainder are comprised of tumor-specific cells, porcine cells, and hematopoietic stem cells. Approximately 16 of the 27 products are under development by companies based in the United States; the remainder are based in Europe and Asia.

15.4.1.2 Commercial Products

A number of first generation cell-based therapeutics have been commercially available for several years. As of November 2009, approximately 44 cell therapy products were marketed in at least one regulated country. This total is *not* limited to products that received formal approval from relevant regulatory authorities; it includes products that have tacit approval in that they are allowed to be marketed by regulatory authorities in their jurisdictions although they have not received formal regulatory approval. The number does *not*, however, include stem cell transplantation as a service even when offered by a company or sold by clinics operating in regulated or unregulated countries. These 40 or so commercial cell therapy products include 10 in the U.S., 21 in Europe (about 9 in Germany and 12 in other European Union countries or across the European Union), about 9 in Korea (Cho 2008; Hong 2008; Kim and Park 2008), 2 in Australia, and 3 in Japan.

Very few of these 40+ commercial products are truly global (Genzyme's Epicel and Carticel). Most are regional, if not limited to one country. The companies behind a few of these products are in such financial distress that the products may not be currently available (York Pharma's CryosSkin and MySkin and Forticell Bioscience's OrCel).

About 35 (or 80%) of the products fall into the categories of cartilage repair and skin (wound, skin ulcer, burn) repair. Marketed cell therapy skin products (Rotislav 2009) include Apligraf (Organogenesis), Bioseed-S (BioTissue Technologies, GmbH), CellSpray (Clinical Cell Culture, also known asC3), Dermagraft (Advanced BioHealing), EPIBASE (Laboratories Genevrier), Epicel (Genzyzme Biosurgery), EpiDex (Modex Therapeutiques), Hyalograft 3D, Laserskin, and TissueTech Autograft System (Fidia Advanced Biopolymers), MySkin (York Pharma), OrCel (Ortec International), PolyActive (HC Implants BV), and Transcyte and Dermagraft TC (Advanced BioHealing).

No commercially available product in Europe has received formal regulatory approval even in its home country with the exception of TiGenix's ChondroCelect—the first cell therapy to receive market approval under the new European Medicines Agency (EMA) Advanced Therapy Medicinal Products (ATMPs) regulation (see Chapter 19 by Bravery). All the commercially available products in the European Union reached the market in a regulatory vacuum or under state regulations that no longer exist. They are now governed (1) as non-medicinal products (even though they are therapeutics) under the European Tissues and Cells Directive (EUTCD) which, like the U.S. Good Tissue Practice rule, governs facilities and processes rather than products—premarket authorization is not required for such products—or (b) via the EMEA's ATMPs (although state regulations also apply within the overarching EU/EMEA frameworks).

ATMPs are medicinal products for human use, based on gene therapy, somatic cell therapy, or tissue engineering. The regulatory framework for ATMPs is established by Regulation EC 1394/2007 (see Chapter 19 by Bravery).

Regenerative medicine products now commercially available in Europe will have to be confirmed as falling within the non-medicinal classification, allowing them to fall outside the ATMPs regulatory framework or be transitioned as ATMPs. Somatic cell therapies (SCTs) and tissue engineering products (TEPs) that were legally marketed before December 30, 2008 will have to comply with the ATMP regulation no later than December 30, 2011 and December 30, 2012, respectively. This means the companies behind these products will have to submit market approval applications (MAAs) just as if their products were new therapies. This is not likely to be a trivial effort because some products have not undergone traditional clinical trials. Their prior market use will carry no weight and the data presented will be judged by the same standards as any other medicine. From a practical standpoint this may eradicate many if not most of the current commercial regenerative medicine products available in Europe.

The cell therapy products commercially available in the U.S. are:

Provenge (Dendreon, Inc.)
Dermagraft (Advanced BioHealing, Inc.)
Prokera (Biotissue, Inc.)
Amniograft (Biotissue, Inc.)
OrCel (Forticell Bioscience)*
Epicel (Genzyme Corporation)
Carticel (Genzyme Corporation)
Osteocel (Nuvasive, Inc.)
Apligraf (Organogenesis, Inc.)
Fusionary (SpineSmith, LP)

Only Dermagraft, Epicel, Carticel, and Apligraf have formal FDA approval (Carpenter 2009). Some would include the Therakos Photopheresis System point-of-care device (Therakos, Inc.) on the list because it delivers a cell-based product manipulated ex vivo. Genzyme's Carticel has now been used to treat over 10,000 patients, Organogenesis's Apligraf surpassed 250,000 patient therapies in 2008 (Parson 2008) and Osteocel (Nuvasive) has treated over 30,000 patients to date. Dendreon's Provenge, a cell-based immunotherapeutic which some would suggest is not a regenerative medicine despite being a cell therapy, is the first cell-based therapy to receive FDA approval in a decade.

15.4.1.3 Supporting Technologies for Regenerative Medicine Products and Companies

In addition to the development of therapeutic products, the supporting technologies sector around such products and the companies developing them has seen significant growth.

---

* Bankrupt as of this writing.

These life sciences "pick-and-shovel" companies include Life, Lonza, Millipore, Thermo Scientific, GE Healthcare, Invetech, and others. They are developing and providing products (tools, reagents, systems, cell lines, bioreactors) and services (process development, manufacturing, engineering) that are not therapeutic but are or will be integral to the development and/or production of regenerative medicines (see Chapter 13 by Goswami). These companies are mainly mature, multinational companies that leverage their experience and expertise from other sectors. Their investment in the industry is a positive harbinger of sector maturity as their presence and commitment provide credibility to the industry's commercial potential and also inevitably help realize its potential.

15.4.2 Non-Cell-Based Regenerative Medicine Companies and Products

Proteus Venture Partners reports tracking approximately 1,000 companies with investments in the regenerative medicine industry. Approximately "300 to 400 [of these] are focused solely on regenerative medicine" (Parson 2008).

Many companies with key products used in non-cellular regenerative medicines are device companies (e.g., Matricel GmbH, 3M) supplying the types of technologies (matrices, growth factors, etc.) discussed earlier in this chapter. Others, however, use stem cells and the accumulated knowledge base of stem cell biology to identify traditional biotech and/or pharmaceutical targets to effect desired cellular responses in vivo.

This is a new class of company—a company defined not so much by the type of product it develops (they are quite traditional) but by the intended mechanisms of action. The dividing lines are not clear since there are companies, certain pharmaceutical companies, for example, that heavily incorporate the use of stem cells in their drug discovery, screening, and/or toxicology testing programs (see Chapter 10 by Walker). This activity is often seen as a sector of the regenerative medicine industry. In our opinion, however, the use of stem cells by a pharmaceutical company in its drug development program does not mean it is properly classified as a regenerative medicine company; it remains a pharmaceutical company developing pharmaceuticals.

Pfizer, for instance, has used stem cells in drug development for a number of years. It was not, however, until the company decided to invest in the use of cells as therapies that it became involved in regenerative medicine. Fate Therapeutics, Inc., Stemgent, Inc., and iPierian, Inc. are notable examples of this emerging class of regenerative medicine companies defined not so much by their use of stem cells ex vivo but by their intent to effect cellular reprogramming and/or directed cellular differentiation in vivo.

15.5 CORE COMPONENTS OF REGENERATIVE MEDICINE

The four core components of regenerative medicine are (1) matrices and scaffolds (classic tools of tissue engineering), (2) growth factors, (3) cells, and (4) other biologic or pharmaceutical agents (e.g., rNAI, genes, antibodies, proteins, small molecules, etc.) which invoke or enhance a therapeutic response within the definition of regenerative medicine.

15.5.1 MATRICES AND SCAFFOLDS

Natural polymers and extracellular matrix materials are widely used as scaffolds for tissue engineering and in regenerative medicine. These materials have shown excellent biocompatibility and safety due to their biological characteristics including biodegradability and weak antigenicity.

Scaffolds provide an appropriate environment for restoration of the intrinsic functions of cells and provide structural components including collagen, elastin, adhesive proteins, proteoglycans, and glycosaminoglycans. Biological matrices such as acellular dermis, amniotic membrane, and small intestine submucosa have been employed in numerous wound healing applications; decellularized whole organs, both simple and complex, are being developed as bio-artificial products for organ replacement. The safety and efficacy of biological matrices have been demonstrated in preclinical models and in clinical work when used for bone, tendon, ligament, vessel, urinary tract, and skin regeneration (Seo 2009).

15.5.1.1 Collagen

Collagen is the most abundant protein within the mammalian extracellular matrix (ECM) and can be isolated easily from animal tissues and organs. Collagen has been widely applied in tissue engineering and to some extent in drug delivery systems because it readily interacts with cells in connective tissues and regulates (via signaling) cell anchorage, migration, proliferation, and differentiation. Injection of collagen solution has been largely limited to skin therapy but applications are now expanding into treatment of damaged bone, cartilage, and muscle by combining collagen with growth factors and cells. It has been demonstrated that bone regeneration is accelerated by injecting collagen in combination with recombinant BMP-2 and mesenchymal stem cells (Okafuji 2006; Alonso 2008). The earliest tissue-engineered product to use collagen gel was bio-artificial skin. Apligraf® (Organogenesis, U.S.) was the first FDA-approved product for treatment of skin ulcers (Guerret 2003).

15.5.1.2 Gelatin

Gelatin is produced by denaturing collagen and has been used as a dressing due to excellent hemostatic activity. It has been investigated as a material for cartilage regeneration; a gelatin sponge was mixed with albumin for enhancement of cell adhesion and proliferation and the combination yielded excellent proliferation of chondrocytes and supported the chondrogenic differentiation of mesenchymal stem cells (MSCs) (Mohan 2008).

15.5.1.3 Silk

Silk matrices have been rediscovered and reconsidered as potentially useful materials in a range of applications in regenerative medicine. Silk has been developed as a scaffolding material for the reconstruction of bone, cartilage, ligament, tendon, and skin. Silk scaffolds have been designed to exhibit complex demanding mechanical properties and fabricated in various structures such as porous form, non-woven form, and film (Sugihara 2000; Kim 2005). Use of a BMP-2 infused porous silk scaffold resulted in effective bone regeneration (Zhao 2009; Kirker-Head 2007) and has effectively been utilized as a substrate for cartilage engineering.

15.5.1.4 Fibrin

Fibrin is a major component of clotted blood and plays a critical role in wound healing. It can be harvested readily from patients and is used as a hydrogel for the delivery of cells and bioactive materials; in addition it is used widely as a bio-adhesive in surgery due to its hemostatic, chemotactic, and mitogenic properties. Fibrin is available for the delivery of growth factors (Jeon 2005) and fibrin containing nerve growth factor was effective in promoting neural regeneration (Taylor 2004). Commercially available fibrin glue products are Tisseel VH® (Baxter, U.S.) and Greenplast® (Greencross, Korea). Commercial kits for harvesting individual patients' fibrin are marketed as Cryoseal® (Thermogenesis, U.S.), Vivostat® (Vivolution, Denmark), and a combination device that harvests both fibrin and platelet-rich plasma under development as a dual processing device (Circle Biologics, U.S.).

15.5.1.5 Laminin

Laminin is one of the major components of the basal lamina along with type IV collagen and proteoglycans. Laminin is widely used in the fabrication of basal lamina substrates along with type IV collagen and in the manufacture of bio-artificial skin. Laminin coating of a dermal substrate resulted in the adhesion of epidermal cells, reduction of MMP-9 production, and enhancement of keratinocyte proliferation (Dowgiert 2004). More recently, laminin has been demonstrated to have utility in nerve regeneration as it plays a critical role in neural cell migration, differentiation, and axonal extension because the laminin protein is continuously produced and secreted after nerve damage (Rangappa 2000).

15.5.1.6 Hyaluronic Acid

Hyaluronic acid (HA) is clinically available and has been utilized in ophthalmologic and joint regions. During skin development in the embryo, HA is highly expressed and during wound healing the material has been shown to be involved in cell migration and differentiation (Alexander 1980). A bio-artificial skin comprised of two layers has been developed in the form of a thin collagen layer covering an HA sponge layer. In addition, a cultured epidermis (Laserskin®, Advanced Biopolymers, Italy) is manufactured by culturing keratinocytes on a film of HA (Kubo 2003). HA is therapeutically employed for lubrication of damaged connective tissue and joints and recently data demonstrated that adding functional groups into the terminal regions of HA increased its ability to support cell adhesion and drive cell proliferation (Park 2003).

15.5.1.7 Biological Acellular Matrices

Biological matrices are now being widely studied and developed for therapeutic use in tissue engineering and regenerative medicine. Acellular matrices are developed by removing the cellular component of a whole organ or tissue either through mechanical agitation or retrograde perfusion using detergents (Ott 2007). Acellular dermis obtained from cadaveric or xenogeneic skin is clinically available.

Applications include temporary wound dressings that permit neovascularization in the wound bed as well as permanent grafts. A permanent graft is manufactured by removing cells after separating the epidermis from the dermis. This acellular dermis has shown little graft vs. host reaction and a very high engraftment rate when kerotinocytes have been cultured (Cuono 1986).

For example, Paolo Macchiarini and colleagues recently used an engineered graft to replace a diseased portion of windpipe in a woman who had failed all previous treatments, and they succeeded in restoring breathing function to this patient (Macchiarini 2008). Their strategy involved use of a simple biologic scaffold in the form of a donated human trachea that was pretreated to remove endogenous cells and immune-stimulatory proteins (Hung 2009). This scaffold was seeded with two populations of patient-derived cells: cultured epithelial cells isolated by a small airway biopsy, and cartilage cells derived in vitro from bone marrow stem cells obtained from a needle aspirate laboratory to patient (Hung 2009). Recent work by Ott et al. (Ott 2007) demonstrated the ability to engineer complex heart tissue based on a decellularized biological matrix. Importantly, Ott was able to show that by using perfusion decellularization, a complex, biocompatible cardiac ECM scaffold with a perfusable vascular tree, patent valves, and a four-chamber-geometry template could be created for biomimetic tissue engineering. In addition, this report shows that by applying perfusion decellularization to a porcine heart, the technology can be scaled to organs of human size and complexity (Ott 2007). Human-implanted engineered tissues comprised of cellularized scaffolds (e.g., skin [Parenteau 1999], cartilage [Ochi 2004], and bladder [Atala 2006]) have also been demonstrated to be effective.

15.5.2 Growth Factors

The growth factor segment of the regenerative medicine market includes natural and recombinant growth factors, used alone or in combination with other materials such as transforming growth factor-beta (TGF-beta), platelet-derived growth factor (PDGF), fibroblast growth factor (FGF), and bone morphogenetic protein (BMP), predominantly for the treatment of orthopedic indications. The growth factor segment is dominated by bone morphogenic proteins that account for around 90% of the market; autologous growth factors (platelet-rich plasma) account for the remainder.

Initially discovered in 1965 by Marshal Urist, BMPs are the only known proteins capable of inducing new bone formation. Several BMPs (BMP-2, BMP-7, and BMP-14) have been shown experimentally or clinically to induce bone formation in spinal fusion. The InFUSE (Medtronic USA) bone graft represents an rhBMP-2 (recombinant human bone morphogenetic protein-2) formulation combined with a bovine-derived absorbable collagen sponge (ACS) carrier. The rhBMP-2 induces the body to grow its own bone, eliminating the need for a painful second surgery to harvest bone from elsewhere in the body. InFUSE bone graft is indicated for use in spinal fusion and for the treatment of certain types of acute, open fractures of the tibial shaft, as well as certain oral maxillofacial indications (Episcom 2008).

Platelets contain various growth factors (also called cytokines). Platelet-derived growth factor (PDGF), transforming growth factor-beta (TGF-B), insulin-like growth factor (IGF), and vascular endothelial growth factor (VEGF) are contained within the platelets. These proteins play a role in bone formation, but none of them, individually or in combination, are capable of inducing new bone formation. Two are of particular interest in dealing with orthopedic injuries due to their roles in recruiting connective tissue, matrix formation, and overall tissue healing: PDGF and TGF-beta-33. While platelet concentrates contain growth factors, they are not osteoinductive because they do not include bone morphogenic protein.

Platelet-rich plasma (PRP) is a platelet concentrate widely used to accelerate soft and hard tissue healing. Post-surgically, blood clots initiate the healing and regeneration of hard and soft tissues. A natural clot contains 95% red blood cells, approximately 5% platelets, and less than 1% white cells. Platelets are primarily involved in wound healing through clot formation and the release of growth factors that initiate and support it. Using PRP involves taking a sample of a patient's blood preoperatively, concentrating autologous platelets, and applying the resultant gel to the surgical site. This technique produces a blood clot that has nearly a reverse ratio of red blood cells and platelets compared with a natural clot (Episcom 2008). Surgical sites enhanced with PRP have been shown to heal at rates two to three times that of a normal surgical site. One advantage of using these is that they are easily removed from the patient's blood with very few complications. Furthermore, the ability to derive this platelet component from the patient's own blood also has the advantage of reducing the risk of cross-infection and disease transmission.

The use of PRP in place of recombinant growth factors has several advantages in that growth factors obtained from platelets exert their own specific actions on tissues and also interact with other growth factors, resulting in the activation of gene expression and protein production. Therefore, the properties of PRP are based on the production and release of multiple growth and differentiation factors upon platelet activation. These factors are critical in the regulation and stimulation of the wound healing process and they play an important role in regulating cellular processes such as mitogenesis, chemotaxis, differentiation, and metabolism.

15.5.3 Cells

Stem cells, progenitor cells, and other cell types that exhibit or can be reprogrammed to exhibit "stemness"-like properties are life's biological repair mechanisms and the structural and functional units of all living organisms. Multipotent cells are integral components of the regenerative medicine industry and used for both therapeutic and drug discovery applications. Almost 3,000 clinical trials using stem cells are listed on clinicaltrials.gov but the majority of these studies continue to focus on hematopoietic disorders and cancer.

While we will focus here on stem cells, primary cells are used in commercially available cell therapy products (e.g., autologous chondrocyte transplants) and/or products under development. Primary cells taken directly from living tissue (biopsy materials) are often, but not always, expanded in vitro before being re-administered. Some believe that because such cells undergo very few population doublings, they are more likely to function like the tissue from which they were derived when compared to cells created from continuous (tumor or artificially immortalized) cell lines.

In 1963, James Edgar Till and Ernest McCullough paired together to study the area of hematopoiesis and leukemia. Till and McCulloch noticed in a laboratory experiment small nodules on the spleens of irradiated mice that they had injected with bone marrow. They reported in *Nature* in 1963, that each of the nodules arose from a single cell that attained the ability to differentiate and self renew—they named this unique newly discovered cell that was responsible for the nodule formation a "stem cell" (Becker 1963). To date, Till and McCullough's defining characteristics of a stem cell still stands true.

A stem cell's natural capacity to self-renew and differentiate into a multitude of cell types has made the study of stem cell biology a cornerstone research area in the field of regenerative medicine and the focus of the cell section of this chapter. The derivation, research, and clinical application of stem cells has been, and continues to be, in the spotlight of international regulators, policy makers, economic development officers, layers, bioethicists, and of course scientists, physicians, investors, and regenerative medicine business leaders. The translation of stem cell science into regenerative medicine therapies offers hope and potential cures to countless diseases and disorders and has therefore become a core research component for many regenerative medicine companies.

Stem cells, progenitor cells, and other cell types that exhibit or can be reprogrammed to exhibit "stemness"-like properties are life's biological repair mechanism and the structural and functional units of all living organisms. Stem cells, as well as other pluripotent and multipotent cells, are an integral component of the regenerative medicine industry and used for both therapeutic, drug discovery, and disease modeling applications.

The two fundamental properties that distinguish stem cells from other somatic cells are self-renewal and potency. The first stem cell property is the capacity to self-renew or undergo mitotic cell division; at least one of the two daughter cells retains the same development potential as the mother cell. The other daughter cell undergoes a finite number of cell divisions and differentiates to become a specified somatic cell. Self-renewal is a necessary cellular function to maintain a sufficient number of stem cells in different tissues and organs of the body to maintain and repair lost or damaged cells and tissues. Depending on the age and type of stem cell, the capacity to self-renew will vary except for the pluripotent embryonic stem cells that maintain the ability to self-renew indefinitely under proper culture conditions.

The second distinguishing property of a stem cell is the ability to differentiate into mature specialized cell types. The number of specialized cell types a stem cell can produce defines its potency (totipotent, pluripotent, multipotent, unipotent). Stem cells that give rise to all cell types of the body including the embryonic components of the trophoblast and placenta are defined as totipotent. They are derived from the pre-implanted embryo at the morula stage of development.

Like totipotent stem cells, pluripotent stem cells also contain the capacity to become all cells of a developed organism but cannot produce trophoblast and placenta to support uterus implantation and organismal development. Naturally existing pluripotent stem cells are found within the inner cell mass of the blastula stage embryo and have the natural capacity to differentiate into all germ layers and cell types. Under normal developmental conditions these cells cannot naturally revert back to the pluripotent state once committed to a specific cell lineage or after they pass the gastrulation stage of embryonic development.

Multipotent stem cells continue to express the defining characteristics of a stem cell—self-renewal and differentiation—but they do not retain the capacity to produce cell types of all three germ layers. Instead, multipotent stem cells give rise to multiple but limited numbers of lineages within a tissue, organ, or physiological system. Hematopoietic stem cells, mesenchymal stem cells, and neural stem cells represent three well defined multipotent, tissue-specific types. Finally, unipotent stem cells maintain the capacity to self-renew but give rise to only one mature cell type and therefore have the lowest potency. Skin cells that maintain the ability to self-renew are examples of unipotent stem cells.

Various types of multipotent stem cells are found throughout the body during all stages of life and they function to produce tissue and act as the body's natural repair mechanism. As the body ages, the number of stem cells and progenitor cells found in adult tissue decreases. This may occur because aging tissue-specific multipotent stem cells lose their propensity to self-renew and instead differentiate into mature cell types.

Cell potency is one of the critical factors for cellular therapeutic companies to consider when developing a regenerative cellular therapeutic. Other important characteristics include ease of procurement, versatility (plasticity), non-immunogenicity, safety (non-teratoma forming), expandability, and the ethics of procurement. Cells at different stages of maturity and different types of stem cells exhibit different characteristics that may be advantageous or detrimental to the cell function needed in terms of ex vivo production and intended in vivo effect.

15.5.3.1 Cell Type

Selection of a cell type for a therapeutic product is a multiparametric exercise based on several factors including but certainly not limited to (1) the intended mechanism of action, (2) the scientific interpretation of how a particular cell type will act in vivo, and (3) the desired propensities of that cell type to act ex vivo as needed. Described below are five major stem cell types (HSCs, MSCs, NSCs, iPS cells, and ESCs) selected to discuss various challenges with these and other types of cell-based therapies.

15.5.3.1.1 Hematopoietic Stem Cells (HSCs)

Hematopoietic stem cells are progenitor cells of platelets, erythrocytes, granulocytes, B and T lymphocytes, monocytes, tissue macrophages, and dendritic cells. Predominantly used in treating autoimmune diseases (ADs), the aim of HSC transplantation (HSCT), for example, is to replace host autoaggressive immune effector cells with donor-derived non-autoaggressive cells as a means of inducing tolerance and sustained remission. Currently, autologous HSCT is the most widely used form. Allogeneic HSCT can result in graft-versus-host disease (GVHD), a potentially severe and life-threatening complication, thus making the allogeneic form less attractive as a treatment for AD. Hematopoietic stem cells are also the source materials for other more manipulated autologous and allogeneic products in development.

15.5.3.1.2 Mesenchymal Stem Cells (MSCs)

Mesenchymal stem cells, the non-hematopoietic progenitor cells found in various adult tissues, are characterized by their ease of isolation and rapid growth in vitro while maintaining their differentiation potential, allowing for extensive culture expansion to obtain large quantities suitable for therapeutic use.

Originally derived from bone marrow, MSCs and MSC-like cells have been found to exist in and may be isolated from a large number of adult tissues, where they are postulated to carry out the functions of replacing and regenerating local cells lost to normal tissue turnover, injury, or aging. MSCs are regenerative; they promote tissue repair in the body by modulating immune responses and protecting damaged tissue. They migrate to inflammation sites to influence inflammation processes and the immune system (Aggarwal 2005).

A very important property of MSCs, especially for their use in rheumatic diseases, is their potent immunosuppressive and anti-inflammatory functions that have been demonstrated both in vitro and in vivo. While the original belief was that MSCs were stem or progenitor cells with the potential to contribute to the regeneration of tissue, more recent data suggest that the principal mechanism of MSC activity is the release of soluble mediators that elicit the observed biological response (Horwitz and Dominici 2008). They serve as cellular factories, secreting mediators to stimulate the repair of tissues or modulate the local microenvironment to foster requisite beneficial effects (Horwitz 2008). Recent research has demonstrated that MSCs are generally considered to have three potential applications: (1) differentiating into mature cells and populating the resident tissues; (2) secreting cytokines or other soluble mediators; and (3) serving as gene delivery vehicles (Horwitz, 2008).

15.5.3.1.3 Embryonic Stem Cells (ESCs)

ESCs have been derived from a variety of animals including humans (hESCs), have the ability to self-renew, and are pluripotent or totipotent, giving rise to cells from all three germinal layers (endoderm, ectoderm, and mesoderm) (Thomson 1998; Ware 2006). In addition, ESC lines can be grown indefinitely in vitro if the correct conditions are met. Importantly, these cells continue to retain their ability to form all specialized cell types after they are removed from the conditions that keep them in an undifferentiated or unspecialized state. The ability to expand and yet control differentiation, however, is an art not yet mastered with any standard level of success.

A critically important characteristic of any embryonic stem cell line slated for therapeutic or transplantation use will be the lack of teratoma formation in vivo, a current challenge associated with ESCs (Carpenter 2009). Overall, hESCs are difficult to procure due to significant ethical issues, present varying levels of immunogenicity, and are often associated with unsuitably high numbers likely to form teratomas in vivo, thus challenging their inherent safety and driving up cost of development by requiring that they be exhaustively characterized. However, hESCs are totipotent, highly expandable, and scalable.

15.5.3.1.4 Induced Pluripotent Cells (iPS Cells)

In 2006, Yamanaka et al. showed that mouse embryonic and adult fibroblasts acquire properties similar to those of ESCs after they retrovirally introduced genes encoding four transcription factors, namely Oct3 and 4, Sox2, Klf4, and c-Myc (Takahashi and Yamanaka 2006). The first generation iPS cells were similar to ESCs in morphology, proliferation, expression of some ESC marker genes, and formation of teratomas. It was postulated that various types of somatic cells derived from pluripotent stem cells could be used in regenerative medicine and iPS cell technology potentially could overcome two important obstacles associated with hESCs: immune rejection after transplantation and ethical concerns regarding the use of human embryos (Carpenter 2009).

However, the clinical application of iPS cells faces many obstacles, some shared with ESCs and others that are unique (see Chapter 10 by Walker). The first common obstacle is teratoma formation—even a small number of undifferentiated cells can result in the formation of teratomas. Therefore, protocols must be further refined to induce differentiation of hESCs or iPS cells into the required cell type while leaving few undifferentiated cells behind. Additional unique hurdles must also be overcome before iPS cells can be used in the clinic, primarily related to the forced reprogramming of somatic cells. Aberrant reprogramming may result in impaired ability to differentiate and may increase the risk of immature teratoma formation after directed differentiation. Thus, incomplete reprogramming of somatic cells to iPS cells could result in impaired differentiation of iPS cells into the required cell type. Recent evidence supports concerns about the epigenetic match of these cells with hESCs.

Another key issue is the presence of transgenes in iPS cells. Currently, most iPS cells are generated by transduction of somatic cells with retroviruses or lentiviruses carrying transgenes, which are integrated into the host cell genome. This genetic manipulation creates significant commercial hurdles.

iPS technology is developing rapidly. Recent work appears to demonstrate the near-term opportunity to reprogram without any genetic factors as well as to reprogram somatic cells directly to other types of primary cells, skipping the pluripotent cell stage.

15.5.3.1.5 Neural Stem Cells (NSCs)

Neural stem cells are tissue-specific, multipotent cells that maintain the capacity to self-renew and differentiate into all three neuroectodermal lineages in vitro using a variety of well documented techniques. Neuroectoderm is defined as the region of embryonic ectoderm that develops into neuronal and glial cells that comprise the tissue of the central nervous system (astrocytes, oligodendrocytes, and neurons) and peripheral nervous system (Schwann cells, satellite cells, enteric glial cells, and neurons).

Human neural stem cells (hNSCs) and progenitor cells can be obtained from fetal or adult tissue postmortem as cadaveric material by removing tissue from the preventricular zone. hNSCs can then be isolated by fluorescence-activated cell sorting (FACS) after enzymatic dissociation of the source tissue. The sorted hNSCs are then grown in specific growth factor-rich media as free floating neurospheres (Huhn 2009).

Neural stem cells may also be obtained by neural differentiation of human embryonic stem cells. Induction of neural stem and progenitor cells from hESCs has been widely reported by treating hESCs with BMP antagonist noggin for approximately

14 days, resulting in the efficient induction of NSCs expressing neuroectodermal markers Sox2, Pax6, and nestin (Pera et al. 2004; Li and Zhang 2006).

Multiple commercial sources of animal-free feeder cells are readily available for research and clinical application, allowing for rapid expansion and long-term sustainability of NSCs in vitro. In vivo predifferentiation of neural stem cells has been shown to increase engraftment, migration, and differentiation using in vivo functional screens (Davies et al. 2006).

15.5.3.2 Cell Procurement

To obtain the cell of choice, the accessibility of stem cells from a particular source or location (somatic niche) is a fundamental factor with development and commercial implications considering that cells reside in various specific microenvironments (or niches) within tissues and organs that vary in accessibility and surgical invasiveness. Each has the potential to significantly impact patient experience, clinical adoption, process and product design, and cost of goods. Furthermore, cell characteristic and/or function may or may not differ dependent on cell source.

Both HSCs and MSCs, for example, are easily obtained via autologous or allogeneic bone marrow aspiration. These cells are relatively well characterized and can be purified using common and accepted procedures under proper manufacturing conditions. However, bone marrow presents other disadvantages as a source for clinical or commercial users when compared, for instance, with circulating blood or adipose tissue.

NSCs are less readily accessible and are often isolated from cadaveric fetal tissue or obtained from an embryonic stem cell line. They are also relatively well characterized and purification protocols are readily available.

ESCs are derived from the inner cell masses of 3- to 5-day old blastocysts. Several well characterized, easy to access, and ethically obtained research and clinical grade embryonic stem cell lines are available worldwide. New federal guidelines in the United States have and will continue to greatly increase the number of human embryonic stem cell lines available for research funding and/or commercial use.

An unresolved issue in the creation of iPS cells is tissue source optimization due to differences demonstrated in reprogramming consistency between iPS cells generated from fibroblasts, keratinocytes, blood progenitors, and other cells (Carpenter 2009).

Different derivation methodologies are employed, depending on the niche selected and the cell target of choice, ranging from aspiration devices to apheresis systems in addition to different reagents and devices used to separate and purify the desired cell populations. Depending on the extent of the ex vivo processing required and the desired business model, cell processing can be performed via single bedside point of care (POC) devices in local or regional centers employing open and/or closed systems or at centralized manufacturing facilities. The latter is typically more common if the processing involves cell expansion.

15.5.3.3 Expansion

The ability to expand a potentially small population of stem cells into a clinically relevant number of cells is fundamental to clinical efficacy (and current dosing requirements) and certainly meaningful for commercialization of allogeneic products where economies of scale may determine commercial viability (Parson 2008). One of the drivers for the large numbers of cells currently required to achieve clinically meaningful results is the loss of cell viability and function during extensive ex vivo manipulation (separation, purification, expansion, cryopreservation, thawing, etc.). While technologies are being developed to minimize the harshness of such procedures and effects on cell function and viability, there are also technologies in development that may potentially make or help us discover how to make transplanted cells more effective in vivo post-transplant.

Although hematopoietic stem cells are easily obtained and give rise to all cell types in blood, hematopoietic stem cell expansion remains an elusive milestone in achieving what many believe to be a scalable, cost-effective, and commercially viable clinical product. HSC expansion efforts have been underway for over 20 years and while various hematopoietic growth factors have been shown to regulate primitive hematopoietic cells, in vitro HSC expansion has met with limited success (Delaney 2010). Mesenchymal stem cells present a major commercial and clinical advantage in this regard. They are easily obtainable and relatively easily expanded in vitro under well established culture conditions. The advancements in animal-free NSC media in conjunction with well documented and tested expansion protocols mean that NSCs are also capable of rapid expansion and long-term survival.

Embryonic stem cells undergo rapid cell division and self-renew indefinitely in culture and are therefore the most expandable cell type. One of the primary difficulties hESCs present, however, is controlling their differentiation during expansion. This is a hurdle several companies and academic researchers are currently grappling to solve.

15.6 CONCLUSION AND PROGNOSTICATION

Regenerative medicine is a multidisciplinary and heterogeneous industry comprised of several core types of technologies, therapeutic products, and companies. In its broadest sense, it includes products ranging from those seemingly very similar to traditional pharmaceuticals or devices to radically transformative therapeutics based on cells or genes. The one common thread running through these products is that they have the potential to be curative rather than palliative; to augment, repair, replace, or regenerate organs or tissues from damage caused by injury and/or disease. It is difficult to find an indication which is not a target for regenerative medicine. By their very nature they have the potential to revolutionize healthcare from both a medical and commercial perspective.

While until 2008 the industry was largely defined by academic study and small venture-backed firms, post-2008 has seen a significant influx of interest in the sector from multinational biopharmaceutical, life science, and healthcare companies (McKernan 2010). The mere creation of the Alliance for Regenerative Medicine (ARM), a Washington, DC-based non-profit organization founded in late 2009 and comprised of a growing list (45+) of organizations, institutes, investors, and biotechnology and biopharmaceutical companies, is itself a sign of a maturing industry.

However, this interest has yet to translate into significant participation, is still modest in terms of investment, and even more cautious in terms of engaging in clinical testing of cell therapies. McKernan et al. cite the following factors for this pharma reticence: "insufficient demonstration of efficacy, regulatory and safety concerns; a belief that cell therapies will not offer substantial benefit over existing therapies or demonstrate uptake by patients; and lack of familiarity with both the business models for commercializing cell-based products and the complexity of developing a product" (McKernan 2010).

Cell-based therapies have seen a first generation of products brought to market before the current regulatory framework was put in place. While some regenerative medicines which are the most like traditional biologics have reached blockbuster status, the cell-based products met with limited commercial success for a variety of reasons. This lack of commercial viability to date has stained the perception of such therapeutics in the mind of many investors. Despite this tranche of product forerunners, the sector is now enjoying a significant investment spoke from multinational biopharmaceutical, life science, and healthcare companies, all of whom are doing so with an eye to the future potential of these therapeutics to significantly impact—if not revolutionize—the practice of medicine, bringing cures to the currently incurable.

Many scientific questions remain unanswered with respect to the types of regenerative medicines that will prove effective for what conditions or injury. Many technical issues have yet to be resolved, including scalability, reproducibility, and characterization. How such products will be successfully commercialized in a way the healthcare system can afford and return a reasonable profit to investors remains to be seen. Nonetheless, most analysts are encouraged that since Pfizer's 2008 announcement of its Regenerative Medicine unit, the size and sophistication of the companies investing in the sector has significantly matured, and it is assumed that these companies will bring answers to these questions by leveraging their experience and expertise from other predecessor biopharmaceutical sectors.

What is interesting to note is that it may not simply be the unique scientific nature of these products that will make them uniquely challenging to develop and commercialize compared to predecessor medicines but the world in which they are being developed.

Not only is regenerative medicine growing up in an era where the bar for developing new medicines is much higher—in the sense that the products are more complex and the expectation is now for cures rather than palliative treatments—but we propose the world in which they are being developed will inevitably and differently impact how they are developed, received, and commercialized compared to previous therapeutics. This is an era of regulatory jurisdiction shopping, medical tourism, patient activism, and the increasingly crippling burden of a trifecta of factors influencing the delivery of healthcare: skyrocketing healthcare costs, aging populations, and national debts. This will force governments and health insurers to make increasingly difficult critical decisions.

REFERENCES

Alexander, S. A. and R. Donoff. 1980. The glycosaminoglycans of open wounds. *J Surg Res* 29: 422.

Alons, M., S. Claros, J. Becarra et al. 2008. The effect of type I collagen on osteochondrogenic differentiation in adipose-derived stromal cells in vivo. *Cytotherapy* 10: 597.

Atala, A., S. B. Bauer, S. Soker, J. J. Yoo, et al. 2006. Tissue-engineered autologous bladders for patients needing cystoplasty. *Lancet* 367: 1241.

Becker, A. J., McCullouch, E. A., and Till, J. E. 1963. Cytological demonstration of the clonal nature of spleen colonies derived from transplanted mouse marrow cells. *Nature* 197: 452–454.

Carpenter, M. K., J. Frey-Vasconcells, and M. S. Rao. 2009. Developing safe therapies from human pluripotent stem cells. *Nat Biotechnol* 27: 606.

Cho, Y. 2008. Cell therapy market in Korea. *BioSpectrum Asia* 13: 57.

Delaney, C., S. Heimfeld, C. Brashem-Stein et al. 2010. Notch-mediated expansion of human cord blood progenitor cells capable of rapid myeloid reconstitution. *Nat Med* 16: 232.

Davies, J. E., C. Huang, and M. Noble. 2006. Astrocytes derived from glial-restricted precursors promote spinal cord repair. *J Biol* 5: 6.

Dowiert, J., G. Sosne, and M. Kurpakus-Wheater. 2004. Laminin-2 stimulates the proliferation of epithelial cells in a conjunctival epithelial cell line. *Cell Prolif* 37: 161.

Guerret, S., E. Govignon, D. J. Hartmann et al. 2003. Long-term remodeling of a bilayered living human skin equivalent (Apligraf) grafted onto nude mice: immunolocalization of human cells and characterization of extracellular matrix. *Wound Repair Regen* 11: 35.

Horwitz, E. M. et al. 2005. Clarification of the nomenclature for MSC: The International Society for Cellular Therapy position statement. *Cytotherapy* 7: 393.

Huhn, S. 2009. Cellular therapy for CNS disorders: a translational perspective. In *World Stem Cell Rep*, Genetics Policy Institute, p. 50.

Hong, S. 2008. Korean Perspective on Biologics Regulation. Presentation. http://www.pmda.go.jp/2008bio-sympo/file/06.Dr.Hong.pdf (accessed January 11, 2010).

Hung, C. 2009. One breath closer to making engineered tissues a clinical reality. *Cell Stem Cell* 4: 5.

Kim, J. M. and J. K. Park. 2008. TERMIS Newsletter interlink, Oct–Dec III(IV) 4.

Kim, K., L. Jeong, H. N. Park et al. 2005. Biological efficacy of silk fibroin nanofiber membranes for guided bone regeneration. *J Biotechnol* 120: 327.

Kirkerhead, C., Karageorgiou, S. Hofmann et al. 2007. BMP–silk composite matrices heal critically sized femoral defects. *Bone* 41: 247.

Lefevre, F. 2008. Blue Cross and Blue Shield Association TEC Assessment Program. 25. http://www.bcbs.com/blueresources/tec/vols/23/autologous-progenitor-cell.html (accessed January 11, 2010).

Li, X. J. and S. C. Zhang. 2006. In vitro differentiation of neural precursors from human embryonic stem cells. *Methods Mol. Biol* 331: 169.

Lysaght, M. J., A. Jaklenec, and E. Deweerd, E. 2008. Great expectations: private sector activity in tissue engineering, regenerative medicine, and stem cell therapeutics. *Tissue Eng A* 14: 305.

Macchiarini, P., P. Jungebluth, T. Go et al. 2008. Clinical transplantation of a tissue-engineered airway. *Lancet* 372: 2023.

Mason, C. 2007. Regenerative medicine 2.0. *Regen Med* 2: 11.

Mason, C. and P. Dunnhill. 2008. A brief definition of regenerative medicine. *Regen Med* 3: 1.

Mason, C. and P. Dunnhill. 2008. The strong financial case for regenerative medicine and the regen industry. *Regen Med* 3: 351.

McAllister, T. N., N. Dusserre, M. Maruszewski et al. 2008. Cell-based therapeutics from an economic perspective: primed for a commercial success or a research sinkhole? *Regen Med* 3: 925.

McKernan R., J. McNeish, and S. Smith. 2010. Pharma's developing interest in stem cells. *Cell Stem Cell* 6: 517.

Mohan, N., P. D. Nair, and Y. Tabata. 2008. A 3D biodegradable protein-based matrix for cartilage tissue engineering and stem cell differentiation to cartilage. *J Mater Sci Mater Med* DOI: 10.1007/s10856-008-3535-x.

Ochi, M., N. Adachi, H. Nobuto et al. 2004. Articular cartilage repair using tissue engineering technique: novel approach with minimally invasive procedure. *Artif Organs* 28: 28.
Okafuji, N., T. Shimuzi, and T. Watanabe. 2006. Three-dimensional observation of reconstruction course of rabbit experimental mandibular defect with rhBMP-2 and atelocollagen gel. *Eur J Med Res* 11: 351.
Ott, H., T. Matthiesen, S. K. Goh et al. 2007. Perfusion-decellularized matrix: using nature's platform to engineer a bioartificial heart. *Nat Med* 1.
Parenteau, N. 1999. Skin: the first tissue-engineered product. *Sci Am* 280: 83.
Park, S., H. Lee, K. H. Lee, and H. Suh. 2003. Biological characterization of EDC-crosslinked collagen–hyaluronic acid matrix in dermal tissue restoration. *Biomaterials* 24: 1631.
Parson, A. B. 2008. Stem cell biotech: seeking a piece of the action. *Cell* 132: 511.
Pera, M. F., J. Andrade, S. Houssami et al. 2004. Regulation of human embryonic stem cell differentiation by BMP-2 and its antagonist noggin. *J Cell Sci* 117: 1269.
Ramalho-Santos, M. and H. Willenbring. 2007. On the origin of the term "stem cell." *Cell Stem Cell* 1: 35.
Rangappa, N., A. Romero, and K. D. Nelsen. 2000. Laminin-coated poly(L-lactide) filaments induce robust neurite growth while providing directional orientation. *J Biomed Mater Res* 51: 625.
Ryu, S. H., J. H. Chung, and B. S. Kim. 2005. Control of basic fibroblast growth factor release from fibrin gel with heparin and concentrations of fibrinogen and thrombin. *J Control Rel* 5: 249.
Shevchenko, R., S. L. James, and S. E. James. 2010. A review of tissue-engineered skin bioconstructs available for skin reconstruction. *J Roy Soc Interface* 7: 229.
Seo, Y. K., H. H. Youn, and J. K. Park. 2009. *Tissue Eng Regen Med* 6: 1088.
Smith, D. M. 2008. In *World Stem Cell Report*. Genetics Policy Institute, p. 158.
Sugihara, A., K. Sugiara, H. Morita et al. 2000. Promotive effects of a silk film on epidermal recovery from full-thickness skin wounds. *Proc Soc Exp Biol Med* 225: 58.
The Global Orthobiologics Market 2008. Players, Products and Technologies Driving Change. *Espicom Business Intelligence.*
Takahashi, K. and S. Yamanaka. 2006. Induction of pluripotent stem cells from mouse embryonic and adult fibroblast cultures by defined factors. *Cell* 126: 663.
Taylor, S., J. Rosenzweig, and J. MacDonald. 2004. Controlled release of neurotrophin-3 from fibrin gels for spinal cord injury. *J Control Rel* 98: 281.
Yi, J., Y. Yoon, and J. Park. 2001. Reconstruction of basement membrane in skin equivalent; role of laminin-1. *Arch Dermatol Res* 293: 356.
Zhao, J., S. Zhang, S. Wang et al. 2009. Apatite-coated silk fibroin scaffolds to healing mandibular border defects in canines. *Bone* 45: 517.

Section 4

Intellectual Property

16 Regenerating Intellectual Property
*Europe after WARF**

Julian Hitchcock and Devanand Crease

CONTENTS

16.1 A Moving Target 239
16.2 European Patent Environment 240
    16.2.1 Introduction 240
    16.2.2 Morality Problem 240
    16.2.3 European Union Law 241
    16.2.4 European Patent Convention (EPC) 242
    16.2.5 The WARF Case: Background 242
    16.2.6 WARF in the European Patent Organisation 243
    16.2.7 Resolving the "Embryo" Definition 244
    16.2.8 Historical Background 244
    16.2.9 iPS Cells as "Embryos" 245
    16.2.10 Conjoined Twins? 246
16.3 Parallel Routes to Patent Protection in Europe: The National Route 247
16.4 Commercial Landscape 247
    16.4.1 Regenerative Medicine Intellectual Property 248
    16.4.2 Types of Stem Cell Claims in Granted European Patents 250
16.5 Scopes of Stem Cell Claims 251
16.6 Duration of Protection 253
16.7 Future for Stem Cell Patents in Europe 253
References 254

16.1 A MOVING TARGET

The European regenerative medicine patent landscape of 2010 is dense, complex, fragmented, and exploding. Legal developments in connection with human embryos may divide the two dominant institutions: the European Union (EU) and the European Patent Organisation. However, technical progress has reduced the significance of human embryonic cells. Induced pluripotent stem (iPS) cells have moved to the fore, bringing with them new issues. Meanwhile, reprogramming patents

---

* Acronym for the decision of the Enlarged Board of Appeal in Case G 0002/06 of November 2008 on an appeal by the Wisconsin Alumni Research Foundation.

239

that avoid the need for pluripotency appear to be in the ascendant. Advances in cellular engineering are complemented by those in enabling technologies. The mass of regenerative medicine patents brings its own problems, the resolution of which provides a significant opportunity.

16.2 EUROPEAN PATENT ENVIRONMENT

In this section, we examine the legal environment surrounding stem cell patents in Europe.

16.2.1 Introduction

The key difference between the European and United States stem cell patent environments arises from an option to the member states of the World Trade Organisation. Article 27(2) of Trade-Related Aspects of Intellectual Property (TRIPS), the treaty that sets out a common framework for the intellectual property laws of member states, provides:

> Members may exclude from patentability inventions, the prevention within their territory of the commercial exploitation of which is necessary to protect *ordre public* or morality ... provided that such exclusion is not merely because the exploitation is prohibited by their law.

Although the United States has not adopted this morality option, Europe (like China) has. However, the *Europe* designation may suggest more uniformity than actually exists. Aside from differences between the cultural views of member states, the two independent pillars of European patent law, the European Patent Organisation and the European Union (EU) are not bound to agree with one another.

Virtually all countries in Europe, including the entire EU, have harmonized their patent laws to conform to the European Patent Convention (EPC) (European Patent Office 2007). This has a two-fold effect of (1) enabling most European countries to participate within the European Patent Organisation (an institution independent of the European Union), through which coordinated examinations and grants of patents are handled by the European Patent Office (EPO), and (2) providing a level playing field in terms of the uniformity of interpretation of patent law at the national level. Where nation states in Europe diverge from this common system is in the interpretation of patent laws, typically after grant, and particularly in relation to what is and is not patentable. With perhaps the exception of the patentability of computer software, the most heightened debate over patentability in recent times has concentrated on human embryonic stem cells (hESCs).

16.2.2 Morality Problem

The problem arises from the exercise of the TRIPS morality option by the EU and EPC. In the guise of Article 53(a) of the EPC, the option provides that "European Patents shall not be granted in respect of ... inventions, the commercial exploitation of which would be contrary to *ordre public* or morality." Biotechnological inventions are not strangers to Article 53. Indeed, they have served to clarify the meaning of the section. Notably, the EPO's Technical Board of Appeal confirmed that, "*Ordre public* covers the protection of public security and the physical integrity of individuals as part of society" and protection of the environment and that, "The concept of morality is related to the belief that some behaviour is right and acceptable whereas other behaviour is wrong, this belief being founded on the totality of the accepted norms which are deeply rooted in a particular culture" (Kinkeldye, 1995). However, the Technical Board's assumption as to the singularity of European culture appears somewhat fanciful:

> For the purposes of the EPC, the culture in question is the culture inherent in European society and civilisation. Accordingly, under Article 53(a) EPC, inventions the exploitation of which is not in conformity with the conventionally-accepted standards of conduct pertaining to this culture are to be excluded from patentability as being contrary to morality.

The practical effect is that a board of patent examiners has become a forum in which to debate "conventionally accepted standards of conduct."

One of the first cases to consider the issue of patentability of biological material involved the so-called Harvard Oncomouse (European Patent 0169672). The application was filed by scientists based at Harvard University in the mid-1980s and related to a genetically modified animal that had an increased predisposition to develop cancer. The application was immediately considered controversial and underwent an extended period of examination by the EPO. To allow the case, the EPO had to consider both the legal issues of patentability and the complex ethical questions raised by many third party pressure groups.

These issues were considered again after grant, because the EPC system allows third parties to object to granted patents within a 9-month period after the date of grant. Following convoluted opposition and appeal proceedings, the patent was eventually upheld in substantially amended form only a few months before it was due to expire at the end of its 20-year term. The rationale for the decision was that the benefit to mankind outweighed the suffering of the mice and environmental risk. The Technical Board of Appeal added that the moral exception to patentability would apply only in a plain case.

16.2.3 European Union Law

While the Oncomouse case served to clarify and confirm the position over patentability of animals and animal cells under the EPC, it also heightened public interest in issues concerning intellectual property rights to living materials—issues subsequently picked over and expanded upon by the EU Parliament and Council which added a biotech gloss to the morality rule and incorporated it into the law of the EU. Article 6 of the Directive on the Legal Protection for Biotechnological Inventions (Directive 98/44/EC, also known as the Biotech Directive) then provided that, "Processes for modifying the genetic identity of animals which are likely to cause them suffering without substantial medical benefit to man or animal, and also animals resulting from such processes" should henceforward be considered unpatentable as contrary to "*ordre public* or morality." The "animals" term does not include humans, for which a separate exclusion to patentability for cloning and modification of the human germ line exists. The authors suggest that "genetic identity" should be construed narrowly: i.e., to refer to genes per se and not to epigenetic modifications.

Oncomouse was not the only factor to influence the directive. Article 6 also reflected concerns arising from the cloning of the sheep named Dolly and the derivation of hESCs by further immoralizing "processes for cloning human beings" and "uses of human embryos for industrial or commercial purposes."

European directives require the laws of member states to comply with them. While most member states incorporated the Biotech Directive before the implementation deadline, others lagged behind, largely on the grounds of moral objections on the part of their national legislatures. However, since 2006, the national patent law of all EU states has incorporated the directive. As with all such provisions, interpretation of national measures is the preserve of national courts, with an ultimate right to refer the matter to the EU's European Court of Justice (ECJ).

16.2.4 EUROPEAN PATENT CONVENTION (EPC)

Because the European Patent Organisation is entirely independent of the EU, the Biotech Directive has no necessary bearing upon the European Patent Convention. Nevertheless, in 1999 the EPO's Administrative Council chose to incorporate it into the "Implementing Regulations" of a revised European Patent Convention (EPC 2000) that became effective December 13, 2007. In its new EPC garb, the morality exception became Rule 28. Because EPC member states include non-EU states such as Switzerland and Turkey, Rule 28 now affects the patentability of stem cells and human-derived tissues across virtually the entire continent.

Rule 28 applies to patents filed under the EPC, as opposed to those filed at a national level and subject to local implementation of the Biotechnology Directive (discussed below). In principle, therefore, rulings affecting "European patents" may be dealt with differently from national ones, even when the same rule is involved. In practice however, national laws (such as the U.K.'s Patents Act, 1977) require courts to take account of rulings of EPO tribunals. In November 2008, everyone took notice.

16.2.5 THE WARF CASE: BACKGROUND

The value of human embryonic stem cells to regenerative medicine and drug discovery is, of course, their "stemminess"—their ability to beget any cell type of the human body. That is a property of nature, not of a named individual or employer. Indeed, EPC Rule 29 (voluntarily taking its cue from Article 5 of the Biotech Directive) provides that, "the human body, at the various stages of its formation and development, and the simple discovery of one of its elements ... cannot constitute a patentable invention." Assuming, for the sake of argument, that a waste blastocyst left over from an in vitro fertilization (IVF) procedure is a "human body," then, whatever else the law may have to say on the subject, anyone can use hESCs from that blastocyst without paying heed to any patent; after all, they are hardly new. However, Dr. Thomson's patent application on behalf of the Wisconsin Alumni Research Foundation (WARF) (Wisconsin Alumni Research Foundation, 2008) did not seek to claim hESCs or their discovery. Instead, it adhered to a culture of characterized hESCs derived by the method specified in the application. That was quite different. As elements isolated from the human body or otherwise produced by a technical process, cultured hESCs are capable of constituting a patentable invention. EPC Rule 29 says so (and thus adult stem cells are patentable), subject to the basic requirements. In effect, the possibility of invention hangs upon the technical processes of derivation and culture. However, those same processes had to run the gauntlet of the morality exception under Rule 28. Inevitably, the case proceeded to the tribunal that served as the highest level court of the European Patent Organisation, the Enlarged Board of Appeal.

16.2.6 WARF in the European Patent Organisation

European patent law requires claims to be interpreted in the light of the description and drawings. Taking this holistic view, the Enlarged Board of Appeal was of the opinion that, even though the claims were drafted to refer to a cell culture, the destruction of human "embryos" was an indispensable step in the invention. The board observed that clever claim drafting should not equip an applicant to avoid statutory prohibitions, in this case, on commoditization and commercialization of the human embryo. Despite the delay and speculation leading up to it, the board's decision was ultimately as narrow as the case could sustain:

> In view of the questions referred, this decision is not concerned with the patentability in general of inventions relating to human stem cells or human stem cell cultures. It holds unpatentable inventions concerning products (here human stem cell cultures) which can only be obtained by the use involving their [sic] destruction of human embryos (Messerli, 2008).

In effect, the ruling comes down to "kill 'embryo,' kill application." However, this should not be taken to suggest that if the "embryo" survives hESC extraction (prior to sluicing) then so can the application. The mischief that Rule 28 refers to is use. Destruction is merely a consequence of use.

The Enlarged Board of Appeal's statement that the decision did not apply to the patentability in general of inventions relating to human stem cells or stem cell cultures caused confusion within the stem cell community as to the patentability of methods and products relating to existing ESC lines. It is, of course, a non-rule: a statement about what an embryo is not. Some have taken the view that as long as there is no express requirement in the application as a whole to use human embryos, an invention is clear of the WARF decision (PatentWatch 2009).

In tentative steps to operate post-WARF, the EPO Examining Division has taken the view that inventions for products that claim cultures of human ES cells are patentable if at the filing date of the patent application the skilled person would have been able to procure human ES cells from previously existing lines without resorting to the use of human embryos (European Patent Application No. 05806471). Once the root cell is secured, progeny provisions come into play: cells possessing the same characteristics as patented cells from which they are derived are also protected, whether in an identical or divergent form, with equivalent provision as regards production processes (European Union 1998).

16.2.7 Resolving the "Embryo" Definition

In fact, WARF provides significant clues as to future cases in its approach to defining the "embryo" term. This matters because, in the view of many, embryos are not destroyed when hESCs are extracted because the ball of cells from which an embryo is derived is unworthy of the moniker. We will not elaborate on the detailed arguments concerning the meaning of the word. However, given its impact on European patent law in relation to cells derived from blastocysts and those in which pluripotency is induced, we provide a brief history.

16.2.8 Historical Background

As the most world's most advanced IVF state, the United Kingdom can be considered to have pioneered fertilization and embryology laws. During the planning phases of the country's Human Fertilisation & Embryology Act 1990, the great developmental biologist, Anne McLaren, the only scientist on the government inquiry leading up to that Act, advised the Warnock Committee that the word "embryo" would be an inappropriate term for a blastocyst. In her view, the very fact that its cells were undifferentiated meant that nothing in the blastocyst could sensibly be described as being a part of a human body—no organs, tissues, blood, bone, skin, nervous system, nothing coherent to speak of (c.f. Zernicka-Goetz 2009). McLaren proposed that the phases from fertilization up to and including the preimplantation blastocyst should be distinguished from those of the developing embryo.

Unfortunately, the majority view of the Warnock Committee was that human material had a special moral status, on the unqualified basis that it was human. The immediate consequence was that the 1990 Act, a piece of legislation intended primarily to control fertility clinics, expanded the term "embryo" beyond its usual scientific limits to include the earliest stages of development: an egg in the process of fertilization and the two-cell zygote (HFEA 1990). A less direct effect may have been to confuse the public into believing that "embryo research" concerned a developed entity; researchers were interested in such material precisely because it was undeveloped.

The WARF case introduced the same extended definition into the entire European patent system, the Enlarged Board of Appeal considering it in the context of the EPC. The appellant argued, as had McLaren in the early 1980s, that "embryo" should be given its medical meaning: "embryos of 14 days or older, in accordance with usage in the medical field" (Messerli 2008). The Enlarged Board, however, noted that neither the EPC, nor the Biotech Directive that it incorporated, defined "embryo." However, the Enlarged Board found guidance in national laws: the German

Gesetz zum Schutz von Embryonen of December 13, 1990 (which defines "embryo" to include a fertilized egg) and the UK's Human Fertilisation & Embryology Act 1990 (which included an egg in the process of fertilization):

> The EU and the EPC legislators must presumably have been aware of the definitions used in national laws on regulating embryos, and yet chose to leave the term undefined. Given the purpose to protect human dignity and prevent the commercialization of embryos, the Enlarged Board can only presume that "embryo" was not to be given any restrictive meaning in Rule 28 ..., as to do so would undermine the intention of the legislator, and that what is an embryo is a question of fact in the context of any particular patent application (Messerli, 2008).

With respect to the Enlarged Board, the purpose of the directive as regards patent law is not to protect human dignity. Rather, "patent law must be applied so as to respect the fundamental principles safeguarding the dignity and integrity of the person" (European Union 1998). The authors submit that a "person" means a legal person and that a blastocyst is not a person. For example, a blastocyst is not entitled to own property. However, the telling part of the Enlarged Board's statement is that each case will be a question of fact for any particular patent.

16.2.9 iPS Cells as "Embryos"

This is important because the statutory definitions to which the EPO may refer are not static. In particular, just 12 days before the Enlarged Board of Appeal delivered its decision in WARF, the UK enacted legislation that, from the end of 2009, significantly broadened the definition of "embryo" (HFEA 2008). In large part, the amendment simply catches up on the case law arising from the innovation of cell nuclear replacement (UK House of Lords 2003)—an innovation that removed the necessity for fertilization. Today, an "embryo" is "a live human embryo" and "references to an embryo include an egg that is in the process of fertilisation or is undergoing any other process capable of resulting in an embryo."

Though distinctly self-referential, this statement seems to capture entities, however brought about, with the theoretical potential to develop into fully differentiated and organized human beings, even if they are not implanted. Bearing in mind that the EPO will now take this definition on board, the significance of this observation will become apparent when one considers that two Chinese teams have successfully cloned live mice from adult mouse fibroblasts and that human blastocysts have reportedly been cloned from skin donated by Dr. Samuel Wood, a fertility doctor and founder of Stemagen, a California-based company (Anonymous 2009a).

At some point, an iPS cell becomes (at least in principle) an induced totipotent (iTS) cell. When that happens, as seems to have been the case with Dr. Wood, the question emerges as to whether such a cell should have the same legal (and moral) status as an embryo. Plainly, under the new statutory definition, an iPS-dubbed cell that is merely capable of developing into a fully differentiated and organized human being, even if it is never to be implanted in a woman, is an "embryo." In the hands of the EPO, this may mean that the use of such cells for industrial or commercial purposes will be unpatentable under Rule 28(c). The exclusion regarding "processes for cloning human beings" may also apply.

In keeping with the WARF case, claims for cells derived from iTS cells will also apply because "embryos" were involved in their derivation. Indeed, the position may be worse than it was in WARF. While extracted embryonic cells are partly differentiated so as to be incapable of producing trophectoderm, induced cells are not necessarily so limited. They may therefore fall victim to that other import from the Biotech Directive, Rule 29, which states, "the human body, at the various stages of its formation and development … cannot constitute a patentable invention."

Plainly, most iPS cells are only pluripotent and incapable of giving rise to all the tissues of the body. However, examiners will look beyond the form and title to the substance of the claim. In doing so, they may take notice of genetic or epigenetic features that prevent implantation. For example, induced Cdx2 deficiency has been shown to prevent implantation (Hurlbut 2004). Examiners could conceivably look at repression of Elf5, recently identified as "the gatekeeper gene" that is methylated in somatic embryonic stem cells but unmethylated in trophoblast stem cells (Hemberger 2008). It will be observed, however, that even if efforts are taken to ensure the silencing of genes such as Elf5, thereby side-stepping Rule 29 and the broadened embryo definition, epigenetic watermarking for the sake of the law is unlikely to be of any scientific, therapeutic, or even moral benefit.

16.2.10 Conjoined Twins?

The above discussion has focused upon the EPC and the impact of EPO case law. As we noted, such decisions impact upon national practices, for example, the UK Intellectual Property Office (IPO) followed the WARF decision (see below). However, the current travels two ways. It will be recalled that the relevant provisions of the EPC are in effect a voluntary copying of EU law under the Biotech Directive. Flattering as this may be, the EU need not reciprocate. The judges of the ECJ, the EU's top tribunal, are under no obligation to follow the decisions of EPO patent examiners. In reviewing identical provisions, the Grand Chamber is fully entitled to reach entirely different conclusions. How the EPO would respond to such a decision would be its own affair.

At the end of 2009, this emerged as a real possibility. In Germany, where the ethical debate continues along similar lines to that at the EPO, a decision of the German Federal Patents Court (Bundespatentgericht 2006) proved to be influential in the approach subsequently taken by the Enlarged Board of Appeal of the EPO in the WARF decision. This is not coincidental. The headquarters of the EPO is in Munich, Bavaria, one of the most religious and conservative regions of Germany. The subject matter of German Patent 19756864 related to methods for deriving precursor cells with neural or glial characteristics from embryonic stem cells. The Federal Patents Court partially rescinded it following a challenge from Greenpeace. However, in November 2009, the case went on appeal to the German Federal Court of Justice, which referred two key questions to the ECJ: (1) the meaning of a human "embryo" under the Biotech Directive and (2) whether using the "embryos" in accordance with the patent would comprise an "industrial or commercial use" (European Union 2010). These issues were, of course, at the heart of the WARF decision. It is therefore possible that the continent will have divergent authorities on the same points. The authors suggest that, in such event, the EPO will follow the ECJ, possibly with some reluctance. Judgment may be expected before the end of 2012.

16.3 PARALLEL ROUTES TO PATENT PROTECTION IN EUROPE: THE NATIONAL ROUTE

As noted, the EPO does not represent the only route to obtaining national patent rights in Europe. Virtually all of the European Patent Organisation member states run their own parallel national patent systems. Economic imperatives typically favored applicants seeking the centralized EPO route if broad territorial protection was required. However, in the particular area of stem cells and regenerative medicine, the ethical debate at the EPO has motivated some commercial players in the stem cell industry to utilize the national route as well. The historical unease about eugenics and religious concerns in Germany and many Mediterranean countries that cause problems for many stem cell-related technologies are not shared by the more liberal utilitarian ideals of Northern European states such as the UK and Scandinavia (Cornwell 2009)—a difference reflected in the varied ratification of the Oviedo Convention on Biomedicine (Oviedo 1997), Article 18 of which limits the scope of signatories such as Germany by declaring that, "Where the law allows research on embryos in vitro, it shall ensure adequate protection of the embryo. The creation of human embryos for research purposes is prohibited."

While the EPO has adopted a largely cautious approach to patentability of human embryonic stem cells, the UKIPO issued guidelines that perhaps are more generous in scope (the UK is not a signatory to the Oviedo Convention). The UKIPO position is to exclude human totipotent cells and uses of human embryos for industrial purposes from patentability, but not to preclude the patenting of human pluripotent cells that can be grown in culture and the cell lines stored in cell banks provided that at the date of filing the cells may be obtained by means other than the destruction of human embryos (UKIPO 2008). The EPO has shown recent signs of moving toward the UKIPO position in some areas but remains vulnerable to interests in Europe that are averse to stem cell patents in general.

In practice, this has permitted a level of adeptness in claim drafting at the national level, at least in the UK, that might provide a broader scope of protection than would be obtained through the corresponding EPO route. Essentially, if the prohibited acts are not included in the claims and the application is not purely restricted to pluripotent cells obtained from human embryos. a patent is likely to be granted in the UK, assuming the conventional requirements of patentability are also met. In addition, UK patents are not open to post-grant opposition unlike European patents, so the kinds of third party challenge from pressure groups that embroiled the Oncomouse patent and WARF application in controversy are far less likely in the UK.

16.4 COMMERCIAL LANDSCAPE

Within the overall sphere of regenerative medicine, patents that are solely concerned with human embryonic stem cells represent only a fraction of the total. In 2005, a report of the UK Stem Cell Initiative (the Pattison Report) set a course for UK stem cell endeavor for the next 10 years—a course the government adopted (Pattison 2005). Europe's only state-level stem cell strategy, the Pattison Report, set out "to act as a guide for public and charity sector investment in UK stem cell research."

Commercialization was linked less to direct therapies than to the use of cells in drug discovery; indeed, the expression "regenerative medicine" appeared on only 9 of its 122 pages. Nevertheless, the Pattison Report provides a useful comparator for where we are now. Five years ago, "stem cells" was virtually a synonym for the beckoning horizon of regenerative medicine. To get a feel for activity in that distant sector, a search against the term "stem cell" would generally do the job. Today this would appear naïve because the focus has shifted from pluripotency to the mastery of factors that determine phenotype: "cybercytology" might be a better name for the art.

The single most striking aspect of the regenerative medicine intellectual property (RMIP) landscape, in Europe and elsewhere, is its explosive nature. The ignition point undoubtedly concerns human embryonic stem cells. For a time, these provided the only route to pluripotency, in effect presenting a tollgate to anyone contemplating working with such cells. Then induced pluripotent stem cell patents appeared and cut a new route to pluripotency. While this was taking place, reprogramming patent applications flared up, avoiding the need for pluripotency in the first place. Seemingly unrelated areas of technology added fuel to the fire. RM-related applications appeared from sectors such as material sciences, chemistry, physics, engineering, and robotics. As the scale of a research project becomes apparent, enabling technologies grow to meet demand. The dawning realization that the technology might actually work led to a few tentative therapeutic applications.

16.4.1 REGENERATIVE MEDICINE INTELLECTUAL PROPERTY

What is remarkable about this is not the pattern, but the extraordinary productivity and pace of development. At the time of writing, three patents have been granted in respect of induced pluripotent stem cells (to Kyoto University in Japan, to iPierian in the UK, and to Fate Therapeutics in the United States), the last two within a week of each other. At the time of this writing, there were around 75 reprogramming applications and there is no sign of any reduction in the flow (Anonymous 2009b). The sheer volume and rush of intellectual property serve as significant features of the IP landscape in their own right.

Concerns have been expressed that the number of patents and their fragmented ownership present an impediment to research and ultimate commercial deployment. With such a dense patent environment, exploitation of one technology should increasingly mean taking a license to another or risking infringement proceedings. Both are potentially exorbitant courses. In order to produce therapies at a price that the market can afford, royalties to underlying technologies must be squeezed. The process of negotiating such royalty stacks adds to the overall cost burden on a new industry with more than its fair share of challenges, effectively choking it in its cot, at least in principle. A related concern is that the grant of exclusive licenses will literally exclude researchers from key technologies, blocking their development. Anticipating such problems, the 2005 Pattison Report proposed a solution:

> ... It is important that any necessary licensing of stem cell lines or techniques should promote the principles being developed by the Organisation for Economic Cooperation and Development (OECD). Namely, that licensing practices should increase rather than decrease access to inventions for research purposes and that commercial considerations in public research activities should not unduly hinder the academic freedom of researchers. One way to achieve this would be to pool IP rights among patent holders so that patents could be licensed broadly.

Despite this endorsement, little or no effort was made to bring interested parties together to explore the possibility of such cooperation, whether by patent pooling or otherwise, until 2009 when a group of leading stakeholders and policy makers met to discuss and explore the issues under the Chatham House rule with a view to informing policy choices and future cooperation in both research and commercialization. The Vine Street meeting concluded that patent pooling was not an appropriate model for imminent cooperation in regenerative medicine, although there was a demonstrable appetite for other models. In reaching this informal conclusion, participants examined the question of freedom to operate in the shoulder-jostling RMIP environment and reflected on the research exemption (Hitchcock 2009).

The Vine Street participants noted that, despite the clear difficulties of conducting freedom-to-operate searches, remarkably few deals were struck between RMIP owners, perhaps because few patents were essential. They suggested that the patent landscape would be winnowed out, as limited university budgets lead to non-renewals and cases were brought to challenge the validity of certain patents. They especially noted that the violent pace of technology quickly rendered new advances obsolete, notably in the iPS cell field.

Other concerns had been expressed that the research exemption under Section 60(5)(b) of the UK Patents Act 1977 (mirrored in other European jurisdictions without the necessity for harmonization) is too weak to comfort researchers about liability from infringement.

The exemption is certainly narrow. It applies only to work "done for experimental purposes relating to the subject-matter of the invention." Suppose, for example, that the WARF patent on human embryonic stem cells were to be valid in the U.K. Although the use of such cells in a university research project to derive hepatocytes might be "done for experimental purposes," because these purposes would not relate to the subject matter of the patent (hESCs) but to the derivation of hepatocytes, the defence would not be available and a license would be required.

Nor did the Vine Street participants or the respondents to the UKIPO's consultation on the exemption (UKIPO 2009) report any compelling evidence that the weakness of the research exemption was actually deterring research. In practice, universities are rarely sued for using patented products or processes in the course of research. This is because of an unwritten convention that recognizes that universities are net providers of valuable intellectual property and make no commercial profits by such exploitation. Nevertheless, as commercial entities become more involved in research, such concerns may grow. The research exemption was the subject of a recent UKIPO consultation. A report on this is expected to be published in 2010–2011. The UK's *Gowers Review on Intellectual Property* favored clarification of the exemption, for example, along the lines of Swiss patent law (Gowers 2006). However, the emerging view seems to be that the existing position is adequate and that any amendment will be slight.

16.4.2 Types of Stem Cell Claims in Granted European Patents

During the period when the WARF patent was under examination, the EPO placed a large number of cases that may have contained related subject matter "on ice" pending the outcome of the Enlarged Board of Appeal decision. Consequently, a considerable number of stem cell and regenerative medicine-related patent applications exist but comparatively few have actually proceeded to grant.

While it is difficult to provide a paradigm patent claim that would be considered acceptable under the present criteria of the EPO, it is possible to identify examples of patents in the field of regenerative medicine that have proceeded to grant. Neuralstem Inc. obtained a granted patent (European Patent 0915968) for its "core technology" in 2008 (Baltimore 2008). Claim 1 of this patent reads as follows:

> A method for expansion and culture in vitro of stem cells of the central nervous system of a mammal, wherein the stem cells maintain the multipotential capacity to differentiate into neurons, astrocytes and oligodendrocytes, comprising the steps of:
>
> a) dissociating cells from central nervous system tissue by mechanical trituration in the absence of divalent cations;
> b) culturing the dissociated cells adhered onto a plate in a serum-free culture medium;
> c) plating dissociated cells at a density not exceeding 20,000 cells/cm^2 and, in subsequent passages, replating the cultured cells at a density not exceeding 10,000 cells/cm^2;
> d) adding daily to the culture cells a growth factor selected from the group consisting of:
>   i) bFGF at a concentration of at least 10 ng/ml,
>   ii) EGF at a concentration of at least 10 ng/ml,
>   iii) TGF-alpha at a concentration of at least 10 ng/ml, and
>   iv) aFGF at a concentration of at least 10 ng/ml plus 1 mg/ml heparin;
> e) passaging the cultured cells within four days after plating so as not to exceed 50% confluence; and
> f) passaging the cultured cells by treating the cultured cells with saline solution.

In the above case, the patent comprises a fairly specific method claim. However, other claims have been granted that appear more broad in scope such as that for ReNeuron Limited's patent granted in 2004 (European Patent 1161521):

> Use of a hematopoietic stem cell, which is not a human embryonic stem cell, in the manufacture of a medicament for intracerebral transplantation into a damaged brain for the treatment of a sensory, motor and/or cognitive defect.

This claim includes what has become the standard disclaimer to avoid coverage of human embryonic stem cells. It is also adopts the so-called Swiss format of wording (use of X in the manufacture of a medicament for treating Y); this is a claim wording convention to avoid another European prohibition on the patenting of methods of medical treatment.

More examples of granted stem cell patents exist in the UK because of the more permissive environment for such research and also because, unlike the EPO, the UKIPO did not impose a moratorium on the prosecution of patent applications in this area until the outcome of the WARF case was settled. In January 2008, the Singaporean-based Cell Research Corporation was granted a UK patent 2432166 for a method of isolating umbilical cord stem cells that represents a good example of the types of claims that could be obtained in the UK for such technology. Claim 1 of its patent recites:

> A method for isolating stem/progenitor cells from the amniotic membrane of umbilical cord, the method comprising: separating the amniotic membrane from the other components of the umbilical cord in vitro; culturing the amniotic membrane tissue in step (a) under conditions allowing cell proliferation; and isolating the stem/progenitor cells.

The patent also claims stem cells isolated by the method recited in the patent and detailed in claim 1.

To summarize, the de facto situation in Europe and in many European countries permits patents to be granted for human and animal cells and cell lines, isolated or engineered tissues that are the result of technological processes, and substances derived therefrom. Not considered patentable are human body parts, processes for modifying the human germ line, or any invention or method that results in the commercialization or destruction of human embryos. By the nature of the technology, RMIP often sails close to the tense borderline between these two positions.

The growth of the granted (and thus enforceable) patent estate for stem cell and regenerative medicine in Europe has been unduly delayed by the ongoing machinations over morality and ethics discussed previously. How the various European courts will approach and consider stem cell patents when they are eventually enforced represents the next challenge for patent holders in this field.

16.5 SCOPES OF STEM CELL CLAIMS

It is established case law by both the EPO and the UK courts that a patent that claims a product protects the exploitation of that product in its entirety irrespective of how it was made. For simple products such as articles of industrial manufacture, this is straightforward. Few would question that a patent covering a new type of engine spark plug should extend to cover any way of making that spark plug also (UKIPO 1977).

In the field of stem cells there is uncertainty because a stem cell line that we consider the "product" is defined frequently by only a few parameters, typically including how it was isolated or maintained. If a patent claims a neural precursor cell, say, and defines that cell by way of morphology and expression of some key surface markers, should the patent cover that cell and all ways of making that cell? This is particularly important where the cell defined in the patent is isolated from one particular source, such as adult tissue, and inventors from another organization succeed in generating a similar or possibly identical cell via a different route such as by reprogramming of a fibroblast cell line. This type of question has been addressed in the UK in two leading cases: *Biogen v. Medeva* (UK House of Lords 1997) and more recently in *Generics v. Lundbeck* (UK House of Lords 2009). The approach taken in these cases was to differentiate between pure product and so-called product-by-process claims.

In a pure product claim, all the characterizing features of the product are inherent within it; one is primarily concerned with what has been invented, not how it has been invented. By contrast, in a product-by-process claim at least part of the characterization of the product in the patent is by way of a method step, such as the process by which it was isolated. As the method of manufacture is irrelevant to pure product claims, a product that can be envisaged as such, with all characteristics determining its identity together with its properties in use, is non-obvious and claimable in a patent if there is no known way to make it.

The claimed methods for its preparation are therefore the first to achieve this in an inventive manner (Szabo 1993). However, if the method by which the product is made is integral to the definition of what the product is in the patent as claimed, then this would appear to limit the scope of protection available if competing products are obtained using different processes.

In the present era of division between various regenerative medicine technologies (embryonic stem cells, induced pluripotent stem cells, and isolated adult stem cells) that can obtain stem cells with perhaps equivalent, if not identical, phenotypes the extent to which a patent from one of these technologies can dominate the landscape and preclude entry into the market by the others leads to an interesting debate. With a view on current case law and approaches taken in Europe, it is truly a case-by-case analysis.

Patents that claim a stem cell line and owe much to their particular methods of isolating the specific stem cell, such that the method is part of the claim, are likely to be limited to their method of isolation in the coverage of their stem cell lines. More holistically drafted patents that suggest different potential origins for the claimed isolated stem cell and provide a detailed phenotypic characterization of the cell without specifying particular methods of derivation in the claim may benefit from much broader protection.

It is easy to confuse a culture with the cells it comprises. Bluntly, however, not all pluripotent cells from one blastocyst are the same and today's iPS cell cultures are far from completely reprogrammed (Daley 2009) and continuous passaging may affect differential potential (Hochedlinger 2010). A key aim of regenerative medicine is therefore to produce uniform and better-characterized cell cultures. As this is achieved, one might anticipate objections raised against the patentability of the purer lines. After all, we might ask, weren't exactly the same cells around before?

In effect, this is equivalent to the issue addressed recently in the European national patent courts, including the UK's most senior court, in the *Generics v. Lundbeck* litigation. The disputed claim was to a (+) enantiomer, not defined by a class of manufacturing processes that had already been employed in a medicine for treating the same indication, albeit in a racemate (mixture of stereoisomers). If we read "culture" for "racemate" and "highly phenotyped cell type" for "enantiomer," the purer culture may be claimable as a product as such, even if the pure cell type was a constituent of earlier cultures. Not only would the patentee's process of purification be covered, but any method of producing the patented monoculture would be susceptible because it is the monoculture that is the protectable technical contribution to the art. This is not to say that a later inventor could not patent better production methods. However, the exercise of such methods would require a license from the holder of the original culture patent.

16.6 DURATION OF PROTECTION

Although the monopoly conferred by a patent is limited to 20 years from the date of application, medicinal products, of which somatic cell therapies and tissue engineered products are varieties (EU 2007), enjoy a separate period of protection under Supplementary Protection Certificates (SPCs) (EU 2009) that comes into effect upon expiry of the patent. SPCs are granted to compensate for the period between patent application and marketing authorization during which the product cannot be lawfully exploited.

In many cases, the greatest commercial return on a pharmaceutical product occurs during the SPC period. The SPC therefore fulfills an important public policy role by providing an economic incentive for early stage investment in healthcare innovation. However, the SPC period is reduced by 5 years and subject to a 5-year cap. Moreover, with only 30% of pharmaceutical products paying for their research and development costs (Di Masi 2007) and the regulatory time to market likely to be longer in the cases of cell products, investors may reasonably wonder whether there is any point in subsidizing the research in the first place. This has led some to argue for a longer SPC term in the case of stem cell therapeutics.

While the case appears strong, if only to allay investor confusion, regard may be had to a form of the de facto intellectual property protection conferred by data exclusivity provisions in respect of biological medicinal products (European Union 2004) that are effectively more advantageous than would be the case for a small molecule drug because of the need for follow-on manufacturers to supply fresh data.

The EU *Bolar* defense (named after a United States case) enables generic manufacturers to conduct such research before expiry of the relevant patent or SPC. In the absence of case law, it is worth considering U.S. jurisprudence in connection with the equivalent provision. In *Merck v. Integra* (U.S. Supreme Court 2005) the court held that the use of patented compounds in studies is permissible if there is a reasonable basis to believe that the compound tested could be the subject of an FDA submission or if experiments will produce the types of information relevant to an Investigational New Drug or New Drug Application.

16.7 FUTURE FOR STEM CELL PATENTS IN EUROPE

The sheer volume and flow of patent applications filed in the fields of stem cells and regenerative medicine have become significant features of the European RMIP landscape. An exceptionally diverse number of stakeholders from academia and industry also characterize a unique IP landscape. However, comparatively few patents have been granted as a result of years of sometimes ferocious debate over the morality of allowing patents to certain aspects of this technology. As Europe now emerges from a period of intense discussion, the RMIP estate will begin to mature and with it further challenges of enforcement and opportunities to consolidate will present themselves to those with active interests in this rapidly developing field. Besides consideration of an extended patent term for cellular inventions, some may feel that the time has come for the creation of a new form of IP right that is as specific to cells as plant variety rights are to new breeds of plant.

REFERENCES

Anonymous. 2009a. Doctor clones human embryo from his own skin cell. *Nature* July 23.
Anonymous. 2009b. The gold rush for induced pluripotent stem cells. *Nat Biotechnol* 27.
Baltimore Daily Record. 2008. Rockville-based Neuralstem's European Patent. April 29, 2008.
Bundespatentgericht (German Patents Court). December 5, 2006.
Cell Research Corporation, UK Patent 2432166.
Chan, E. M., S. Ratanasirintrawoot, I. H. Park et al. 2009. Live cell imaging distinguishes bona fide human iPS cells from partially reprogrammed cells. *Nat Biotechnol* 27: 1033.
Cornwell, J. January 26, 2009. The dilemma on the tip of a needle. *New Statesman* p. 36.
Daley, G. Q. et al. 2009. Live cell imaging distinguishes bona fide human iPS cells from partially reprogrammed cells. *Nature Biotechnology*, 27(11).
DiMasi, J. and H. G. Grabowski. 2007. The cost of biopharmaceutical R&D: is biotech different? *Manag Decision Econ* 28.
European Patent Office. 2007. *European Patent Convention*, 13th Ed. Munich.
European Union. July 30, 1998. Biotechnology Directive 98/44. http://eur-lex.europa.eu/LexUriServ/LexUriServ.do?uri=CELEX:31998L0044:EN:HTML
European Union. 2004. Directive 2001/83/EC of the European Parliament and Council of November 6, 2001 on the Community code relating to medicinal products for human use. *Official Journal L* 67/128, 311.
European Union. 2007. Regulation 1394/2007 of the European Parliament and Council for Advanced Therapy Medicinal Products. *Official Journal L* 324/121.
European Union. 2009. Regulation 469/2009 of the European Parliament and Council for Pharmaceutical Products. *Official Journal L* 152/1.
European Union. 2010. Reference for a preliminary ruling lodged on 21 January 2010–Oliver Brüstle v Greenpeace (Case C-34/10).
Gowers Review of Intellectual Property. December 2006. http://www.hm-treasury.gov.uk/d/pbr06_gowers_report_755.pdf
Hemberger, M. et al. 2008. Epigenetic restriction of embryonic cell lineage fate by methylation of Elf5. *Nat Cell Biol* 10(11): 1280–1290.
Hitchcock, J. 2009. Report on the Vine Street Meeting on Patent Pooling in Regenerative Medicine held at Field Fisher Waterhouse LLP on September 29, 2009. www.ffw.com/pdf/Patent-Pooling-Report.pdf
Hochedlinger, K. et al. 2010. Cell type of origin influences the molecular and functional properties of mouse induced pluripotent stem cells. *Nature Biotechnology Online* 19 July 2010.
Human Fertilisation & Embryology Act (HFEA). 2008. http://www.opsi.gov.uk/acts/acts2008/ukpga_20080022_en_1.
Human Fertilisation & Embryology Act (HFEA). 1990. http://www.opsi.gov.uk/Acts/acts1990/ukpga_19900037_en_1.

Hurlbut, W. B. 2004. *Altered Nuclear Transfer as a Morally Acceptable Means for the Procurement of Human Embryonic Stem Cells*. Washington: President's Council on Bioethics. http://www.bioethics.gov.reports/white_paper/index.html

Kinkeldye, U. M. 1995. European Patent Office Technical Board of Appeal. Case T 356/93, European Patent Office.

Messerli P. November 2008. European Patent Office Enlarged Board of Appeal. Case G 0002/06, European Patent Office.

Ng, R. K., W. Dean, C. Dawson et al. 2008. Epigenetic restriction of embryonic cell lineage fate by methylation of Elf5. *Nat. Cell. Biol.* 10: 1280.

Oncomouse.

Oviedo. 1997. Convention for the Protection of Human Rights and Dignity of the Human Being with Regard to the Application of Biology and Medicine: Convention on Human Rights and Biomedicine. http://conventions.coe.int/Treaty/en/Treaties/Word/164.doc

PatentWatch. 2009. European Patent Office's stem-cell decision. *Nat Rev Drug Disc* 8: 12.

Pattison Report. November 2005. UK Stem Cell Initiative: Report and Recommendations. http://www.advisorybodies.doh.gov.uk/uksci/uksci-reportnov05.pdf

ReNeuron, European Patent 1161521.

Szabo, G. S. A. 1993. European Patent Office Technical Board of Appeal. Case T595/90, European Patent Office.

UK House of Lords. 1997. *Biogen Inc. v. Medeva* Plc. RPC 1.

UK House of Lords. 2003. *Quintavalle v. Secretary of State for Health*. UKHL 13.

UK House of Lords. 2009. *Generics (UK) Limited et al. v. H Lundbeck A/S*. UKHL 12.

UK Intellectual Property Office (UKIPO). 2008. Patentability of human embryonic stem cells. http://www.ipo.gov.uk/pro-types/pro-patent/p-policy/p-policy-biotech/p-policy-biotech-stemcell.htm

UK Intellectual Property Office (UKIPO). 2009. The Patent Research Exception Consultation: Summary of Responses. http://www.ipo.gov.uk/response-patresearch.pdf

UK Patents Act. 1977. UK Intellectual Property Office. http://www.ipo.gov.uk/patentsact1977.pdf

United States Supreme Court. 2005. *Merck v. Integra*, 545 US 193.

Wisconsin Alumni Research Foundation (WARF). May 2, 1998. European Patent Office Publication 0770125.

Zernicka-Goetz, M., S. A. Morris, and A. W. Bruce. 2009. Making a firm decision: multifaceted regulation of cell fate in the early mouse embryo. *Nat Rev Genet* 10: 467.

17 Protecting Regenerative Medicine Intellectual Property in the United States
Problems and Strategies

David Resnick, Ronald I. Eisenstein, and Joseph M. McWilliams

CONTENTS

17.1	Introduction	258
17.2	Patent System: Background	260
17.3	Approaches for Claiming Regenerative Technology: Highlight on Stem Cells	260
17.4	United States Patent Climate: Why Obtaining Patent Protection for Regenerative-Related Technology Is Problematic	262
	17.4.1 Early Stage Technologies and Speed	262
	17.4.2 Scrutiny	262
	17.4.3 Competition	263
17.5	Strategies	267
17.6	Regenerative Medicine Patent Landscape	268
17.7	Strategic Alliances and Partnership in Regenerative Medicine Patent Landscape	270
17.8	Wisconsin Alumni Research Foundation (WARF) Patents	270
17.9	Stem Cells	272
17.10	Human Embryonic Stem Cell Therapy Landscape	273
17.11	Non-Embryonic Stem Cell Landscape	273
17.12	Induced Pluripotent Stem Cell Patent Landscape	274
17.13	Tissue Engineering and Replacement Organ Patent Landscape	275
17.14	Conclusions	278
References		279

17.1 INTRODUCTION

In the United States, when one thinks of regenerative medicine, the first things that typically come to mind are embryonic stem cells and cloning. According to the National Institutes of Health (NIH), regenerative medicine is the process of creating living, functional tissues to repair or replace tissue or organ function lost due to disease, damage, congenital defects, or age (Regenerative Medicine 2006). Accordingly, the subject area is much broader than that initial impression and can include artificial skin, organ scaffolds, organ replacement and regeneration, etc. However, due to the politics and the public arguments surrounding the subject, the focus on embryonic stem cells and their use has exerted an effect that goes beyond the ability to obtain federal funding.

While the most significant impact of this debate has yet to be determined, the issue of federal funding of research, particularly academic research, and its impact are becoming apparent. After the Bush administration announced its limited ban on funding such research, many states decided to get involved. The impacts of such state involvement can be seen in many areas. As will be discussed below, the economic effect has dramatically influenced the origins and directions of related intellectual property. The public involvement in "good" science versus "bad" science may be at the heart of public attacks on the patent system, such as the American Civil Liberties Union's (ACLU's) attack on gene patents in the *Myriad* lawsuit (Association for Molecular Pathology, et al. v. United States Patent and Trademark Office, et al., 2010).

In contrast to the politics concerning the research, the patent landscape has been relatively apolitical. Unlike the situation in the European Patent Office (EPO) when a ban was imposed on the use of materials from an embryo (see Chapter 16 by Hitchcock and Crease), there was no ban on using such materials in the United States. Thus, in the U.S., Thompson was granted three patents that generally related to embryonic stem cells, whereas the EPO concluded such subject matter was prohibited by statute. However, all the politics and accompanying scrutiny produced an effect. Over time, patent practitioners believed that examiners became increasingly hesitant to grant broad claims that could subject the U.S. Trade and Patent Office (USPTO) to criticism, and more individuals in the USPTO became involved in deciding whether a patent should be issued.

In this section, we examine whether perception matches reality. Is it harder to obtain a patent now? Has the scope been affected? Where are the applications coming from?

It is important to recognize that advancements in this subject area have been spectacular. For example, Dr. Anthony Atala reported in 2006 that he had grown human bladders in his lab and that the patients who received them were still healthy 5 years later. Today, Atala and his staff at North Carolina's Wake Forest Institute for Regenerative Medicine are working to replicate their success with other body parts. They are currently growing 22 different tissues, from heart valves to muscles to fingers. From stem cells to tissue engineering and organ regeneration, regenerative medicine technologies have the potential to address major unmet clinical needs across a wide range of areas (Table 17.1). Thus, the public interest and potential for commercial investment are easy to see.

TABLE 17.1
Promise of Regenerative Medicine

Medical Need	Example
Novel methods of insulin replacement and pancreatic islet regeneration for diabetes	Bone marrow stem cell transplantation for pancreatic regeneration
	Microencapsulation for immunoisolation of transplanted islets
	Cultured insulin-producing cells from embryonic stem cells, pancreatic progenitor cells, or hepatic cells
	Genetically engineered cells to stably express insulin and contain glucose-sensing mechanism
Autologous cells for regeneration of heart muscle	Myocardial patch for cardiac regeneration
	Direct injection of autologous bone marrow mononuclear cells for cardiac repair
	Stromal cell injection for myocardial regeneration
	Localized angiogenic factor therapy through controlled release systems or gene therapy
Immune system enhancement by engineered immune cells and novel vaccination strategies for infectious disease	Genetically engineered immune cells to enhance or repair immune function
	Single-injection DNA vaccines
Tissue-engineered skin substitutes, autologous stem or progenitor cells, intelligent dressings, and other technologies for skin loss to burns, wounds, and diabetic ulcers	Bilayered living skin constructs (e.g., Apligraf)
	Engineered growth factors applied in conjunction with topical treatments
	Intelligent dressings composed of slow-releasing growth hormone polymers
	Epithelial cell sprays
Biocompatible blood substitutes for transfusion requirements	Polyhemoglobin blood substitutes for overcoming blood shortages and contamination issues
Umbilical cord blood banking for future cell replacement therapies and other applications	Preserved umbilical cord blood stem cells to provide future cell replacement therapies for diseases such as diabetes, stroke, myocardial ischemia, and Parkinson's disease
	Pooled cord blood for treatment of leukemia
Tissue-engineered cartilage, modified chondrocytes, and other tissue engineering technologies for traumatic and degenerative joint disease	Matrix-induced autologous chondrocyte implantation for cartilage repair
	Tissue-engineered cartilage production using mesenchymal stem cells
Gene therapy and stem cell transplants for inherited blood disorders	Genetically engineered hematopoietic stem cells to restore normal blood production in beta-thalassemic patients
Nerve regeneration technologies using growth factors, stem cells, and synthetic nerve guides for spinal cord and peripheral nerve injuries	Synthetic nerve guides to protect regenerating nerves
	Stem cell therapy for spinal cord regeneration
	Growth factor-seeded scaffolds to enhance and direct nerve regeneration

continued

TABLE 17.1 (continued)
Promise of Regenerative Medicine

Medical Need	Example
Hepatocyte transplants for chronic liver diseases or liver failure	Microencapsulation of hepatocytes to prevent immunological reaction
	Derivation of hepatocytes for transplantation from embryonic stem cells
	Transdifferentiation of hepatocytes for transplantation from bone marrow cells

Source: Greenwood, H. L., P. A. Singer, G. P. Downey, et al. 2006. *PLoS Med* 3: e381. DOI: 10.1371/journal.pmed.0030381.

Intellectual property (IP) rights are important aspects of this arena because they provide a means for the owners of technology to recover their investments in the technology and, in some cases, make a profit (Atala et al. 2008). Patent protection of biological technologies encourages investors and pharmaceutical companies to continue research and development for the treatment of disease. Patents enable the development of critical technologies to help those who need them the most. According to Alan Lewis, former CEO of Novocell, a diabetes cell therapy company, intellectual property is a cornerstone of the biotechnology industry. "We wouldn't get any investors if we weren't filing IP [rights] or getting IP [patents] issued," he said. Indeed, without intellectual property rights, it is exceedingly difficult to obtain the money necessary to see a product or process reach the clinic level to benefit patients. Without the patent system, the business of drug development and medical treatments, as we know it, would not exist.

17.2 PATENT SYSTEM: BACKGROUND

A patent is a grant by the government to a patent applicant that permits the owner of the patent to exclude others from making, using, or selling the technology described in the patent. This right of exclusion is given in exchange for providing the information necessary to make and use the technology to the public. A patent uses a set of "claims" that define the scope of what the owner has the right to exclude.

Historically, the patent system evolved around technologies relating to machinery or devices with moving parts, but, since the 1980s, it has become increasingly involved in the area of biotechnology. However, defining biological processes, biological products, and methodologies for production of, for example, induced pluripotent stem cells (iPS) results in a number of unique challenges.

17.3 APPROACHES FOR CLAIMING REGENERATIVE TECHNOLOGY: HIGHLIGHT ON STEM CELLS

The U.S. Patent Law, Chapter 35 of the United States Code (USC), §101, sets forth the scope of what types of inventions can be patented. For example, the U.S. statute permits a patent to cover a process, machine, manufacture, or composition of matter. The process (or method), machine, manufacture, or composition must be novel (35 USC §102), non-obvious (35 USC §103), and useful. Where stem cell technology is concerned, the areas of primary focus are likely to be the cells themselves (compositions), methods of making or preparing stem cells (processes or machines), methods of differentiating stem cells to useful phenotypes (processes), and methods of using the cells or their differentiated progeny, especially to treat disease (processes). Each of these avenues has advantages and disadvantages, to the extent that a combination of them will most often be the prudent approach. Herein, we highlight some of the common approaches applicable to the protection of a stem cell technology.

The general wisdom is that a patent that claims a composition, such as a stem cell or stem cell preparation, tends to be of the greatest value, because it usually does not matter how the composition is made. For example, an induced pluripotent stem cell can be defined in a claim as a "composition." An example of a claim relating to a cell that can be induced to be pluripotent would read: "An isolated human cell expressing at least three stem cell-specific markers selected from the group consisting of SSEA-3, SSEA-4, TRA-1-60, TRA-1-8, TRA-2-49/6E, and Nanog." This claim has not yet been granted; however, if granted in this form, it would not matter how the cell was made or for what it would be used. Anyone who made a cell with all of the characteristics described in the claim, used the cell, or sold or imported it would be infringing the granted patent (35 USC §271(a)). [*Note:* To date, there has not been a patent granted on an iPS cell composition in the United States.]

One of the major drawbacks of protecting a natural stem cell itself is that, as a policy matter, cells and natural phenomena *as they exist in nature* are not considered to be patentable subject matter under U.S. law (*Diamond v. Chakrabarty* 1980). However, one can claim such material if it has been modified, for example, isolated or purified. In addition, one can claim processes using such material. Because it is preferred that cells used to treat a disease differ as little as possible from naturally occurring cells, one must look at a variety of claim types. Thus, your claim is dependent on what you are using. Accordingly, a composition-of-matter claim may not be the correct claim to provide the protection you seek.

Another approach for claiming stem cell compositions is to describe the compositions in terms of the process used to make them via a product-by-process claim, for example, "an induced pluripotent stem cell *obtained by the method* of reprogramming a somatic cell, comprising contacting a nuclear reprogramming factor with the somatic cell to obtain an induced pluripotent stem cell." This approach will avoid the problem of the claim not coming within the patent statute, but it will not overcome a prior-art problem. An applicant must show that the claimed product differs from the prior-art product. Further, the Federal Circuit ruled that process terms in product-by-process claims serve as limitations in determining infringement. In other words, the claim is limited to the process set forth in the claim and not all processes for obtaining the product (*Lupin Ltd. and Lupin Pharmaceuticals v. Abbott Laboratories and Astellas Pharma* 2009). Thus, while this type of claim can solve certain problems in how to describe the invention, it does not solve all the problems. The product-by-process approach allows a competitor to avoid infringement by using a different process to make the product from the process set forth in the claim.

Stem cell technologies can also be protected using a method-style claim, where the cell itself is not patented, but rather the method used to produce the cell or a method of using it is the subject of the claim. To date, one iPS cell patent has been granted for a method of making a stem cell. Dr. Shinya Yamanaka obtained a patent in Japan for a method for producing an induced pluripotent stem cell (Japanese Patent 2008-131577); the English translation of the allowed claim recites: "A method for producing an induced pluripotent stem cell from a somatic cell, comprising the step of introducing the following four kinds of genes into the somatic cell: Oct3/4, Klf4, c-Myc, and Sox2." While a method claim has the advantage of being more easily defined than a stem cell composition, the drawback is that, if another party can produce the same stem cell by a different method, for example, the patent is not enforceable against that party.

17.4 UNITED STATES PATENT CLIMATE: WHY OBTAINING PATENT PROTECTION FOR REGENERATIVE-RELATED TECHNOLOGY IS PROBLEMATIC

Since about 2000, when the USPTO and the courts were viewed as pro-patent, the climate has changed to one where it is much more difficult to obtain broad patents. For example, citing concerns about pre-emption of a field, recent court decisions made it easier for the patent office to reject an invention as obvious. While policies differ worldwide, this section highlights some difficulties with capturing intellectual property related to regenerative medicine technologies and provides strategies for dealing with these difficulties.

17.4.1 Early Stage Technologies and Speed

It is often difficult to obtain meaningful patent protection in an area of research that is evolving intensely, such as the regenerative medicine field. For example, new or improved technologies relating to safer, faster, and more reliable methods for the use of stem cells in treatment and for the induction of stem cells are revealed almost daily. Thus, one can imagine that a method that works well today may well be obsolete in another week or month as the field evolves, let alone in the 2 to 5 years common for the patent application process. The speed at which this technology moves begs the question: do we gain any commercial advantage by filing on this technology?

17.4.2 Scrutiny

A rapidly changing field also poses problems for the USPTO. The sheer number of new discoveries prompts the filing of dozens of applications; however, despite scientific promise, new fields (such as stem cells) lack clear guidelines for granting patents. Moreover, 1 million U.S. patent applications are pending examination today, waiting an average of 3 years to be examined and ruled upon via an antiquated USPTO technology system. The office has only recently started to take advantage of the option of sharing search results with other patent offices around the world that now redundantly examine identical applications filed in their respective jurisdictions (Weiss 2009).

The problems caused by the relative lack of precedent for patents in this field and the large number of patent applications pending examination are compounded by a reluctance of the USPTO to grant patents that might completely dominate or "preempt" a field. This reluctance has led to enhanced levels of scrutiny in all cases, making it harder to get applications allowed and thus increasing costs. This is particularly true in the case of stem cells, given both the ethical issues and public opinion (Figure 17.1) associated with embryonic stem cell research and the publicity of the criticism aimed at the USPTO for the broad nature of the WARF patents (see Section 17.8 below). Many in the stem cell community perceive these patents to be preemptive and pose high barriers to progress in the stem cell field, impeding investment in U.S. stem cell companies.

One way the USPTO approaches this issue is requiring patent applications in new fields to provide greater detail to satisfy the "enablement" requirement. In particular, specifications "shall contain a 'written description' of the invention, and of the manner and process of making and using it, in such full, clear, concise, and exact terms as to 'enable' any person skilled in the art to which it pertains … to make and use the same, and shall set forth the best mode contemplated by the inventor of carrying out his invention." Thus, to satisfy the requirements of enablement, a patent applicant may need to provide more data or more complete studies to demonstrate that the claimed invention works as taught. Thus, while it may be prudent to file an application early and attempt to prevent the worldwide competition from gaining patent rights, better developed technologies may provide more support to satisfy the enablement requirement. This can pose a significant challenge for academic and non-profit research.

17.4.3 COMPETITION

Another issue affecting the current U.S. patent climate is competition. Since 2001, nearly a dozen states have introduced local initiatives to augment federal funds for stem cell research (Figure 17.2). The California Institute for Regenerative Medicine (CIRM), established in 2004, is the largest of these initiatives, with $3 billion in initiatives to be funded over 10 years. Massachusetts established the Massachusetts Life Sciences Center in 2008 as part of a $1 billion initiative. Other states, such as New York, New Jersey, Connecticut, Wisconsin, Maryland, Wisconsin, Illinois, and Texas, have also established initiatives to fund stem cell and other regenerative research. Since 2003, state initiatives have pledged nearly $6 billion in funding.

States are also competing fiercely with each other to realize the economic benefits of their investments: retaining and forming companies, creating new jobs, encouraging investors with new patent applications and issued patents, and advancing scientific discoveries toward effective therapeutic products. As a result of these funding initiatives, the sheer number of new discoveries in regenerative medicine has led to a significant increase in the filing of patent applications in these states (Figure 17.2).

California, funded by the unprecedented $3 billion CIRM since 2004, has experienced the most significant rise in the number of patent applications compared to any

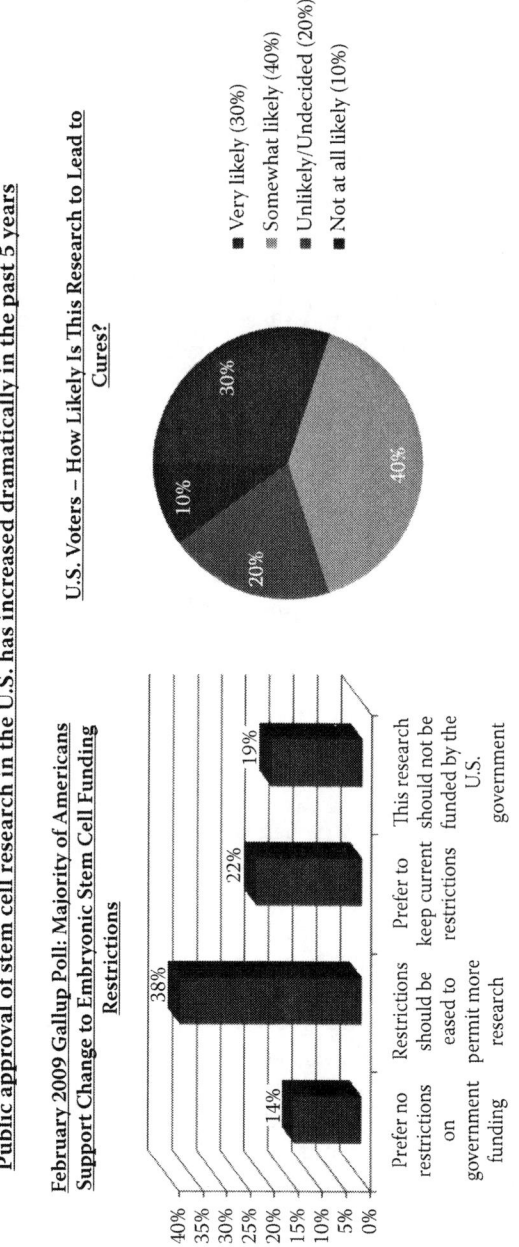

FIGURE 17.1 Public approval of stem cell research in the U.S. has increased dramatically. (From Sharma Group, LLC (TSG). Advances in Replacement Organs and Tissue Engineering. http://www.tsg-partners.com.)

Since 2001, nearly a dozen states have introduced local initiatives to augment federal funds for embryonic stem cell research. The California Institute for Regenerative Medicine (CIRM) is the largest of these initiatives at $3 billion.

State Stem Cell Initiatives have Pledged ~ $6Bn towards Stem Cell Research since 2003

State	Amount Pledged
California	$3 billion
Connecticut	$100 million
Illinois	$10 million
Maryland	$60 million*
Massachusetts	$1 billion
New Jersey	$150 million**
New York	$600 million
Texas	$50 million
Wisconsin	$750 million

Funding for Stem Cell Research by States: 2003-2007

- State Embryonic: 55%
- State Non-Embryonic: 30%
- State General Stem Cell: 12%
- State Undesignated: 3%

National stem cell policy since 2001 has forced states to spend additional money on stem cell research

A Significant Increase in Related Patent Applications is Evidenced in States Funding Regenerative Stem Cell Research

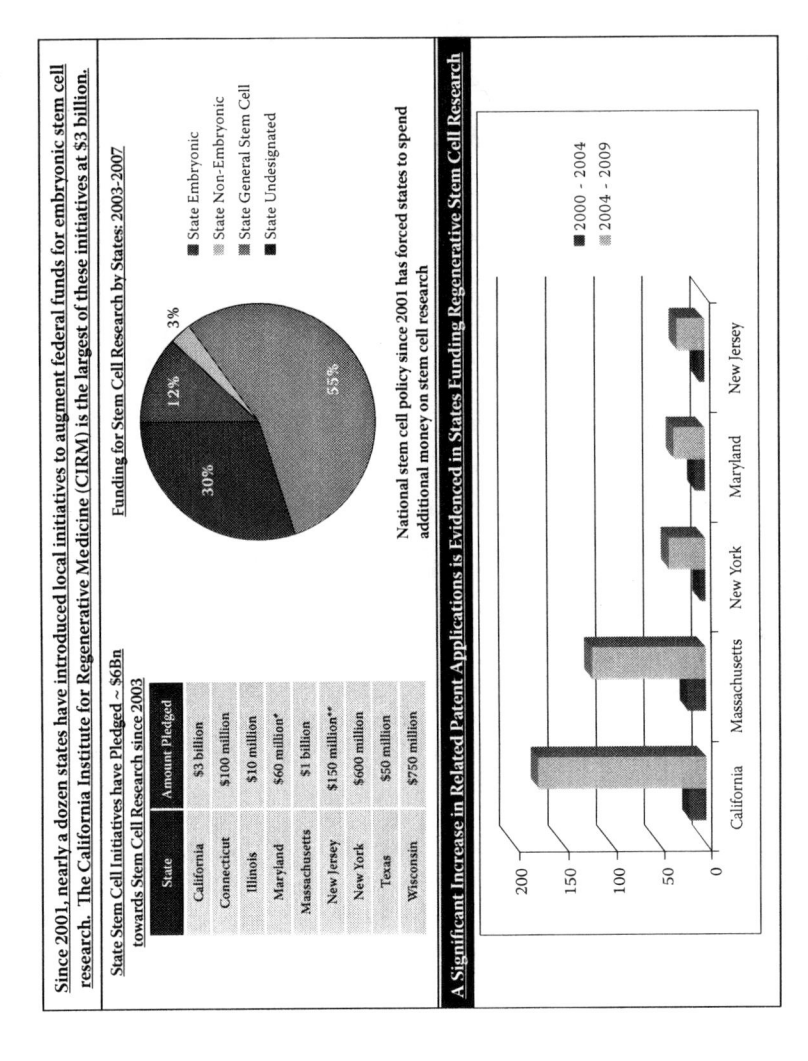

(Bar chart showing patent applications 2000-2004 vs 2004-2009 for California, Massachusetts, New York, Maryland, New Jersey)

FIGURE 17.2 State stem cell funding initiatives and increases in patent applications. * = Maryland stem cell funding determined year-to-year; ** = New Jersey stem cell construction funds on hold. (From Sharma Group, LLC (TSG). Advances in Replacement Organs and Tissue Engineering. http://www.tsg-partners.com.)

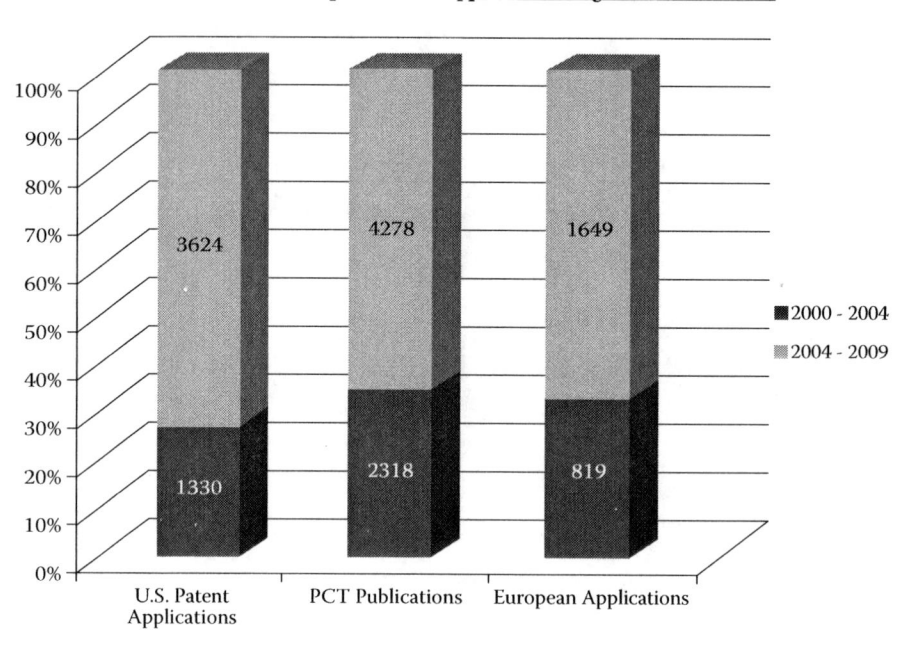

U.S., PCT, and European Patent Applications: Regenerative Medicine

```
         U.S. Patent      PCT Publications    European Applications
         Applications
  2000-2004:  1330             2318                  819
  2004-2009:  3624             4278                  1649
```

FIGURE 17.3 Regenerative medicine patent applications. (From http://www.uspto.gov.)

other state. It is interesting to note that, in 2004, California and Massachusetts—both leaders in life sciences research and regenerative medicine—were in a virtual tie for the number of patent applications produced in regenerative medicine. However, 5 years into CIRM funding in 2009 (and only 1 year after the establishment of the Massachusetts Life Sciences Center), California had 33% more patent applications directed to regenerative medicine than Massachusetts. As states continue to fund regenerative medicine research, this trend is predicted to continue, as new intellectual property is translated into new patent applications and potential economic benefits.

In addition to the increase in patent applications in U.S. states with funding program, the first decade of the 21st century has also experienced a worldwide explosion in the number of regenerative medicine patent applications (Figure 17.3). Interestingly, this is in stark contrast to the number of granted patents in regenerative medicine worldwide, which has increased at a far slower pace (Figure 17.4). As a consequence of this worldwide increase in applications and in scientific publications, an abundance of prior art has flooded the system, leading to narrower claim scopes as a result of rejections for lack of novelty and obviousness.

Viewed from a slightly different angle, the constantly expanding scientific literature and the recent increase in the number of patent applications in the U.S. that relate to pluripotent stem cells, for example, may be helpful for those in the field. Increased publications in a field provide an index of the level of "skill in the art" for the USPTO to use as a gauge to determine whether a patent application accurately represents a technological advance in the field and also whether the technology described in

Patent Applications vs. Granted Patents: U.S., PCT, European, Japan, 2004–2009

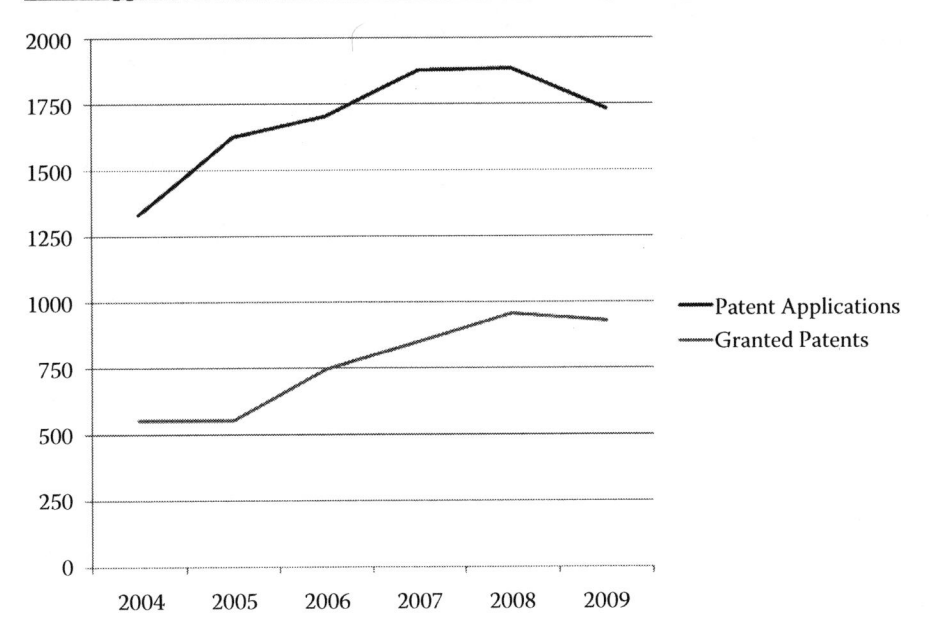

FIGURE 17.4 Regenerative medicine patent applications versus granted patents. (From http://www.uspto.gov.)

the application could be understood by a scientist working in the stem cell field. For example, before September 2008, no published applications relating to iPS cells were pending with the USPTO. However, a basic patent search now yields at least 30 applications published worldwide that relate directly to iPS cells and several others that relate indirectly (e.g., cell culture methods, etc.). At this time, we are seeing only the tip of the iceberg in terms of patent applications directed to iPS cells or other stem cell technologies. This is, in part, because a patent application does not publish in the U.S. until 18 months after filing and technologies pursued for patent protection are not known to the public during that time. This delay can make it difficult to predict whether a competitor is filing on a similar technology and may encourage investigators and/or institutions to file multiple applications relating to a family of technologies such as iPS cells.

17.5 STRATEGIES

In this rapidly changing field, patent protection of regenerative technologies is pursued aggressively by many research institutions and companies. However, the reluctance of the USPTO to allow broad patent coverage suggests that patenting stem cell technologies will continue to be a remarkably tedious and slow process compared to the scientific discoveries in the field. The changing landscape of medical patent law in the U.S. amplifies the difficulty of patent prosecution and necessitates a strategic approach to achieve patent protection.

As law and policy continue to change, it is wise to protect regenerative medicine technologies such as stem cells using a combination of claim approaches. For example, method claims relating to the differentiation of stem cells to a desired cell phenotype may be important to protect, as are methods related to controlling the level of pluripotency of a stem cell. A method that permits stem cells to be used therapeutically in a safe manner (e.g., with minimal risk of producing a cancerous tumor) may be particularly important to further research and development in the stem cell arena. Furthermore, methods that relate to effective stem cell (or differentiated progeny) delivery to an individual for treatment of a disease will also be important advances in harnessing stem cell technology for therapeutic use. Stem cells are also important in vitro tools for screening therapeutic agents, and may benefit from patent protection to encourage development of diagnostics or drug discovery platforms.

It is equally important to consider protecting more narrow advances, such as the use of a particular combination of growth factors that exhibits an unexpected advantage in use. A series of narrow patents can provide commercially valuable exclusivity. A patent or patent portfolio with a variety of claim styles can be the first step in building business platforms and also for encouraging future research and development in the emerging field of stem cell technology, with the ultimate goal of getting therapies into the clinic.

17.6 REGENERATIVE MEDICINE PATENT LANDSCAPE

The regenerative medicine patent landscape is dominated by the USPTO—by far, the largest granter of patents. This dominance is reflective of the commercial importance of this market and perhaps the relative speed with which U.S. patents are granted, particularly compared with the EPO, and the uncertainty as to the patentability of some inventions, particularly in Europe. From 2000 to 2009, the trend has consistently been that more patents are granted each year.

Certainly, patent grant activity does not fully explain the patent landscape in regenerative medicine. Therefore, it is helpful to analyze the number of patent applications. The number of regenerative medicine patent applications each year is consistently around two to three times the number of patents granted, although the variance by country is considerable. The U.S. publishes only slightly more patent applications each year than granted patents. In contrast, European publications represent around six times more than the number of granted patents.

To further understand key dynamics affecting the regenerative medicine industry, it can also be instructive to look at the locations of priority filings to gain insight into the locations of key research institutions and companies. It should be noted, however, that many non-U.S. applicants will file first in the U.S. for commercially beneficial reasons. Globally, the U.S. dominates, although China, Japan, and Australia show significant levels of priority applications. In its entirety, Europe closely mirrors China, although priority filings are split among countries. While strong in regenerative medicine research, Canada shows relatively few filings, likely because Canadian applicants file priority applications in the U.S. This is less common for European companies.

Government and public bodies dominate the list of organizations that filed significant numbers of patents from 2000 to 2009, although a number of specialized

TABLE 17.2
Top Regenerative Medicine Patent Assignees, 2000 to 2009

Genentech, Inc.
Regents of University of California
General Hospital Corporation
Isis Pharmaceuticals, Inc.
Board of Regents of University of Texas System
Regents of University of Michigan
Human Genome Sciences, Inc.
Lexicon Genetics, Inc.
Merck & Co., Inc.
Massachusetts Institute of Technology
Geron Corporation
Osiris Therapeutics, Inc.
Elan Pharmaceuticals, Inc.
United States of America (Department of Health and Human Services)
Amgen, Inc.

Source: http://www.uspto.gov.

regenerative medicine companies and existing biopharmaceutical companies are present. The current predominance of government and public bodies is further influenced by the funding initiatives in a number of U.S. states, the numerous publicly funded stem cell institutes around the world, and the recent lifting of some federal restrictions on stem cell research in the U.S. Future patent filings are likely to be more heavily influenced by government and public bodies, most particularly until investors feel more confident that regenerative medicine technologies such as stem cell will provide good returns. Today, this confidence is aided by the progress of leading private companies in regenerative medicine, such as Organogenesis and Tengion, both leaders in regenerative tissues and organs. However, pending healthcare reform in the U.S. could temper investment (see Chapter 21 by Meurgey and Wille and Chapter 22 by Faulkner). Leading patent holders (assignees) in regenerative medicine are listed in Table 17.2.

In addition to analyzing the major patent holders in regenerative medicine, analysis of the rate of patent filing and the position of a company's intellectual property position relative to its competitors provides insight into promising organizations in the field. These companies may not yet have built sufficient patent portfolios to appear on the list of top patent holders, but can represent a wide range of cutting edge technologies and companies with strong future commercial potential. For example, Medistem Laboratories, Inc. is a U.S.-based biotechnology company focused on the development and commercialization of adult stem cell-based technologies used in the treatment of inflammatory and degenerative diseases. Medistem holds a number of patents in the "endometrial regenerative cell" (ERC) area. Its intellectual property position is in its infancy and will not expire soon.

However, patents on the first generation technology that leads stem cell research are set to expire in the near to mid future. For example, some of the composition-of-matter patents covering CD34+ stem cells (e.g., U.S. Patent 4,714,680 assigned to Johns

Hopkins and licensed to Baxter) have already expired. Other composition-of-matter patents on mesenchymal stem cells, such as Osiris's main mesenchymal stem cell patent (5,486,359), expire in 2013 (Medistem, Inc. 2009).

17.7 STRATEGIC ALLIANCES AND PARTNERSHIP IN REGENERATIVE MEDICINE PATENT LANDSCAPE

Regenerative medicine companies can benefit from strategic alliances and partnerships. In the case of large pharmaceutical companies, such deals can be mutually beneficial. For regenerative medicine companies, large pharma operations bring cash, infrastructure, and expertise to the table (see Chapter 5 by Prescott, Chapter 6 by Benjamin, and Chapter 12 by Vertès). Creating a successful regenerative medicine product requires a convergence of multiple intellectual property rights and technologies that can be obtained only from an entity with financial resources. Because the current regenerative medicine industry is fragmented from an intellectual property and technology perspective, large pharma could provide the access to intellectual property, delivery systems, cells, and manufacturing expertise necessary to create a successful product. This would be extremely difficult for a single regenerative medicine company to accomplish.

These deals also benefit large pharmaceutical companies that now face slowing growth due to patent expiries and a declining number of new drug approvals. Patent expiries from 2009 to 2012 will account for $65 billion in today's revenues. By 2014, large pharma will see more than 50% percent of their current revenues decline due to patent expirations. FDA approvals are now 50% of the rates seen in the 1990s. Large pharma is in need of new platforms that can drive revenue growth (see Chapter 12 by Vertès). These companies are both actively acquiring and licensing high-value products and investing in early-stage technologies. Therefore, regenerative medicine technologies and products are attractive to them.

The most lucrative partnership with a large company to date was the Genzyme deal with Osiris for Prochymal and Chondrogen, with total deal terms of more than $1 billion (Osiris Therapeutics, Inc. 2008). Other deals include Johnson & Johnson's funding of the Coulter Foundation Translational Research Partnership (2008); Pfizer's partnership with University College London (UCL) to develop stem cell-based therapies for ophthalmic conditions; and the partnership of AstraZeneca, Roche, GSK, and the UK government to develop stem cell technology for safety testing of investigational compounds (2007) (DataMonitor 2009). Table 17.3 lists additional strategic alliances and partnerships in the stem cell field.

17.8 WISCONSIN ALUMNI RESEARCH FOUNDATION (WARF) PATENTS

In the mid-1990s, Dr. James Thomson and co-workers made a breakthrough discovery when they isolated and cultured primate and human embryonic stem cells (hESCs) at the University of Wisconsin Madison. As a result of these significant accomplishments, the Wisconsin Alumni Research Foundation (WARF), the university's commercialization branch, filed a series of patents that continue to affect the intellectual

TABLE 17.3
Notable Strategic Alliances and Partnerships in Stem Cell Field

Date	Development	Date	Development
Apr 24 2009	Pfizer Regenerative Medicine enters collaboration with University College London to advance development of stem cell therapies	Aug 07 2008	Orthofix International signs development and commercialization agreements with Musculoskeletal Transplant Foundation
Mar 02 2009	Epistem announces R&D collaboration with Novartis	Jul 29 2008	ALS Therapy Development Institute and California Stem Cell Inc. announce long-term collaboration
Jan 15 2009	VistaGen Therapeutics signs commercialization agreement with Capsant Neurotechnologies for stem cell technology	Jul 29 2008	GlaxoSmithKline and Harvard Stem Cell Institute partner in stem cell research
Jan 02 2009	CHA Biotech forms joint venture with Advanced Cell Technology	Jun 19 2008	Sigma-Aldrich announces collaboration with University of California San Francisco to develop monoclonal antibodies
Dec 19 2008	VistaGen and WARF sign licensing agreement	Jun 18 2008	Invitrogen signs letter of intent with International Regulome Consortium to develop tools for stem cell research
Dec 02 2008	Neuralstem collaborates with China Medical University Hospital to develop additional ALS stem cell clinical trials	May 22 2008	MediStem Laboratories and OrcistBio enter into collaboration agreement focused on augmenting stem cell activity
Nov 25 2008	Sigma-Aldrich enters research collaboration program with D-Finitive Cell Technologies Inc.	Apr 14 2008	Buck Institute for Age Research and Q Therapeutics form stem cell collaboration to combat Parkinson's disease
Nov 06 2008	Aldagen, Inc. and UC Davis Health System form collaboration related to neural diseases	Apr 09 2008	International Stem Cell Corporation announces collaboration with University of Cambridge to study parthenogenetic stem cells
Nov 04 2008	Genzyme and Osiris partner to develop and commercialize first-in-class adult stem cell products	Apr 03 2008	BioTime and Embryome Sciences partner with International Stem Cell Corporation and Lifeline Cell Technology
Oct 23 2008	Novo Nordisk signs R&D agreement for stem cell diabetes treatment	Mar 13 2008	Vistagen Therapeutics forms broad ESC alliance with University Health Network and McEwen Centre for Regenerative Medicine

continued

TABLE 17.3 (continued)
Notable Strategic Alliances and Partnerships in Stem Cell Field

Date	Development	Date	Development
Oct 09 2008	Power3 Medical Products in collaboration with StemTroniX Inc. to identify and monitor autologous adult stem cells for use in stem cell therapy using protein biomarkers	Mar 06 2008	Boston Scientific and Osiris Therapeutics announce stem cell alliance
Aug 26 2008	Cyntellect Inc. collaborates with Burnham Institute for Medical Research	Mar 04 2008	Millipore and Guava Technologies join to create new solutions for fast growing cell biology market

Source: Sharma Group, LLC (TSG). Advances in Replacement Organs and Tissue Engineering. http://www.tsg-partners.com.

property landscape. These patents (5,843,780, 6,200,806, and 7,029,913) concerning embryonic stem cells are controversial because, on their face, they cover all human embryonic stem cell lines in the U.S. regardless of genotype, method or culture medium used, and who produced the line. Since then, WARF has become the major "gatekeeper" in determining which companies have been allowed to conduct research and create commercial products using hESCs.

Since obtaining its patents in 1998, WARF has entered into licensing agreements with 27 commercial partners for its intellectual property portfolio related to research in isolating and differentiating hESCs, of which only 15 have been disclosed. In addition, 300 academic laboratories have received agreements for using the technology without charge. Geron has an exclusive license for the development of therapeutic and diagnostic products. While Invitrogen (Life Technologies), Becton Dickinson, and Chemicon (Millipore/Merck KGaA) have disclosed licensing agreements with WARF, many research products companies have had to explore less costly alternatives, such as circumventing the claims, conducting research offshore, exploring the potential use of abnormal karyotypes, developing embryonic stem cell products for other species, and pursuing strategic collaborations (BioInformant Worldwide 2009).

The WARF patents were challenged in a July 2006 re-examination at the USPTO by the Foundation for Taxpayer and Consumer Rights in Santa Monica and the New York-based Public Patent Foundation. While WARF's patents were upheld by the USPTO, the claim for 7,029,913 (hES cells) was narrowed to require the cells be isolated from *pre-implantation* human embryos. Thus, the re-examination clarified that the scope of these controversial patents did not extend to iPS cells.

17.9 STEM CELLS

The ability of scientists to tap into the processes of cell differentiation and development represents one of the greatest potential areas in regenerative medicine. These efforts can provide us with unprecedented insights about human development and diseases and potentially transformative clinical therapies. At the forefront of this work is stem cell research.

Stem cells are characterized by their properties of self-renewal and potency. Self-renewal is defined as the ability to complete numerous cycles of cell division while maintaining an undifferentiated state. Potency is defined as the capacity to differentiate into specialized cell types. The two broad types of mammalian stem cells are embryonic and adult. Embryonic stem cells are pluripotent cells isolated from the inner cell masses of blastocysts. Adult stem cells, also known as somatic stem cells, are the most common multipotent cells found in adult tissues. A more nascent discovery involves induced pluripotent stem cells—a type of pluripotent stem cell artificially derived from a non-pluripotent cell, typically an adult somatic cell.

Stemcellpatents.com identified 1,434 stem cell patents across multiple classes including 3D cultures and scaffolds, administration, differentiation, expansion, mobilization, type, un-classified use, and extraction/preservation. The number of patent applications for stem cell technologies rose tremendously from 2000 to 2009. Based on increased state and federal government funding for stem cell research, it is expected that the number of patents will rise in coming years (Stemcellpatents.com 2009).

17.10 HUMAN EMBRYONIC STEM CELL THERAPY LANDSCAPE

Embryonic stem cells have been hailed as the most versatile of the stem cells and have the ability to differentiate into almost any type of cell in the body. This gives them great therapeutic potential. The first embryonic stem cell lines were established in 1998, putting embryonic stem cell research, compared to adult stem cell research, in an earlier phase of discovery. It is also immersed in controversy. In May 2008, the FDA suspended the start of clinical trials for Geron Corp.'s GRN0PC1, which was expected to be the first human embryonic stem cell product to start clinical trials. In November 2009, Advanced Cell Technology applied for a license to carry out a human embryonic stem cell research clinical trial on patients in the U.S. suffering from a type of macular degeneration (Frost and Sullivan 2008). Table 17.4 describes leading patent holders for tools and technologies involved in regenerative neuronal human embryonic stem cell therapy.

17.11 NON-EMBRYONIC STEM CELL LANDSCAPE

Adult stem cells can turn into multiple types of body cells, but are only a fraction of the 200+ body cells that pluripotent stem cells can turn into. Adult multipotent stem cells can be procured from bone marrow, body fat, and other sources. Scientists have begun to demonstrate how these may be useful for vascular or muscle therapies. Recent research developments include adult stem cells aiding in the healing of bone fractures in mice and the conversion of human fat stem cells into smooth muscle cells (Center for American Progress 2009). An example of a recent patent claim directed to an amniotic stem cell is Claim 1 of U.S. Patent 7,569,385,

TABLE 17.4
Leading Regenerative Medicine Patent Holders:
Neuronal Human Embryonic Stem Cells

Tools	IP Owners
IVF	Geron/Wisconsin Alumni Research Foundation
Cell fusion	
Media and factors	
Culture supports	
Genetic engineering	
Media & factors	University of California
Culture supports	
Genetic engineering	
Media and factors	University of Cambridge
Culture supports	
Genetic engineering	
Drugs	Merck & Co., Inc.
Scaffolds	
Tissue engineering	

Source: Sharma Group, LLC (TSG). Advances in Replacement Organs and Tissue Engineering. http://www.tsg-partners.com.

issued August 4, 2009, to the University of California Oakland. The claim reach is as follows:

> A composition comprising isolated, mortal, epithelioid multipotent stem cells derived from amniotic fluid, wherein said stem cells are characterized by expression of SSEA3, SSEA4, Tra1-60, Tra1-81, Tra2-54, Oct-4, and CD105, and by less than 5-percent expression of SSEA1.

17.12 INDUCED PLURIPOTENT STEM CELL PATENT LANDSCAPE

Induced pluripotent stem (iPS) cells are recently discovered types of pluripotent stem cells artificially derived by reprogramming somatic cells (see Chapter 10 by Walker). iPS cells are morphologically similar to embryonic stem cells and are capable of differentiating into a variety of different somatic cell types. iPS cells were first produced in 2006 from mouse cells (Takahashi et al. 2006) and in 2007 from human cells (Takahashi et al. 2007). This technology allows researchers to obtain pluripotent stem cells for use in a research setting. iPS cells may also have therapeutic uses for the treatment of disease without the need for stem cells derived from an embryonic source. Table 17.5 and Table 17.6 illustrate U.S. iPS patent applications and iPS PCT publications through 2009. No patent has been issued yet in the U.S. for iPS cells. A published claim (PCT Publication WO 08/151058, Applicant: The General Hospital Corporation) directed to iPS methodology covers a method of selecting induced pluripotent stem cells, the method comprising:

TABLE 17.5
Induced Pluripotent Cells: U.S. Patent Applications

Assignee/Inventor	U.S. Patent Application Number	Application Title
Kyoto University	20090227032A1	Nuclear Reprogramming Factor and Induced Pluripotent Stem Cells
Ma, Yupo	20090191171A1	Reprogramming of Differentiated Progenitor or Somatic Cells Using Homologous Recombination
Kyoto University	20090047263A1	Nuclear Reprogramming Factor and Induced Pluripotent Stem Cells
Izumi Bio, Inc.,	20090299763A1	Methods of Cell-Based Technologies
Biodontos, LLC	20090291496A1	Neural Stem Cell Isolates from Dental Papillary Annulus of Developing Teeth
Kubo, Atsushi; Bonham, Kristina; Stull, Robert; Snodgrass, H. Ralph	20090280096A1	Pancreatic Endocrine Progenitor Cells Derived from Pluripotent Stem Cells
Kyoto University	20090246875A1	Efficient Method for Nuclear Reprogramming
Cleveland Clinic Foundation	20090246179A1	Method of Treating Myocardial Injury
Conklin, Bruce R.; Aalto-Setala, Katriina	20090227469A1	Cells and Assays for Use in Detecting Long Qt Syndrome
Yamanaka, Shinya	20090068742A1	Nuclear Reprogramming Factor

Source: http://www.uspto.gov.

a) providing a female cell that is heterozygous for a selectable marker on the X chromosome, wherein the selectable marker is mutant on the active X chromosome and wild-type on the inactive X chromosome, and wherein the cell does not express Nanog mRNA when measured by RT-PCR;
b) re-programming said cell to a pluripotent phenotype;
c) culturing the cell with a selection agent, wherein the reactivation of the inactive X chromosome permits the expression of wild-type selectable marker and permits cell survival in the presence of the selection agent, whereby surviving cells are induced pluripotent stem cells.

17.13 TISSUE ENGINEERING AND REPLACEMENT ORGAN PATENT LANDSCAPE

Tissue engineering is among the foremost technologies that occupy the best minds in the biological sciences today. Most companies in the industry today focus on biomaterial applications, followed in equal measure by organ regeneration technologies. Cell therapy and organ regeneration represent the future for tissue engineering. Table 17.7 and Table 17.8 list the major industry and academic patents for tissue engineering and replacement organs. A published claim (U.S. Patent 7569076,

TABLE 17.6
Induced Pluripotent Stem Cell Patent Publications

Assignee/Inventor	PCT Publication Number	Publication Title
Helmholtz Zentrum München–Deutsches Forschungszentrum für Gesundheit Und Umwelt (GMBH), Germany	WO2009115295A1	Vectors and Methods for Generating Vector-Free Induced Pluripotent Stem (ips) Cells Using Site-Specific Recombination
Max-planck-gesellschaft Zur Förderung Der Wissenschaften EV, Germany	WO2009144008A1	Generation of Induced Pluripotent Stem (ips) Cells
Kyoto University, Japan	WO2009133971A1	Method of Nuclear Reprogramming
Kyoto University, Japan	WO2009118928A1	Efficient Productiona and Use of Highly Cardiogenic Progenitor and Cardiomyocytes from Embryonic and Induced Pluripotent Stem Cells
Nevada Cancer Institute	WO2009092042A1	Reprogramming of Differentiated Progenitor or Somatic Cells Using Homologous Recombination
Childrens Medical Center	WO2009061442C1	Method to Produce Induced Pluripotent Stem (ips) Cells from Non-Embryonic Human Cells
Childrens Medical Center	WO2009061442A8	Method to Produce Induced Pluripotent Stem (ips) Cells from Non-Embryonic Human Cells
Childrens Medical Center	WO2009061442A1	Method to Produce Induced Pluripotent Stem (ips) Cells from Non-Embryonic Human Cells
Vistagen Therapeutics, Inc.	WO2009137844A2	Pancreatic Endocrine Progenitor Cells Derived from Pluripotent Stem Cells
Mirae Biotech Co., Ltd.	WO2009131262A1	Method of Manufacturing Induced Pluripotent Stem Cells Originated from Human Somatic Cells
Scripps Research Institute	WO2009117439A2	Combined Chemical and Genetic Approaches for Generation of Induced Pluripotent Stem Cells
J. David Gladstone Institutes	WO2009114133A1	Cells and Assays for Use in Detecting Long QT Syndrome
President and Fellows of Harvard College	WO2009102983A2	Efficient Induction of Pluripotent Stem Cells Using Small Molecule Compounds
Mirae Biotech Co., Ltd.	WO2009096614A1	Method of Manufacturing Induced Pluripotent Stem Cells Originated from Somatic Cells
Australian Stem Cell Centre, Ltd.	WO2009055868A1	Process and Compositions for Culturing Cells
Whitehead Institute for Biomedical Research	WO2009032194A1	Wnt Pathway Stimulation in Reprogramming Somatic Cells
General Hospital Corporation	WO2008151058A3	Methods of Generating Pluripotent Cells from Somatic Cells
General Hospital Corporation	WO2008151058A2	Methods of Generating Pluripotent Cells from Somatic Cells

Source: http://www.uspto.gov.

TABLE 17.7
Leading Industry Patents: Tissue Engineering and Replacement Organs

Assignee	Patent Number	Title
Warsaw Orthopedic	7,341,601	Woven orthopedic implants
Stick Tech OY	7,354,969	Dental and medical polymer composites and compositions
De Paois; Potito; Slehmoghaddam; Saleh	7,128,836	Dialysis device (artificial liver)
Vital Therapies	7,390,651	C3A serum-free clonal cell line and method of use
Tapic International	7,084,082	Collagen material and its production process
Boston Scientific Scimed	7,326,571	Decellularized bone marrow extracellular matrix
Dentigenix	7,309,232	Methods for treating dental conditions using tissue scaffolds
RenaMed Biologics	7,332,330	Device for maintaining vascularization near implant
St3 Development Corporation	7,410,792	Instrumented bioreactor with material property measurement capability and process-based adjustment for conditioning tissue engineered medical products
Medtronic Vascular	7,387,645	Cellular therapy to heal vascular tissue
Scimed Life Systems	7,384,786	Aligned scaffolds for improved myocardial regeneration
Encelle	6,992,062	Method of stimulation of hair growth
Aastrom Biosciences	6,835,566	Human lineage committed cell composition with enhanced proliferative potential, biological effector function, or both; methods for obtaining same; and their uses
Cryolife	7,318,998	Tissue decellularization
CryoLife	7,129,035	Method of preserving tissue
Tepha	7,025,980	Polyhydroxyalkanoate compositions for soft tissue repair, augmentation, and viscosupplementation

Source: Frost and Sullivan. 2008. Advances in Replacement Organs and Tissue Engineering. http://www.frost.com.

issued to Children's Medical Center Corporation) directed to organ replacement methodology says:

> A method for the replacement or repair of a bladder, or portion of a bladder, in a human patient in need of such treatment, comprising the steps of:
> a) providing a biocompatible synthetic, or natural, polymeric matrix structure in the shape of a bladder, or portion of a bladder, wherein said polymeric matrix is coated with a biocompatible and biodegradable shape-setting material comprising a liquefied poly-lactide-co-glycolide copolymer;
> b) depositing at least one first cell population on or in said polymeric matrix; and

TABLE 17.8
Key Academic Patents: Tissue Engineering and Replacement Organs

Assignee	Patent Number	Patent Title
University of South Carolina	7,338,517	Tissue scaffold having aligned fibrils and artificial tissue comprising the same
Japan Science and Technology Agency	7,399,634	Method of preparing basement membrane, method of constructing basement membrane specimen, reconstituted artificial tissue using the basement membrane specimen and process for producing the same
Trustees of University of Pennsylvania	7,396,537	Cell delivery patch for myocardial tissue engineering
Regents of University of California	7,326,570	Induction of tubular morphogenesis using pleiotrophin
Board of Regents of University of Oklahoma	7,344,712	Urinary tract tissue graft compositions and methods for producing same
General Hospital Corporation	7,371,400	Multilayer device for tissue engineering
Children's Medical Center Corporation	7,049,057	Tissue engineered uterus
William Marsh Rice University	7,393,687	Biomimetic 3-dimensional scaffold with metabolic stream separation for bioartificial liver device
University of Hong Kong	7,393,437	Photochemically crosslinked collagen scaffolds and methods for their preparation
Board of Trustees of University of Illinois	7,375,077	In vivo synthesis of connective tissues
Regents of University of Michigan	7,368,279	Three-dimensional bioengineered smooth muscle tissue and sphincters and methods thereof
Aderans Research Institute	7,198,641	Porous, bioabsorbable scaffolds for tissue engineering of human hair follicles, methods for their manufacture and methods of their use in creating new hair
University of Washington	7,300,962	Hydrogels formed by non-covalent linkages

Source: Frost and Sullivan. 2008. Advances in Replacement Organs and Tissue Engineering. http://www.frost.com.

c) implanting the shaped polymeric matrix of step (b) into the human patient at a site of a bladder for the formation of a laminarily organized functional bladder structure.

17.14 CONCLUSIONS

Regenerative medicine will be a vital component of cutting edge life sciences in the 21st century. It is a disruptive technology that offers the potential to replace a number of significant molecular pharmaceuticals and medical prostheses (see Chapter 1 by Polak). The potential therapies range from transforming the pancreatic cells of diabetics so they can produce insulin to regenerating vital organs such as the liver. With the recent decision by President Obama to lift the ban on federal funding for embryonic stem cell research, a further rise in innovative regenerative technologies is anticipated.

In this rapidly changing field, the complexity of patent prosecution is a hurdle faced by many universities, governments, and companies worldwide. Yet, intellectual property rights are also a cornerstone of the industry, relied on by numerous biotechnology companies for successful business models and closely analyzed by investors to identify innovative research and commercialization opportunities. The regenerative medicine sector is quite dynamic, with a prominent role played by government and public bodies. This is likely to increase in the future as individual states and the U.S. government and other nations develop additional funding programs and research centers.

The recent discovery of induced pluripotent stem cells has the potential to be disruptive and transformative as ethical concerns surrounding embryonic stem cell research are mitigated and much promise has been shown. Growth in the regenerative medicine sector will continue, bringing with it the discovery of novel technologies, promising therapies, new business models, and intellectual property challenges. As the extraordinary pace of scientific advancement in regenerative medicine continues, protection of intellectual property rights will ensure that technologies can proceed from scientific discovery to therapeutic products benefiting humankind.

REFERENCES

Association for Molecular Pathology, et al. v United States Patent and Trademark Office, et al. 2010. http://lp.findlaw.com/
Atala, A., R. Lanza, J. Thomson et al. 2008. *Principles of Regenerative Medicine*. San Diego: Academic Press.
BioInformant Worldwide. 2009. http://www.bioinformant.com
Center for American Progress. 2009. http://www.americanprogress.org/
DataMonitor. 2009.http://www.datamonitor.com/
Diamond v. Chakrabarty, 447 U.S. 303, 1980. http://lp.findlaw.com/
Frost and Sullivan. 2008. Advances in Replacement Organs and Tissue Engineering. http://www.frost.com.
Greenwood, H. L., P. A. Singer, G. P. Downey, et al. 2006. Regenerative medicine and the developing world. *PLoS Med* 3: e381. DOI: 10.1371/journal.pmed.0030381.
Lupin Ltd. and Lupin Pharmaceuticals Inc. v. Abbott Laboratories and Astellas Pharma, Inc., No. 2007-1446 (Fed. Cir. May 18, 2009). http://www.cafc.uscourts.gov/
Medistem, Inc. 2009. http://www.medisteminc.com/
Osiris Therapeutics, Inc. 2008. http://investor.osiris.com/releases.cfm?Year=2008
Regenerative Medicine. Department of Health and Human Services. 2006. http://www.stemcells.nih.gov/info/scireport/2006report.htm
Sharma Group, LLC (TSG). Advances in Replacement Organs and Tissue Engineering. 2009. http://www.tsg-partners.com.
Stemcellpatents.com. 2009. http://www.stemcellpatents.com

Takahashi, K. and S. Yamanaka. 2006. Induction of pluripotent stem cells from mouse embryonic and adult fibroblast cultures by defined factors. *Cell* 126: 663.

Takahashi, K., K. Tanabe, M. Ohnuki et al. 2007. Induction of pluripotent stem cells from adult human fibroblasts by defined factors. *Cell* 131: 861.

35 USC §101 et seq. Manual of Patenting and Examining Procedure, 8th Ed., Rev. 7. http://www.uspto.gov

35 USC §271(a), Manual of Patenting and Examining Procedure, 8th Ed., Rev. 7. http://www.uspto.gov

Weiss, R., February 9, 2009. The "patent pending" problem. The Boston Globe. http://www.boston.com.

18 Impacts of Indian Policies and Laws on Regenerative Medicine Patent Applications

Prabuddha Ganguli

CONTENTS

18.1 Introduction 281
18.2 Indian Patent Landscape...282
18.3 Patent Act 1970 and 1995, 1999, and 2002 Amendments 282
    18.3.1 1972 to 1995..282
    18.3.2 1995 to 2005 283
    18.3.3 2002: Second Amendment to Patent Act 1970 284
    18.3.4 2005 to Present: Third Amendment to Patent Act 1970..................286
18.4 Impacts of Policies and Laws on Regenerative Medicine Patent Applications..288
References..289

18.1 INTRODUCTION

It has been well documented over the centuries that traditional Indian systems of medicine (Ayurveda, Siddha, and Unani) have embraced diverse aspects of regenerative medicine. Regenerative medicine has been a priority area for research and development in India for the past two decades and as a result, has seen major investments by the Indian government. More recently, there has been a renewed interest to integrate the learnings from traditional medicine with those of modern science to deliver innovations in regenerative medicine and thereby enhance interdisciplinary research and healthcare in India. The development of appropriate national policies with a balanced intellectual property right (IPR) system underpins and promotes the national innovation processes and industrial growth leading to sustainable development.

India, as a developing nation and a member of the World Trade Organisation (WTO), is obliged to comply with all the requirements of the Trade-Related Aspects of Intellectual Property (TRIPS) agreement. Significant progress has been made by way of introducing new legislation and also through amendments to existing IPR Laws (Ganguli 1999 and 2003).

TABLE 18.1
Chronological Account of Indian Patent Legislation

Act VI of 1856
Act IX of 1857
Act XV in 1859
Patterns and Designs Protection Act 1872
Protection of Inventions Act (XVI) 1883
Inventions and Designs Act (V) 1888
Indian Patents and Designs Act (II) 1911
Justice Tek Chand Report, August 4, 1949
Amendment of 1911 Act in 1950 (Act XXXII) followed by further amendment in 1952
Bill in Parliament in 1953 (Bill 59 of 1959), lapsed due to dissolution of Lok Shaba
Justice N. Rajagopala Aayangar Committee appointed 1957, submitted report September 1959
Patents Bill 1965 introduced September 21, 1965, subsequently lapsed; amended bill introduced in 1967
Indian Patents Act 1970 (Bill 39 of 1970) became effective from April 20, 1972
Repealing and Amending Act 1974 (Bill 56 of 1974)
Delegated Legislation Provisions (Amendment) Act 1985 (Bill 4 of 1986)
Ordinance to amend Patents Act 1970 on December 31, 1994, effective from January 1,1995, ceased to operate after 6 months; subsequent ordinance issued in 1999
Patents (Amendment) Act 1999 (Bill 17 of 1999) retroactively effective from January 1, 1995
Patents (Amendment) Act 2002 (Bill 38 of 2002) effective May 20, 2003
Patents (Amendment) Ordinance 2004 (Ordinance 7 of 2004) effective October 1, 2005
Patents (Amendment) Act 2005 (Bill 15 of 2005), deemed in force from January 1, 2005; with Patent (Amendment) Rules of 2005 notified December 28, 2004 amending Patent Rules of 2003, which were further amended by Patent (Amendment) Rules of May 5, 2006

18.2 INDIAN PATENT LANDSCAPE

Patent law in India has been evolving since 1856, and in more recent times, has seen multiple amendments to earlier patent Acts, that impact on the patenting of regenerative medicine and related products. Table 18.1 is a chronological list of Indian legislative measures. Recently, a third amendment was introduced by way of the Indian Patents Act Amendment (2005) and the Patent Amendment Rules (2005). The major transition was the deletion of Section 5 of the Indian Patents Act 1970, thereby re-introducing patents for products of chemical reactions and substances that are medicines, drugs, pharmaceuticals, etc. Processes continue to remain a subject matter of patents based on the Patents Act 1970.

18.3 PATENT ACT 1970 AND 1995, 1999, AND 2002 AMENDMENTS

18.3.1 1972 TO 1995

The 1970 Indian Patents Act introduced significant changes to the 1911 Patent Act, with limitations related to the patenting of inventions in the area of medicines, drugs, chemicals, and pharmaceuticals. The patenting of products (substances capable of

Impacts of Indian Policies and Laws

being used as drugs and pharmaceuticals) was discontinued although processes for making such products continued to be subject matters of patents. Furthermore, it was not possible to patent inventions related to a method, any process for the medicinal, surgical, curative, prophylactic, or other treatment of human beings or any process that renders animals or plants free of disease, or increases their economic values (or the values of their products).

According to the 1970 Act, process patents relating to inventions of drugs, medicines, and pharmaceuticals that used the terms "medicines" or "drugs" were shortened to 7 years from the date of filing of the complete specification or 5 years from the date of sealing the patent, whichever was shorter. For patents in other fields of technologies, the patent term was reduced from 16 to 14 years.

The conditions for compulsory licensing were also made fairly liberal and introduced the concept of "license of right" for patents related to drugs and pharmaceuticals. The immediate and long-term impact of the 1970 act and the similarly restrictive industrial policies of the Indian government did not encourage activities related to patenting of inventions, especially in the field of pharmaceuticals, medicines, and methods of treatments. The patenting system did, however, encourage the filing of patents for new processes and technologies leading to the production of drugs and medicines.

18.3.2 1995 TO 2005

The Patent Ordinance 1994 (supplementing the Indian Patents Act 1970) was intended to amend the 1970 act in order to provide a means for the filing and handling of patent applications for pharmaceutical products as required by Subparagraph (a) of Article 70.8 of the TRIPS agreement. The amendment also enabled the grant of exclusive marketing rights with respect to the products that are the subjects of patent applications. The ordinance became effective January 1, 1995.

The 1995 amendment was intended to give permanent legislative effect to the provisions of the ordinance, and was introduced in the Lok Sabha (lower house of parliament) in March 1995. Lok Sabha passed the amendment but it was rejected subsequently in the Rajya Shaba (upper house of parliament). The Bill was referred to a Select Committee for re-examination. The committee started its work but was unable to present its report before the dissolution of the Lok Sabha on May 10, 1996 and therefore, the 1995 amendment lapsed with the dissolution of the tenth Lok Sabha.

This was followed by a dispute against India, initiated by the United States, and involving the WTO Dispute Settlement Board (DSB) in July 1996 and December 1997. The U.S., supported by the European Union (EU), alleged that India had not complied with the obligations of TRIPS (particularly Articles 70.8, 70.9, and 63). In December 1997 the DSB directed India to take necessary steps for compliance before April 1999.

In December 1998, India joined the Paris Convention and became a member of the Patent Co-operation Treaty (PCT). Subsequently a presidential ordinance was proclaimed in January 1999 to satisfy all conditions of the transitional arrangements. The patent amendment bill was passed by Lok Sabha in March that year and received endorsement of the Rajya Sabha by a voice vote, also in March. Rules for the implementation of the amendments were published in April 1999. The 1999 patent amendment bill further established continuity with respect to all actions under the Ordinance of 1994.

As a requirement of Articles 70.8 and 70.9 of the TRIPS agreement, India introduced a mail box provision for the filing of product patents for inventions related to medicines and drugs and introduced a provision for exclusive marketing rights. Such mail box applications were entitled to be substantively examined only after December 31, 2004 or after the Indian Patents Act underwent further amendments, whichever was earlier. The mail box provision made it possible to file for product patents related to drugs and medicines in India. Therefore, in 1999, the 1970 Indian Patent Act was amended to state:

> 1) Inventions where any methods or processes of manufacture are patentable. This meant that no patent shall be granted in respect of claims for the substances themselves but claims for the methods or process of manufacture would be patentable in the case of inventions (a) claiming substances intended for use, or capable of being used, as food or as medicine or drug, or (b) relating to substances prepared or produced by chemical processes (including alloys, optical glass, semiconductors and intermetallic compounds). 2) Not withstanding anything contained in subsection (1), a claim for the patenting of an invention for a substance itself intended for use, or capable of being used, as medicine or drug, except the medicine or drug (specified under subclause (v) of clause (1) of subsection (1) of section (2) may be made and shall be dealt, without prejudice to the other provisions of the act in manner provided by Chapter IVA.

Under the mail box provision, several patent applications involving products in biotechnology and regenerative processes were filed in India between 1995 and 1999.

18.3.3   2002: Second Amendment to Patent Act 1970

The second amendment, as a continuation of the first amendment, took further steps toward making the 1970 Indian Patents Act TRIPS compliant. The key aspects of the amended system that would impact the patenting of inventions in biotechnology were:

- Redefining "invention" to mean a new product or process involving an inventive step and capable of industrial application.
- "Inventive step" means a feature that makes the invention not obvious to a person skilled in the art.
- Extending the patent term to 20 years irrespective of the field of technology.
- Formation of an appellate board with a technical member to deal with appeals related to decisions by the controller under specific circumstances.
- In the case of infringement of process patents, reversing the burden of proof on the defendant.

A specific section (107A) was introduced to provide a safe harbor for any act of "making, constructing, using or selling a patented invention solely for uses reasonably related to the development and submission of information." This section regulated the manufacture, construction, use, or sale of any product to be considered exempt from being an act of infringement and the importation of patented products by any person duly authorized by the patentee to sell or distribute the product. The amended Act (Section 3) listed inventions that would not patentable in India following the second amendment of the 1970 Act. These related to products, formulations, dosage forms, medicine, surgery, and so forth and included:

- An invention which is frivolous or which claims anything obviously contrary to well established natural laws.
- An invention, the primary or intended use or commercial exploitation of which could be contrary to public order or morality or which causes serious prejudice to human, animal, or plant life or health or to the environment.
- Mere discovery of a scientific principle or the formulation of an abstract theory or discovery of any living thing or non-living substance occurring in nature.
- Mere discovery of any new property or new use for a known substance or of the mere use of a known process, machine, or apparatus unless such known process results in a new product or employs at least one new reactant.
- A substance obtained by a mere admixture resulting only in the aggregation of the properties of the components thereof or a process of producing such substance.
- Mere arrangement or re-arrangement or duplication of known devices each functioning independently of one another in a known way.
- Any process for the medicinal, surgical, curative, prophylactic, diagnostic, therapeutic, or other treatment of human beings or process for a similar treatment of animals to render them free of disease or to increase their economic value or that of their products.
- An invention which, in effect, is traditional knowledge or which is an aggregation or duplication of known properties of a traditionally known component or components.

A new requirement was also introduced (Section 10) into the 1970 Indian Patents Act related to the contents of a patent specification using biological materials. The amendment states:

"The specification shall be accompanied by an abstract to provide technical information on the invention. If the applicant mentions a biological material in the specifications which may not be described in such a way as to satisfy the clauses requiring the disclosure with the best method of performing the invention, such that anyone trained in the art can reproduce the invention, and if such material is not available to the public, the application shall be completed by depositing the material to an authorized depository institution as may be notified by the Central Government in the Official Gazette and by fulfilling the following conditions, namely: (a) The deposit of the material shall be made not later than the date of the patent application in India. (b) All the available characteristics of the material required for it to be correctly identified or indicated are included in the specification including the name and the address of the depository institution and the date and number of the deposit of the material at the institution (as per the Budapest Treaty of which India is a signatory). (c) Access to the material is available in the depository institution only after the date of the application for patent in India or if a priority is claimed after the date of the priority. (d) Disclose the source and geographical origin of the biological material in the specification when used in an invention."

India continued to follow the system of pre-grant opposition. Following examination and acceptance by the patent controller a patent would then be advertised in the *Gazette of India*. Any opposition to the accepted application could be registered with the patent office within 4 months of the date of publication of the specification. If any party was dissatisfied with the decision of the controller, an appeal could be filed with the appellate board set up under the second amendment to the 1970 Patents Act. In addition to the existing grounds for opposition (Section 25), the second amendment included new provisions:

- The complete specification does not disclose or wrongly mentions the source or geographical origin of biological material used for the invention.
- The invention so far as claimed in any claim of the complete specification is anticipated having regard to the knowledge, oral or otherwise, available within any local or indigenous community in India or elsewhere.

Any interested party could revoke a patent by petition to the high court or the central government or by counter claim in a suit for infringement of the patent (Section 64). In addition to the conventional grounds (listed in Section 64), two further provisions were included:

- The complete specification does not disclose or wrongly disclose the source or geographical origin of the biological material used for the invention.
- The invention claimed in any claim of the complete specification was anticipated having regard to the knowledge, oral or otherwise, available within any local or indigenous community in India or elsewhere.

During this period, several patent applications related to processes, devices, and methods of influencing cell growth and tissue regeneration were filed in India and a number of these have been recently granted (Table 18.2).

18.3.4 2005 to Present: Third Amendment to Patent Act 1970

The key feature of the 2005 Patents (Amendment) Act and the Patent (Amendment) Rules (following the abolition of Section 5) was India's re-entry into the product patents regime, providing for product patents across all fields of technology including drugs, chemicals, and microorganisms. The lack of any definition of *inventive step* in earlier Indian legislation resulted in the inclusion of the following definitions (Section 2):

- "Inventive step" means a feature of an invention that involves technical advance as compared to the existing knowledge or having economic significance or both and that makes the invention not obvious to a person skilled in the art.
- "New invention" means any invention or technology which has not been anticipated by publication in any document or used in the country or elsewhere in the world before the date of filing of the patent application with

TABLE 18.2
Recently Granted Indian Patents on Subjects Related to Regenerative Systems

Application Number	Filing Date	Title	Applicant	International Classification
IN/PCT/2001/361/CHE	March 3, 2001	Method for preparing conditioned medium	Geron Corporation	C12N5/06
397/MAS/2001	May 16, 2001	Medicinal compositions containing neurons made from pluripotent stem cells	Geron Corporation	A61K 35/30
2755/DELNP/2005	June 21, 2005	Culture medium composition, culture method, and myoblasts obtained and their uses	Celogos	C12N 5/06
2544/CHENP/2005	Oct. 5, 2005	Expansion of renewable stem cell populations using modulators of P13-kinase	Gamida Cell Ltd.	C12N
395/MAS/2002	May 23, 2002	Method of isolating progenitor liver cells and their trans-differentiation	Centre for Liver Research and Diagnostics	A 61 K 35/407
240/MUM/2004	Feb. 26, 2004	Self renewing embryonic-like pluripotent stem cells from corneal limbus for use in regenerative therapy and method of producing same	Reliance Life Sciences Private Ltd.	C12N5/06
00545/CHENP/2004	March 12, 2004	Prolactin-induced increase in neutral stem cell numbers and therapeutic use thereof	Stem Cell Therapeutics Inc.	C12N5/06,5/08
IN/PCT/2001/1131/CHE	Aug. 9, 2001	Method and apparatus for maintenance and expansion of hemopoietic stem cells and/or progenitor cells	Technion Research and Development Foundation Ltd.	C12N 5/00
IN/PCT/2002/311/CHE	Feb. 28, 2002	Multipotent adult stem cells and methods for isolation	MCL LLC	C12N
IN/PCT/2002/1297/CHE	Aug. 19, 2002	A method of processing non-fetal donor tissue	University of North Carolina at Chapel Hill	C12N 5/06
582/CHENP/2005	April 8, 2005	Devices and methods for improving vision	Ocular Sciences, Inc.	A61F 2/14
2332/CHENP/2005	Sept. 20, 2005	Methods of using adipose tissue-derived cells in treatment of cardiovascular conditions	Macropore Biosurgery, Inc.	C12N

complete specification, i.e., the subject matter has not fallen into the public domain or that it does not form part of the state of the art.
- "Pharmaceutical substance" means any new entity involving one or more inventive steps.

Section 3, which deals with the limitations on inventions for the purposes of patenting, included Paragraph (d):

- The mere discovery of a new form of a known substance which does not result in the enhancement of the known efficacy of that substance or the mere discovery of any new property or new use for a known substance or of the mere use of a known process, machine or apparatus unless such known process results in a new product or employs at least one new reactant.

For the purposes of this clause, salts, esters, ethers, polymorphs, metabolites, pure form, particle size, isomers, mixtures of isomers, complexes, combinations and other derivatives of known substance shall be considered to be the same substance, unless they differ significantly in properties with regard to efficacy.

The introduction of (d) above has major implications on the patentability of inventions related to new chemical entities in pharmaceuticals and in regenerative medicine, and indeed has sparked controversies including several court cases in India. One case in particular tested in an Indian court involved living organisms (*Dimminaco AG v. Controller of Patents*), potentially impacted regenerative medicine patents. A patent titled "Process for preparation of infectious bursitis vaccine" was filed in 1998. The Indian Patent Office rejected the application on the basis that the end product of the process involved a living organism and thus was not patentable under Section 2. The act states that the statutory definition of *manufacture* does not include a process that results in a *living organism*. On appeal, the court stated that there was no statutory bar in the act to accept a manner of manufacture as patentable even if the end product contained a living organism, and directed the patent office to reconsider the patent application in light of the court's observations. The patent was finally granted.

18.4 IMPACTS OF POLICIES AND LAWS ON REGENERATIVE MEDICINE PATENT APPLICATIONS

The 1970 Indian Patents Act had short-term implications for processing patents related to pharmaceuticals and drugs: 5 years from the date of grant or 7 years from the date of filing, whichever was shorter. This was a major disincentive for patent applicants and was reflected in a sharp fall in the number of patent applications in India after the 1970 Patents Act took effect. However, in recent years, India's science and technology (S&T) policies (especially the policy of 2003) regarding R&D and commercialization of biotechnology, coupled with the second and third amendments of the Indian Patents Act (extending the term to 20 years from date of filing and reintroducing product patents) and improvements to the patenting infrastructure have all led to an increase in the confidence of potential patentees to file patents in India, including those relating to regenerative medicine.

The government of India initiated a project under which a traditional knowledge digital library (TKDL) was set up to document traditional knowledge including methods of treatment, practices, and formulations in Ayurveda, Siddha, and Unani systems of medicine. The library now contains more than two hundred thousand entries. The government granted the EPO and USPTO access to the TKDL to facilitate prior art searches.

While it is generally possible to obtain patent protection for stem cell-related inventions, some jurisdictions place restrictions on the patentability of embryonic stem cells as such or methods using such cells. For example, in the U.S., human embryonic stem cells and their uses are patentable (see Chapter 17 by Resnick, Eisenstein, and McWilliams). Fetal and adult stem cells such as hematopoietic stem cells may be patented in Europe as they fall under the EPC's Rule 23C, which explicitly allows the patenting of isolated elements of the human body as long they as do not constitute a mere discovery. However, the patenting of embryonic stem cells per se is not permitted by the EPO. The Indian patent law is aligned with the European provision, in respect to the patenting of products derived from embryos. Under Section 3(b) of the Indian Patents Act, the patenting of human or animal embryos for any purpose is not permitted because it is against public order or morality (Indian Patent Office: *Manual for Patent Practice and Procedure* 2005).

The patenting of inventions related to regenerative medicine is in its relative infancy in the Indian patenting system and those aspects of the system that could significantly impact regenerative patents have yet to be tested in the Indian judicial system. Several stem cell applications relating the derivation of stem cells, cell lines, methods, and media (Table 18.3) have been filed in India.

The *Patent Journal* provides a list of published patent applications and patents granted by the Indian Patents Office (www.ipindia.nic.in). A search of the *Patent Journal* for patents in the last decade in India relating to regenerative medicine, tissue engineering, tissue growth, tissue reengineering, stem cell, cell growth, tissue culture, and cell engineering shows an increasing number of both published patent applications and granted patents, most of which were filed by foreign applicants and entered India through the PCT route.

REFERENCES

Ganguli, P. 1999. Towards TRIPs compliance in India: the Patents Amendment Act 1999 and implications. *World Patent Inform* 21: 279–287.

Ganguli, P. 2003. Indian path towards TRIPS compliance. *World Patent Inform* 25: 143–149.

Indian Patent Office. 2005. *Manual for Patent Practice and Procedure (Draft)*. New Delhi. http://www.ipindia.nic.in

Kumar. S. and Tiku. 2009. *D. Dimminaco AG v. Controller of Patents*. In *The Conservative Path to Biotech Protection: Managing IP*, 6th Ed. http://www.managingip.com/Article/2282163/The-conservative-path-to-biotech-protection.html

TABLE 18.3
Titles of Some India Patent Applications Pending Examination

Application Number	Filing Date	Title	Applicant	International Class
IN/PCT/2001/01108/DEL	Dec. 3, 2001	Differentiated human embryoid cells and method for producing them	Yissum Research Development Company of the Hebrew University of Jerusalem	C12M5/00
395/MAS/2002[a]	May 23, 2002	Method of isolating progenitors liver cells and their trans-differentiation	Centre for Liver Research and Diagnostics	A61K35/407
0022/CHENP/2004	Jan. 6, 2004	Mesenchymal cells and osteoblasts from human embryonic stem cells	Geron Corporation	C12N
1795/DELNP/2004	June 23, 2004	Islet cells from human embryonic stem cells	Geron Corporation	A61K45/12
1793/DELNP/2004	June 23, 2004	Cells from human embryonic stem cells	Geron Corporation	C12N
00799/KOLNP/2005	May 4, 2005	Composition for culturing multipotent cells and utilization of same	Riken; Niwa; Hitoshi	C12N5/02, 5/06
1138/CHENP/2006	April 3, 2006	Method of inducing the differentiation of stem cells into myocardial cells	Fukuda; Keiichi	C12N5/06
1178/MUM/2006	July 25, 2006	Dopaminergic neurons derived from corneal limbus methods of isolation and uses thereof	Reliance Life Sciences Private Limited	C12N5/06
2373/KOLNP/2006	Aug 22, 2006	Improved modalities for treatment of degenerative diseases of retina	Advanced Cell Technology, Inc.	A01N63/00; A01N65/00; A61K48/00; A01N63/
190/CHENP/2007	Jan. 17, 2007	Method for making high purity cardiomyocyte preparations suitable for regenerative medicine	Geron Corporation	C12N5/06

190/CHENP/2007	Jan. 17, 2007	Methods for identifying stem cells based on nuclear morphotypes	Massachusetts Institute of Technology	G01n 33/50
437/CHENP/2007	Feb. 1, 2007	Hepatocyte lineage cells derived from pluripotent stem cells	Geron Corporation	C12N5/00
652/MUM/2007	March 30, 2007	Cardiomyocyte-like cells from human embryonic stem cells	Reliance Life Sciences Private Limited	C12N5/08
1905/KOLNP/2007	May 28, 2007	Platelets from stem cells	Wisconsin Alumni Research Foundation	C12N5/06
6598/DELNP/2007	Aug. 24, 2007	Use of nuclear materials to therapeutically reprogram differentiated cells	Primegen Biotech. Ltd.	C12N5/00
9209/DELNP/2007	Nov. 29, 2007	Method for producing stem cells or stem cell-like cells from mammalian embryo	Innovative Dairy Products PTY Ltd.	C12N5/06
1538/CHENP/2008	March 28, 2008	Activating agent of stem cells and/or progenitor cells	Tobishi Pharmaceutical Co.	C12N9/64
6264/DELNP/2008	July 17, 2008	Isolated liver stem cells	Universite Catholique de Louvain	C125N5/08
4551/CHENP/2008	Aug. 28, 2008	Method for purifying cardiomyocytes or programmed cardiomyocytes derived from stem cells or fetuses	Asubio Pharma Co. Ltd.	C12N5/00
3853/KOLNP/2008	Sept. 22, 2008	Compositions comprising human embryonic stem cells and their derivatives, methods of use, and methods of preparation	Shroff, Geeta	A61K35/48,C12N5/06
6526/CHENP/2009	Nov. 28, 2008	Method for inducing differentiation of pluripotent stem cells into cardiomyocytes	Asubio Pharma Co. Ltd.	C12N5/00
1239/MUMNP/2009	July 1, 2009	Composition and method for enabling proliferation of pluripotent stem cells	Tryggvason Karl	C12N5/06

[a] Granted.

Section 5

Regulatory Landscape

19 A CATalyst for Change
Regulating Regenerative Medicines in Europe

Christopher A. Bravery

CONTENTS

19.1 Introduction 296
    19.1.1 Historical Perspective 296
19.2 Definitions and Legal Classifications in EU/EEA 296
    19.2.1 What Is Regenerative Medicine? 296
    19.2.2 Medicinal Products 297
        19.2.2.1 Definition 297
        19.2.2.2 Advanced Therapy Medicinal Product (ATMP) 300
        19.2.2.3 Biological Medicinal Products 305
    19.2.3 Medical Devices 306
        19.2.3.1 Definition 307
        19.2.3.2 Medical Device Classes 307
        19.2.3.3 Combination Products 308
    19.2.4 Tissues and Cells 308
    19.2.5 Blood 310
    19.2.6 Cosmetics 311
19.3 Licensing Medicines in EU/EEA 311
    19.3.1 Medicines 311
        19.3.1.1 National Procedures 312
        19.3.1.2 Centralized Procedure 312
        19.3.1.3 Certification Procedure for ATMP 314
    19.3.2 Exemptions from Marketing Authorization 314
        19.3.2.1 Article 5 315
        19.3.2.2 Hospital Exemption 315
    19.3.3 Regulatory Guidance 316
        19.3.3.1 International Convention on Harmonisation 316
        19.3.3.2 EMA/CHMP 317
        19.3.3.3 National Scientific Advice 317
        19.3.3.4 Regulatory Advice 318
19.4 Registering Medical Devices in EU/EEA 318
19.5 Conclusion 318
Acknowledgment 319
References 319

19.1 INTRODUCTION

19.1.1 HISTORICAL PERSPECTIVE

Regulation of regenerative medicine in the European Union (EU) can be understood best when viewed in the historical context of the evolution of the EU. In this chapter, reference will be made to key events and legislation over the last 52 years that led to the situation we have today (Figure 19.1). The history starts with the Treaty of Rome, March 25, 1957 (EEC 1957) that established the predecessor to the current EU, the European Economic Community (EEC). In May 1960, the European Free Trade Association (EFTA) was established as a trade bloc alternative for European countries not then part of the EEC (EFTA 1960). The Stockholm Convention, establishing EFTA, was signed on January 4, 1960 in Stockholm by seven countries (Austria, Denmark, Norway, Portugal, Sweden, Switzerland, and the United Kingdom). The EFTA still exists although 3 of the original 7 are now part of the EU.

Most of the legislation described in the chapter applies not just to the EU, but to the European Economic Area (EEA). The EEA includes the EU along with Iceland, Liechtenstein, and Norway (EEA 1994). A detailed discussion of the EEA is outside the scope of this chapter and the expertise of the author, but simply speaking the EEA agreed on free movement of goods, persons, services, and capital, but those outside the EU do not bear the financial burden of membership. It is significant to note that while Switzerland is a part of the EFTA, it is not a part of the EEA (or the EU).

19.2 DEFINITIONS AND LEGAL CLASSIFICATIONS IN EU/EEA

19.2.1 WHAT IS REGENERATIVE MEDICINE?

Regenerative medicine has been defined numerous ways and since the introduction to this book provides a discussion of the definitions by the editor, there is no need to elaborate here. Needless to say, recombinant proteins and small molecules may play their part, but since the regulation of these products is tried and tested over decades there is no value in discussing them further here. Consequently, this chapter will focus on cell-based healthcare products—this term chosen carefully to encompass both true medicinal products and cell transplants.

The distinction between these two categories has vexed regulatory authorities the world over, since transplantation as a discipline has been well established from the early 1980s and no authority wishes to suddenly classify transplantation as a medicine and thus require transplant centers to submit marketing authorizations. Consequently, defining the boundary requires linguistic and regulatory gymnastics, inclusions, exclusions, and caveats that try the comprehension of the most adept brain. Legislation drafted by lawyers is interpreted by regulators, who are by and large scientists, not lawyers. The problem is wording classifications to clearly exclude transplantation from the somewhat broad definition of a medicinal product; consequently, this discussion will start with what constitutes a medicinal product.

19.2.2 Medicinal Products

The first pan-European definition for a medicinal product was provided by Regulation 65/65/EEC (European Council 1965), when the EEC included only 6 member states. This first definition was less precise than the one we have today, but following various amendments and revisions it eventually approximated that of its successor, Directive 2001/83/EC (European Parliament and Council 2001). This directive is the current core directive on the regulation of medicinal products for the EU/EEA. Like its predecessor, the current medicines directive has undergone numerous revisions and amendments. One is Regulation (EC) 1394/2007 (European Parliament and Council 2007), the so-called Advanced Therapy Medicinal Product (ATMP) regulation that will be discussed in some detail later.

19.2.2.1 Definition

Article 1 of Directive 2001/83/EC (European Parliament and Council 2001b) provides various definitions starting with the one covering medicine. The directive cites two possibilities, but for all practical purposes, the one that counts is Article 2(b):

> "Any substance or combination of substances which may be used in or administered to human beings either with a view to restoring, correcting or modifying physiological functions by exerting a pharmacological, immunological or metabolic action, or to making a medical diagnosis."

Testing this definition with the situation of transplantation, we have to conclude that since a transplant will have metabolic action (it contains living cells; see clarification of this point in Regulation 1394/2007) (European Parliament and Council 2007) and may also act through a pharmacologic mode of action; and of course a transplant restores or corrects a physiological function if it is to replace a defective organ. Thus transplantation could meet the definition of a medicinal product. To frame this definition, it is therefore necessary to refer to the scope of the directive; it does not apply to everything medical; it is only intended to apply to medicines.

The scope is provided by Articles 2 to 5. If you assume they are ordered in priority, the first caveat of the scope is that the product is "intended to be placed on the market in Member States and either prepared industrially or manufactured by a method involving an industrial process." There is no definition for an industrial process but a lawyer may argue that if you are able to make the same product by the same method repetitively, the process is industrial because it is repeatable. A regulator, of course, will argue you will not receive a marketing authorization if you cannot make the product reproducibly. However you look at the argument, unless the product comes directly from nature it is manufactured. In the case of transplantation, this does provide a piece of evidence to support the claim that a transplant is not a manufactured medicinal product, it came directly from nature.

This discussion is not complete without mentioning that minimally manipulated cells can become subject to the medicines directive; this is mentioned later in Section 19.2.2.2. If cells are not used "for the same essential function," then it can be argued they are not performing their usual function, which is in essence another form of manipulation. In these cases, it is necessary to evaluate the safety and risk–benefit factors, so the cells are subject to the medicines directive.

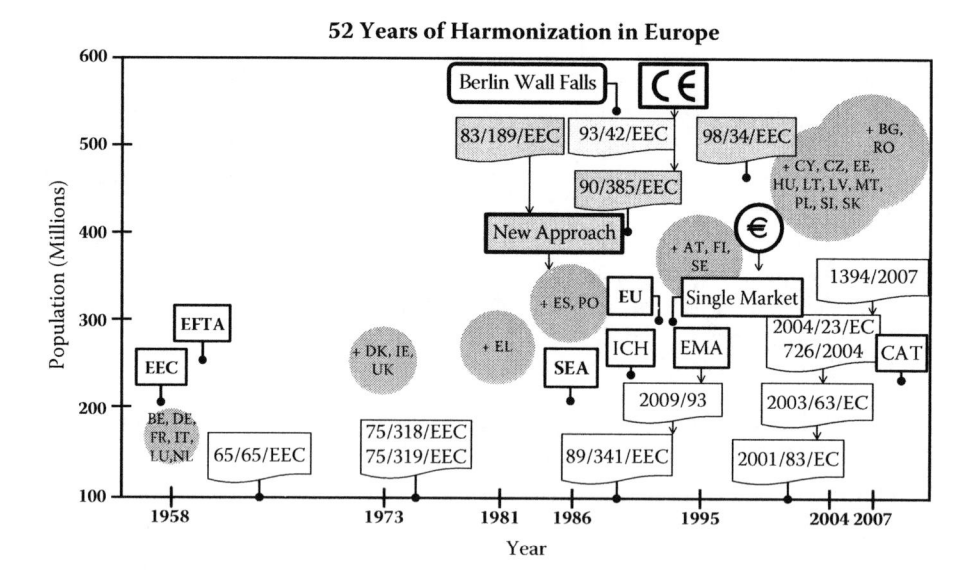

FIGURE 19.1 Fifty-Two Years of Harmonization in Europe

Axis–Population (Conference Board 2009) is plotted against date for the seven phases of EEC/EU enlargement. The area of the circles represents the number of member states involved.
1957–March 25, Treaty of Rome is signed establishing the European Economic Community (EEC 1957) to come into force January 1, 1958.
1958–January 1, EEC is born.
1960–May 3, European Free Trade Association (EFTA) established as a trade bloc alternative for European countries that were not then part of the EEC. Stockholm Convention establishing EFTA signed January 4 in Stockholm by seven countries (Austria, Denmark, Norway, Portugal, Sweden, Switzerland, and United Kingdom). Most of the original seven are now members of the EU; the remaining members today are Iceland, Norway, Switzerland, and Liechtenstein.
1965–Council Directive 65/65/EEC of January 26, on the approximation of provisions laid down by law, regulation, or administrative action relating to medicinal products (European Council 1965).
1975–European Council Directive 75/318/EEC of May 20, 1975 on the approximation of the laws of member states relating to analytical, pharmaco-toxicological, and clinical standards and protocols for testing of proprietary medicinal products (European Council 1975a). This was the predecessor to today's Annex I (European Parliament and Council 2001b) and was repealed by it.
1975–European Council Directive 75/319/EEC on the approximation of provisions laid down by law, regulation, or administrative action relating to medicinal products. This directive together with Directive 65/65/EEC and numerous amendments governed the rules for market authorization until it was repealed by Directive 2001/83/EC (European Parliament and Council 2001b).
1981–Greece joins the EEC.
1983–Directive 83/189/EEC of 28 March 1983 laying down a procedure for the provision of information in the field of technical standards and regulations (European Council 1983).
1985–Council resolution of May 7, 1985 concerning a new approach to technical harmonization and standardization (European Council 1985).
1986–February, the Single European Act (SEA) was signed in Luxembourg on February 17 and in The Hague on February 28 and came into force July 1, 1987 (EEC 1987). The SEA was the first major revision of the 1957 Treaty of Rome and initiated the establishment of a Common Market (by December 31, 1992). Two additional countries join the EEC.
1989–Directive 89/341/EEC of May 3, 1989 amending Directives 65/65/EEC, 75/318/EEC, and 75/319/EEC on the approximation of provisions laid down by law, regulation, or administrative action relating to proprietary medicinal products (European Council 1989).

A CATalyst for Change 299

1989–November 9, the fall of the Berlin Wall followed by reunification of Germany in October 1990, increasing the population of Germany and therefore the EEC by over 16 million people.
1990–April, the birth of ICH at a meeting in Brussels attended by representatives of the regulatory agencies and industry associations of Europe, Japan, and the U.S.
1990–Directive 1990/385/EEC of June 20, 1990 on the approximation of the laws of member states relating to active implantable medical devices (European Council 1990). Amended later by Directive 93/68/EEC (European Council 1993b) introducing CE marking to active implantable medical devices.
1992–Treaty on European Union (European Union 1992), signed in Maastricht on February 7 and coming into force on November 1, 1993.
1993–Directive 93/42/EEC of June 14 (European Council 1993a) and Directive 93/68/EEC (amending Directive 90/385/EEC) of July 22 (European Council 1993b) introduce CE marking for medical devices and active implantable medical devices, respectively, applicable after January 1, 1995.
1993–Council Regulation (EEC) No 2309/93 of July 22 laying down Community procedures for the authorization and supervision of medicinal products for human and veterinary use and establishing a European Agency for the Evaluation of Medicinal Products (European Council 1993c).
1995–Inauguration of the European Medicines Evaluation Agency (EMEA) and the start of the centralized procedure (HMA 2007).
1995–Further EU expansion, three new member states join.
1998–Directive 98/34/EC of the European Parliament and Council of June 22 laying down a procedure for the provision of information on technical standards and regulations (European Parliament and Council 1998).
1999–January, 11 EU Member States (Austria, Belgium, Finland, France, Germany, Ireland, Italy, Luxembourg, Netherlands, Portugal, Spain) adopt the Euro on financial markets. The 11 are joined by Greece in 2001, Slovenia in 2007, Cyprus and Malta in 2008, and Slovakia in 2009. Notes and coins came into use on January 1, 2002.
2001–Major revision of medicines regulations, Directive 65/65/EEC (European Council 1965) and various amendments repealed by Directive 2001/83/EC (European Parliament and Council 2001b).
2003–Technical annex (Annex I) of Directive 2001/83/EC replaced by Directive 2003/63/EC (European Commission 2003). This revision introduces the Advanced Therapy Medicinal Product (ATMP) term and defines gene therapy (GT) and somatic cell therapy (sCT) medicinal products. Amended by Directive 2009/120/EC (European Commission 2009a) to update technical requirements following Regulation (EC) No 1394/2007 and the introduction of Tissue Engineered Products (TEP) as a new category.
2004–EU expansion; an additional 10 countries join.
2004–Publication of Directive 2004/23/EC, the European Tissues and Cells Directive (EUTCD) (European Parliament and Council 2004a) setting standards of quality and safety for the donation, procurement, testing, processing, preservation, storage, and distribution of human tissues and cells. Member states must comply by April 7, 2007. Two technical annexes follow, Directive 2006/17/EC (European Commission 2006a) and Directive 2006/86/EC (European Commission 2006b), the provisions of which must be enacted by November 1, 2006 and September 1, 2007, respectively.
2004–Regulation (EC) No 726/2004 repealing Regulation (EC) No 2309/93 and defining products for which the centralized authorization procedure is obligatory (European Parliament and Council 2004c). It is later amended by Regulation (EC) 1394/2007 adding ATMP to the list of biotechnological products and also adding a new Committee for Advanced Therapies (CAT) to EMEA's committees.
2005–January, the EMEA is renamed the European Medicines Agency. The EMEA abbreviation and logo continue in use (HMA 2007) until December 2009 when a new logo is introduced and the acronym shortened to EMA.
2007–Further EU expansion as two more countries join.
2007–The so-called ATMP regulation, Regulation (EC) No 1394/2007 (European Parliament and Council 2007) is published, introduces Tissue Engineered Products as additional ATMPs (Article 2), and initiates the CAT (Article 17).
2009–January 15–16, the first meeting of the CAT (CAT 2009).

Article 3 excludes various situations, none of which have any real relevance to this discussion except perhaps blood, but it must be remembered Article 3(6) cites unmodified or unprocessed blood (e.g., transfusion). If the blood cells are manipulated, Article 3(6) does not apply. Article 5 allows member states to develop provisions allowing unlicensed medicines to be used for special situations; this is discussed in Section 19.3.2. Other types of medicinal products are classified but only one is directly relevant to regenerative medicine.

Article 4(a) states,

"Advanced therapy medicinal product: A product as defined in Article 2 of Regulation (EC) No 1394/2007 of the European Parliament and of the Council of 13 November 2007 on advanced therapy medicinal products."

Cross referencing Regulation 1394/2007 (European Parliament and Council 2007) brings the definition of an ATMP into Article 1. Before this regulation, ATMPs had been defined within Annex I (European Commission 2003) but had not included tissue engineered products (TEPs). ATMPs are defined as including somatic cell therapy (sCT), TEP, and gene therapy (GT). Article 1 now tells us to look for those definitions within Annex I or in the case of TEP within the ATMP regulation.

Based on the core definitions, we can conclude that cell-based human treatments can fall within both the definition and scope of the directive. As previously mentioned, it should be possible to argue that a transplant is not manufactured and can thus be excluded by Article 2.1 of the scope. However, in case of doubt, we are cross-referenced to the definitions of sCT and TEP (see Sections 19.2.2.2.3 and 19.2.2.2.2). It is also worthy of note that putting the individual definitions of the subcategories of ATMP in Annex I along with the dossier requirements allows for future changes to be made more easily to these definitions to keep pace with scientific advancements (European Council 1999). To understand where the boundary between manipulated and minimally manipulated lies, it is necessary to turn to Annex I of the medicines directive where sCT and TEP are defined.

19.2.2.2 Advanced Therapy Medicinal Product (ATMP)

The recent ATMP regulation is probably the most significant legislation to impact regenerative medicine in Europe since the term was coined. Much of industry perceives this as simply filling a regulatory gap by providing a legal classification for a TEP; this is a misnomer. Leaving aside the interpretation and enactment of the medicines directive by individual member states, since the rewriting of Annex I in 2003 (European Commission 2003), we have had a definition for ATMP. The annex provided a definition for somatic cell therapy (sCT), and included the concept of manipulation (although the phrase used was *substantially altered*) to decide whether a cell was or was not a medicine.

For those of you familiar with this wording, you will agree that it was far from elegant and the recent revisions achieved a far clearer definition (European Commission 2009). The earlier definition for sCT easily encompassed TEPs in most forms. It is true the definition did not specifically allow for non-viable cells, but to constitute sCT a product must first be medicinal and thus must exert a *pharmacological, immunological, or metabolic action*. Non-viable cells having their effect by one of these mechanisms

A CATalyst for Change

would still be medicinal but would not be sCT. Any difference introduced by the ATMP regulation was marginal, although admittedly the wording is far clearer.

The problem of national transposition, before the recent ATMP regulation, may in part be because ATMPs as a group were not required to follow the centralized route (except where they qualified for other reasons); again the ATMP regulation fixes this. The other common misunderstanding is that non-viable human cells (not exerting a *pharmacological, immunological, or metabolic action*) are not medicines and since they cannot be medical devices (they are specifically excluded) are thus unregulated. This is what the European Tissues and Cells Directive (EUTCD) (European Parliament and Council 2004a) is for (see Section 19.2.4), although it does not enforce standardization and conformity of the end product. This may be unnerving for industry sectors more used to comprehensive regulation, but the point goes back to *industrial manufacture*: non-viable human cells are naturally produced in vivo, not manufactured. Consequently there is no gap and the level of regulation is commensurate with the risk.

As a group, all ATMPs are also biological medicinal products, although this was not specifically stated in the earlier directive (European Commission 2003) in which they were described as medicinal products "based on manufacturing processes focused on various gene transfer-produced biomolecules, and/or biologically advanced therapeutic modified cells as active substances or part of active substances." It may be helpful during the following discussion to refer to Table 19.1 for examples of how different product categories are classified.

19.2.2.2.1 *Gene Therapy (GT) Medicinal Product*

"Gene therapy medicinal product means a biological medicinal product which has the following characteristics: (a) it contains an active substance which contains or consists of a recombinant nucleic acid used in or administered to human beings with a view to regulating, repairing, replacing, adding or deleting a genetic sequence; (b) its therapeutic, prophylactic or diagnostic effect relates directly to the recombinant nucleic acid sequence it contains, or to the product of genetic expression of this sequence. Gene therapy medicinal products shall not include vaccines against infectious diseases."

Curiously, GT is now defined as a biological medicinal product which, prior to the above amendment it was not. Whether this provides any loophole for chemically synthesized GT is doubtful; it is almost certain that the same European experts would be called upon to assess such products. Since gene therapy is not core to this chapter it will not be discussed further.

19.2.2.2.2 *Tissue Engineered Product (TEP)*

Article 2(b) (European Parliament and Council 2007) states that

> "'Tissue engineered product' means a product that: contains or consists of engineered cells or tissues, and is presented as having properties for, or is used in or administered to human beings with a view to regenerating, repairing or replacing a human tissue. A tissue engineered product may contain cells or tissues of human or animal origin, or both. The cells or tissues may be viable or non-viable. It may also contain additional substances, such as cellular products, bio-molecules, biomaterials, chemical substances, scaffolds or matrices. Products containing or consisting exclusively of

TABLE 19.1
Product Classifications

Example	Devices Definition	Devices Scope	Medicines Definition	Medicines Scope	EUTCD Definition	EUTCD Scope	Blood Definition	Blood Scope	Conclusion
Heart transplant	Yes[a]	No	Yes	No	Yes	Yes	No	No	Solid organ transplantation, subject to EUTCD only
Lymphocytes (minimally manipulated)	No	No	Yes	No	Yes	Yes	Yes	Yes[b]	Cell treatment, subject to EUTCD only
Blood stem cells (manipulated, e.g., expanded)	No	No	Yes	Yes	Yes	Yes[c]	Yes	No	Medicinal product, also subject to EUTCD
Blood stem cells (not manipulated)	No	No	No	No	Yes	Yes	No	No	Cell treatment, also subject to EUTCD only
Whole blood	No	No	Yes	No	Yes	No	Yes	Yes	Blood transfusion subject to blood directive only
Blood plasma	No	No	Yes	No	No	No	Yes	Yes	Transfusion product, subject to blood directive only
Blood plasma component, e.g., FVIII or fibrin	No	No	Yes	Yes	No	No	Yes	Yes	Medicinal product, also subject in part to blood directive
Human tissue-derived scaffold (physical action only)	Yes	No	No	No	Yes	Yes	No	No	Human-derived scaffold, subject to EUTCD but excluded from medical devices
Animal tissue-derived scaffold (physical action only)	Yes	Yes	No	No	No	No	No	No	Medical device
Lipoaspirate breast reconstruction (same procedure)	Yes[a]	No	Yes	No	Yes	No	No	No	Unregulated surgical procedure
Adipose-derived stem cell breast reconstruction (minimally manipulated)	No	No	Yes	No	Yes	Yes	No	No	Cell treatment, subject to EUTCD only

A CATalyst for Change

| Adipose-derived stem cell breast reconstruction (manipulated, e.g., expansion) | No | No | Yes | Yes | Yes | Yes[c] | No | No | Medicinal product (TEP), also subject in part to EUTCD |

[a] Slightly dubious interpretation, but some may agree.
[b] In practice, excluded following Commission decision.
[c] Donation, procurement, and testing only.

non-viable human or animal cells and/or tissues, which do not contain any viable cells or tissues and which do not act principally by a pharmacological, immunological or metabolic action, shall be excluded from this definition."

Note that this definition is based on mechanism of action, not the nature of the product. A TEP restores, corrects, or modifies a physiological function and does this through a *pharmacological, immunological or, metabolic action* (general definition of a medicine) in order to *regenerate, repair, or replace a human tissue*. If it was not for the fact that a TEP can consist completely of non-viable cells, it would fall into a subcategory of sCT.

Article 2(c) states that

"Cells or tissues shall be considered 'engineered' if they fulfil at least one of the following conditions: the cells or tissues have been subject to substantial manipulation, so that biological characteristics, physiological functions or structural properties relevant for the intended regeneration, repair or replacement are achieved. The manipulations listed in Annex I, in particular, shall not be considered as substantial manipulations; the cells or tissues are not intended to be used for the same essential function or functions in the recipient as in the donor."

It should be pointed out that to be *engineered*, cells must meet one or both of those criteria. Even among European regulators the term *homologous use* is commonly used; this is actually a term of the U.S. Food & Drug Administration (FDA). In Europe, this concept is defined as *same essential function*. This is an important point because the probable intention was to refer to function, whereas *homologous* more accurately refers to form than to function; apparently this has led to heated debate at the FDA.

19.2.2.2.3 Somatic Cell Therapy (sCT) Medicinal Product

"Somatic cell therapy medicinal product means a biological medicinal product which has the following characteristics: (a) contains or consists of cells or tissues that have been subject to substantial manipulation so that biological characteristics, physiological functions or structural properties relevant for the intended clinical use have been altered, or of cells or tissues that are not intended to be used for the same essential function(s) in the recipient and the donor; (b) is presented as having properties for, or is used in or administered to human beings with a view to treating, preventing or diagnosing a disease through the pharmacological, immunological or metabolic action of its cells or tissues. For the purposes of point (a), the manipulations listed in Annex I to Regulation (EC) No 1394/2007, in particular, shall not be considered as substantial manipulations."

It is of note that the wording of the first paragraph is remarkably similar to that for an *engineered* cell. Indeed, it may as well have described sCT as composed of *engineered* cells and saved space. This definition also borrows from the helpful annex (Section 19.2.2.2.4), providing a list of items not considered substantial manipulation.

19.2.2.2.4 Substantial Manipulation

The following are listed in Annex I of the ATMP regulation (European Parliament and Council 2007) as examples of manipulations not considered substantial: cutting, grinding, shaping, centrifugation, soaking in antibiotic or antimicrobial solutions, sterilization, irradiation, cell separation, concentration or purification, filtering, lyophilization, freezing, cryopreservation, vitrification. Naturally this is not exhaustive and although the examples give some indication of what constitutes substantial manipulation, it is not defined.

19.2.2.2.5 Hierarchy

Within the ATMP regulation we are provided with a classification hierarchy. To sum up, where a product meets more than one definition, it is classified in the order GT > TEP > sCT. For example an immunotherapy that involved transfection of the cells with a gene would be a called a GT even though it is also a sCT. There is also clarification with respect to products that may contain both allogeneic and autologous cells; these are considered allogeneic products.

The medicines directive could be described as the prime directive, at least in part because of Article 2(2) which states: "In cases of doubt, where, taking into account all its characteristics, a product may fall within the definition of a "medicinal product" and within the definition of a product covered by other Community legislation the provisions of this Directive shall apply." This is the "get out of jail free card" for authorities in difficult borderline classifications; anyone having this cited to them should make sure it is not used simply because the authority does not understand the product and has decided in doubt to apply Article 2(2).

Another important clarification (or impact) is provided in Article 2.2 of the ATMP regulation (European Parliament and Council 2007) that states: "Where a product contains viable cells or tissues, the pharmacological, immunological or metabolic action of those cells or tissues shall be considered as the principal mode of action of the product." This will presumably overrule any arguments that might be made that the cells are ancillary to the medical device component.

19.2.2.3 Biological Medicinal Products

There are notable omissions among the various core definitions, particularly in defining what a biological medicinal product is. This is important since there are fundamental differences in the regulatory approaches to chemical and biological medicines. Despite that, this vital definition did not appear until 2003 (European Commission 2003) within Annex I. Part I: 3.2.1.1 (General Information on Active Substances) gives this definition:

> "A biological medicinal product is a product, the active substance of which is a biological substance. A biological substance is a substance that is produced by or extracted from a biological source and for which a combination of physico-chemical-biological testing and the production process and its control."

According to this definition, the substance should be of biological origin and, due to its complexity, a combination of physico-chemical and biological testing together with testing and control of the production process is required to determine its quality. The legislation makes it explicit that as a result, the following shall be considered as biological medicinal products: recombinant proteins, monoclonal antibodies, blood products, allergens, and advanced technology products such as gene and cell therapy products. It goes on to say:

"The following shall be considered as biological medicinal products: immunological medicinal products and medicinal products derived from human blood and human plasma as defined, respectively in paragraphs (4) and (10) of Article 1; medicinal products falling within the scope of Part A of the Annex to Regulation (EEC) No 2309/93; advanced therapy medicinal products as defined in Part IV of this Annex."

Note that Regulation (EEC) No 2309/93 (European Council 1993c) was repealed by Regulation 726/2004 (European Parliament and Council 2004c); in such cases we have to look to the successor legislation. These products are therefore:

"Medicinal products developed by means of one of the following biotechnological processes: recombinant DNA technology; controlled expression of genes coding for biologically active proteins in prokaryotes and eukaryotes including transformed mammalian cells; hybridoma and monoclonal antibody methods."

19.2.3 MEDICAL DEVICES

Prior to what was termed the New Approach, each European member state had its own standards and systems to ensure conformity to them (e.g., British Kite Mark, German DIN). The New Approach can be said to have started in 1983 with Directive 83/189/EEC (European Council 1983) that set down a procedure for the provision of information in the field of technical standards and regulations. This directive was necessary to start the process of harmonization of technical standards and requirements within Europe to allow for the free movement of goods.

The first directive to employ the New Approach that might be applicable to regenerative medicine was the active implantable medical devices directive of 1990 (European Council 1990), followed in 1993 by a new medical devices directive (European Council 1993a). This new directive, along with an amendment to the active implantable medical devices directive in the same year, introduced CE marking to all medical devices with effect after 1995. Technical harmonization was also a key commitment (Articles 28 and 30 of the Treaty on the EU) (European Union 1992) toward a single market and led to the replacement of the 1983 directive by the current directive on technical harmonization (European Parliament and Council 1998).

The principle of the New Approach is to lay out the "essential requirements" within the directives and, where possible, develop appropriate standards. A product that conforms with all the appropriate standards can be assumed to have met the essential requirements. Where standards are not available, or for some reason not appropriate, the manufacturer must demonstrate by other means that the essential requirements are met. The applicable standards are published in the *Official Journal of the European Communities*.

The New Approach was not applied to foodstuffs, chemical products, pharmaceutical products, and some others for which community legislation prior to 1985 was well advanced or where the hazards related to such products could not be detailed. It is interesting to note that under the New Approach it is not generally acceptable for member states to maintain or introduce more stringent measures than foreseen in the relevant directive. This point is raised by way of contrast to the European tissues and cells directives (and indeed the blood directive) which lay down "minimum standards"

A CATalyst for Change

and allow all member states to add additional requirements as they see fit. This at first glance seems counter to the treaty since this could form barriers to free trade. The author can only assume that this is allowable under public health grounds since the treaty also allows member states to impose barriers on grounds of health risk.

19.2.3.1 Definition

Directive 93/42/EEC (European Council 1993a) gives us the following definition for a medical device in Article 1(2a):

> "'Medical device' means any instrument, apparatus, appliance, material or other article, whether used alone or in combination, including the software necessary for its proper application intended by the manufacturer to be used for human beings for the purpose of: diagnosis, prevention, monitoring, treatment or alleviation of disease; diagnosis, monitoring, treatment, alleviation of or compensation for an injury or handicap; investigation, replacement or modification of the anatomy or of a physiological process; control of conception, and which does not achieve its principal intended action in or on the human body by pharmacological, immunological or metabolic means, but which may be assisted in its function by such means."

It is reassuring to see that it excludes medicines, although the qualification of "principal intended action" allows medicinal products to be ancillary, e.g., anticoagulant-coated blood tubing or arterial stents with incorporated vasoactive compounds. The directive also defines an "in vitro diagnostic medical device" in Article 1(2c) as:

> "any medical device which is a reagent, reagent product, calibrator, control material, kit, instrument, apparatus, equipment or system, whether used alone or in combination, intended by the manufacturer to be used in vitro for the examination of specimens, including blood and tissue donations, derived from the human body, solely or principally for the purpose of providing information: concerning a physiological or pathological state, or concerning a congenital abnormality, or to determine the safety and compatibility with potential recipients, or to monitor therapeutic measures."

Under this article,

> "specimen receptacles are considered to be in vitro diagnostic medical devices. 'Specimen receptacles' are those devices, whether vacuum-type or not, specifically intended by their manufacturers for the primary containment and preservation of specimens derived from the human body for the purpose of in vitro diagnostic examination."

Conversely,

> "Products for general laboratory use are not in vitro diagnostic medical devices unless such products, in view of their characteristics, are specifically intended by their manufacturer to be used for in vitro diagnostic examination."

19.2.3.2 Medical Device Classes

There are three classes of medical devices and type II devices are further subdivided. The rules on classification are protracted, so for the sake of brevity I have attempted to broadly summarize the key criteria for each class (Table 19.2).

TABLE 19.2
Criteria for Medical Device Classifications

Class	Invasive	Surgically Invasive	Implantable	Biological Effect	Transient (<30 min)	Short-term (<30 days)	Long-term (>30 days)
I	×(√)	×	×	×	(√)		
IIa	√(×)	√	×	×	√	√	×
IIb	√(×)	√	√	×	√	√	√
III	√	√	√	√	√	√	√

√(×) specific criteria may alter general rule.

19.2.3.3 Combination Products

It should be noted that where a device intended to deliver a medicinal product is clearly separate from the medicine, it is subject to the devices directive.

> "If, however, such a device is placed on the market in such a way that the device and the medicinal product form a single integral product which is intended exclusively for use in the given combination and which is not reusable, that single product shall be governed by Directive 65/65/EEC [read Directive 2001/83/EC]. The relevant essential requirements of Annex I to the present Directive shall apply as far as safety and performance related device features are concerned."

This language again emphasizes that the medicines directive is the prime directive. In a clear-cut case, the device component would be totally inert and simply there to "deliver" the medicine and serve no other purpose, for example, an Epipen.

A more complex example might be beta-tricalcium phosphate (TCP) particles soaked in platelet-derived growth factor (PDGF; INN, becaplermin) and placed in a bone void during dental surgery. Many will say that the use of TCP is well established for this procedure and known to be effective; furthermore, the PDGF will only enhance this efficacy, such that the TCP should be considered the principal mode of action. However, the relevant authorities will not always rule in your favor. This product is available in the U.S. on this basis, but in Europe the commission decided PDGF had the principal mode of action and should be regulated as a medicine. The manufacturer recently discovered that the same data heralded by the dental community for being the biggest trial of its kind did not impress the Committee for Human Medicinal Products (CHMP) that gave the application a negative opinion (CHMP 2009a).

19.2.4 Tissues and Cells

Regulation of tissues and cells and blood has already been discussed so there is little more to be said here about the interface. Broadly speaking, the purpose of the directive (European Parliament and Council 2004a) and its technical annexes (European Commission 2006a; European Commission 2006b) is to assure the safety of tissues and cells removed from a human, with the intent to return them to the same or a different human at another time (different procedure). It excludes tissues or cells that are immediately returned to the donor within the same procedure, e.g., plastic surgery. It covers both safety testing and the principles of ethical donation (informed consent) and also ensures traceability and surveillance of safety (reporting of serious adverse events and reactions) and also covers systems to accredit facilities and assure only appropriately trained staff are involved.

As with other directives, the member states had to transpose these directives into national law; as a result, Europe can understandably appear overly complicated on first inspection. In this respect, it is fortunate that the EUTCD does not require licensing of processes; it requires licensing of facilities. Many specialist hospital centers will have more than one purpose for their facilities to be accredited, such that procurement of a new cell or tissue on behalf of a manufacturer of an autologous cell therapy is unlikely to require more than a minor amendment. Where a manufacturer also requires a tissue establishment license, some additional burden is involved. If that establishment also requires a manufacturing license for medicines, it is likely that most of the requirements are already met. The manufacturer has the concern that where a large number of tissue establishments are required the burden in time and cost is likely to fall heaviest on them, with commensurate impact on the cost of the final treatment. It seems probable that this cost will be greatest for autologous cell therapies, a product class that already carries the cost of being tailor-made (and probably hand-made) for each patient.

These directives assure that the cells pose no infectious risk to the recipient and that they were obtained ethically. What they do not do is assure that the cell use will be efficacious nor do they confirm that the cells will not harm the patient in some way, e.g., form a carcinoma or secrete a hormone or growth factor that leads to unwanted effects. However, the reporting requirements should at least identify such problems to allow the competent authority to intervene.

The EUTCD does not assure that the tissues or cells are of a suitable quality to use in the manufacture of a medicinal product beyond data provided through the obligatory testing. For the assessors of cell-based medicinal products the EUTCD provides few shortcuts for assessing—the result is somewhat less useful than a pharmacopeial monograph since it provides only part of the required information. This is a challenge for a manufacturer because autologous products are likely to require "tissue establishment" (the term in the directive) licenses for every EU/EEA member state within which the manufacturer will collect patient cells or tissues in addition to licenses for each manufacturing site that receives them. This is in addition to manufacturing licenses from all member states where manufacturing takes place and a centralized marketing authorization (EMA).

Tissues and cells that are only minimally manipulated benefit from some regulation, some assurance to the public that if their use results in unexpected complications the authority can step in. The absence of a requirement to demonstrate any kind of efficacy may mean that such products can be put into use and potentially generate revenue very quickly, but the lack of evidence will probably mean the manufacturer cannot obtain reimbursement from healthcare providers (see Chapter 21 by Meurgey and Wille and Chapter 22 by Faulkner). Probably for this reason, but no doubt also from a sense of responsibility, many such products are undergoing clinical trials.

These trials do not need authorizations because the tissues and cells are not medicinal products and therefore not subject to the clinical trials directive (European Parliament and Council 2001a). Whether this is a good idea or a bad idea is subjective and the consequential risk is that commercial pressures will influence the quality of these trials.

19.2.5 BLOOD

The blood directive may impact the regulation of some regenerative medicine products, although not cell-based products since nucleated cells of blood fall under the EUTCD. The blood directive covers red blood cells, plasma, and blood derivatives (European Parliament and Council 2003). Since serum and its components are widely used in medicines (e.g., human serum albumin, fibrin, and clotting factors), they are worth mentioning here. The blood directive is completely analogous to the EUTCD; in fact this statement should go the other way around because this directive came earlier.

The purpose of the blood directive, like the EUTCD, is to set minimum standards for the collection, testing, processing, storage, and distribution of human blood and blood components. As with the EUTCD, compliance with the directive does not assure that blood or a blood component is suitable to use as a starting material for the manufacture of a medicinal product. It assures only that the blood or component has been collected, tested, and handled in a way to pose minimum risk to a human recipient. "Risk" primarily means infection. Like the EUTCD, the blood directive specifies ethical standards for donation and requirements for inspection and accreditation of facilities. Each blood establishment should employ a responsible person who takes responsibility for releasing the blood product.

Having been in force a little longer than the EUTCD, the blood directive is well established now, so there are many lessons to be learned from this experience that may be useful in resolving some of the issues encountered with enactment of the EUTCD. Indeed the current thinking is that, as with plasma, the testing requirements chosen for a centrally authorized cell-based medicinal product should encompass the most stringent criteria imposed by all member states.

There is a formal procedure to have a plasma master file (PMF) authorized centrally, thus allowing member states to agree on the final pooling and testing procedure and opening the way for that plasma to be used in centrally authorized medicinal products. One could speculate on whether there is value in having a cell master file system to mirror the procedure for blood. One area where this could be a major advantage would be for feeder cells that will be used by many companies. The ability to have feeder cells certified in some way would then allow companies to buy such cells "off the peg," with all the testing already done. This is, of course, possible without a master file system, but each company would have to maintain a copy of the entire master file and supply it along with its marketing authorization application (MAA).

There is no doubt assessors would be very happy to know they must assess such data only once. Simply checking the master file is valid thereafter. Companies would benefit from the assurance that some objection to the cells per se will not be raised during assessment of their product, although the master file covers only the inherent quality of the source, not the intended use. However, to the author's knowledge, there are no plans for such a system.

19.2.6 Cosmetics

Some regenerative medicine products are being developed not to treat human disease, but to reverse aging or improve the look of the human body. Various terms have been coined for such products including cosmeceuticals and aesthetic medicine, both arising from the view that the products are basically cosmetic. However, in a legal sense, it is unlikely that regenerative medicine products meet the definition of a cosmetic product (European Council 2009) (Article 1.1):

> "Any substance or mixture intended to be placed in contact with the various external parts of the human body (epidermis, hair system, nails, lips and external genital organs) or with the teeth and the mucous membranes of the oral cavity with a view exclusively or mainly to cleaning them, perfuming them, changing their appearance and/or correcting body odours and/or protecting them or keeping them in good condition."

Article 1.2 refers to Annex I of this directive that provides a list of products considered to be cosmetics. Article 4.1 lists exclusions and includes "substances" listed in Annex II that are prohibited. From this list of exclusions, the most relevant to this discussion is Item 416 (of 1,371): "cells, tissues or products of human origin."

However regenerative medicine is defined, its name suggests medicinal products and regenerative products are likely to satisfy the legal definition of a medicine and, if not excluded from the cosmetics directive, will at least be subject to the medicines directive (Article 2.2) (European Parliament and Council 2001b). Unlike the other directives discussed in this chapter, this directive has not been repealed by a newer directive. That does not mean it has not changed. The consolidated version to May 2009 is subject to 62 amending directives, 2 amending acts, and 11 corrigenda.

19.3 LICENSING MEDICINES IN EU/EEA

19.3.1 Medicines

The legal basis for market authorization (MA) within the EU is Article 8 of the medicines directive (European Parliament and Council 2001b) defining the requirements to be met for a full application. There are also provisions in Article 10 for abridged applications, e.g., generics (Article 10.1) or the more recent *similar biological medicinal product* (or biosimilar, Article 10.4). It is of note that an MA is granted *without prejudice to the law relating to the protection of industrial and commercial property* (Article 8.1). This means that the competent authority will not consider the intellectual property situation; this is in stark contrast to the U.S. Where such applications rely on data from another licensed product (e.g., those under Article 10), they will heed the data protection period, which is now harmonized to 10 years, applications being allowed after 8 years but not granted before the 10 years.

Although, as previously acknowledged, some regenerative medicine products may be small molecules or recombinant proteins, this chapter will focus on cell-based products for which abridged applications are not possible; further discussion of abridged applications is beyond the scope of this chapter.

Article 8 provides a high level overview of the nature of the evidence required for any medicinal product to be licensed. For specific product categories, it must be read in conjunction with the relevant section of Annex I. Where a product falls within more than one category as defined in Annex I, it must meet the requirements of both. This chapter deals with the legal bases and procedures only. A discussion of the data requirements would fill a chapter.

19.3.1.1 National Procedures

In general, medicinal products can be authorized within the EU at a national level or through the centralized route (see below) administered by the European Medicines Agency (EMA). There are also two in-between procedures which are very similar, the mutual recognition procedure (MRP) and the decentralized procedure (DCP), that allow an application to be assessed by one member state (the reference member state, RMS) and then mutually recognized by the others (concerned member states, CMS) included in the procedure.

Naturally the CMS have the right to raise concerns during the procedure, but the intention is that they not perform independent assessments. Since these procedures essentially result in national licenses, there is no centralized committee like the CHMP that can reach a consensus. Disputes arising among the member states in the procedure are eventually resolved through the Coordination Group for Mutual Recognition and Decentralised procedures (CMDh). The lower case *h* indicates human to distinguish the group from its veterinary equivalent. However, for cell-based medicinal products, these procedures are now disallowed by Regulation (EC) No 726/2004 which requires them to be authorized through the centralized procedure (European Parliament and Council 2004c).

19.3.1.2 Centralized Procedure

Those products mandated by Regulation (EC) No. 726/2004 must use the centralized procedure, administered by the EMA. The final scientific opinion is the responsibility of the CHMP (European Parliament and Council 2004b), the successor to the Committee for Proprietary Medicinal Products (CPMP) set up in 1977, enacted by Directive 75/318/EEC (European Council 1975a). Once granted by the European Commission, a centralized (or Community) marketing authorization is valid in all European Union (EU) and EEA/EFTA states (Iceland, Liechtenstein, and Norway), a combined population currently around 500 million people.

All licensing procedures run on a 210 day timetable. However, due to clock stops during which an applicant can prepare responses to questions, the overall time is more typically 12 to 24 months before a final opinion is reached by the CHMP. After the opinion is issued, the European Commission usually takes around 3 months to endorse the license. Only one ATMP (for ChondroCelect) has been authorized at the time of going to press (CHMP 2009b), so it is not possible to comment on whether this ATMP will be typical. ChondroCelect was submitted on June 1, 2007 and the final opinion was issued on June 25, 2009; the European Commission granted the license on October 5, 2009, a combined time of 2 years and 4 months. Although the Committee for Advanced Therapies (CAT) did not exist at the start of this procedure, its existence will not alter the time required for scientific assessment.

Centralized applications have been submitted to the EMA for four ATMPs (Table 19.3). The first was for Apligraf in April 2001, but the application by Novartis

TABLE 19.3
Centralized Applications for ATMPs

Product	INN	Company	Submission Date	Opinion	Finalization Date
Apligraf	None assigned	Novartis	April 2001	Unknown	Withdrawn 2002
Cerepro	Adenovirus-mediated herpes simplex virus–thymidine kinase gene	Ark Therapeutics	October 2005	Negative	Withdrawn July 2007
ChondroCelect	None assigned	TiGenix	June 2007	Positive	June 2009
Advexin	Contusugene ladenovec	Gendux Molecular Limited	September 2007	Negative	Withdrawn December 2008
Contusugene Ladenovec Gendux	Contusugene ladenovec	Gendux Molecular Limited	July 2008	Negative	Withdrawn June 2009
Cerepro	Adenovirus-mediated herpes simplex virus–thymidine kinase gene	Ark Therapeutics	January 2009	Negative	December 2009

was withdrawn for business reasons before the end of the procedure. This product predates European Public Assessment Reports (EPARs) so it is not known what concerns the CHMP may have had.

The next application was not until 2005 when Ark Therapeutics submitted Cerepro (INN; adenovirus-mediated herpes simplex virus–thymidine kinase gene) (CHMP 2009c). Cerepro is a gene therapy product for high-grade glioma; it received a negative opinion issued by the CHMP on April 26, 2007. The company requested a re-examination but then withdrew its application on July 16, 2007, before the CHMP reached a second opinion. The withdrawal letter acknowledged that the company needed to complete a larger ongoing study to address the outstanding issues. Cerepro was resubmitted on January 7, 2009, and received a second negative opinion from the CAT and CHMP in December 2009.

Another gene therapy, contusugene ladenovec (its INN), has been submitted twice by Gendux Molecular Limited. The first submission was as Advexin on September 6, 2007 (CHMP 2009d) for the treatment of Li-Fraumeni syndrome, but was withdrawn December 17, 2008 (day 179 of the procedure). On July 2, 2008 the same active substance was submitted as contusugene ladenovec Gendux, this time for squamous cell carcinoma of the head and neck, but again withdrawn on June 12, 2009. Both withdrawals implicated financial limitations (CHMP 2009e). Interestingly, the main concerns of the CHMP were broadly (based on nontechnical summaries) the same for both applications: lack of evidence for efficacy, and potential harmful effects to the environment and close contacts of the patient (Table 19.3).

19.3.1.3 Certification Procedure for ATMP

The ATMP regulation (European Parliament and Council, 2007) also introduced a new procedure for SMEs whereby quality and nonclinical data can be evaluated and certified by the CAT, thus providing a kind of regulatory due diligence service. To understand the thinking of the Commission one should turn to the recitals of the regulation:

> "Studies necessary to demonstrate the quality and nonclinical safety of advanced therapy medicinal products are often carried out by small and medium-sized enterprises. As an incentive to conduct those studies, a system of evaluation and certification of the resulting data by the Agency, independently of any marketing authorisation application, should be introduced. Even though the certification would not be legally binding, this system should also aim at facilitating the evaluation of any future application for clinical trials and marketing authorisation application based on the same data."

During 2009, the Commission published the implementing legislation (European Commission 2009) and the EMA published draft procedural (CAT 2009) and scientific guidance (CAT 2009). The implementing regulation states that the application for certification should be "conducted in accordance with the same scientific and technical requirements as those applicable to a market authorisation application." Since this procedure is available at a relatively early stage, the implementing regulation specifies the minimum set of data required to undertake certification; and this is expanded upon in the (draft) scientific guideline issues by the CAT. As an aside, this guideline may also be of use to any ATMP manufacturer that is unsure where in the CTD dossier certain data should be included since it refers to the content of each main subsection.

Another potentially useful aspect of the certification procedure is the possibility of a site visit that may provide a valuable opportunity to discuss specific manufacturing issues with EU experts. Such a possibility may be of particular interest to those introducing novel manufacturing processes, for instance, automation for autologous cell products.

At the time of writing this chapter, only one application has been submitted to the CAT, so little more can be said about the procedure here. The procedure requires 90 days and is available for all categories of ATMPs, including combined ATMPs (including devices). As with many other regulatory processes there is the possibility of a clock stop, in this case at day 60, to allow an applicant to prepare responses to a day 60 request for supplementary information (RSI) and to allow time for a site visit and oral explanation (if required). Whether SMEs will take up this procedure will depend on whether it proves useful for its presumed intent, which is to reassure investors and business partners to date a product meets current regulatory expectations.

19.3.2 Exemptions from Marketing Authorization

The medicines directive provides two routes by which unlicensed medicinal products may be supplied to patients, one of which is specific for ATMPs (European Parliament and Council 2001b).

19.3.2.1 Article 5

Article 5 of the medicines directive (European Parliament and Council 2001b) allows Member States to allow some products to be excluded from the requirements of the directive. But only where the medicine is "supplied in response to a bona fide unsolicited order, formulated in accordance with the specifications of an authorised health-care professional and for use by an individual under his direct personal responsibility." It also allows member states to "temporarily authorise the distribution of an unauthorised medicinal product" in response to some kind of national emergency (e.g., bioterrorism or pandemic).

Article 5 could be used on a named patient basis to provide ATMPs, but since there is another provision, the so-called hospital exemption that imposes additional obligations above and beyond those of Article 5, it is probably incumbent upon member states not to allow this article to be used for ATMP, or, if it is allowed, to apply the pharmacovigilance and traceability requirements of the hospital exemption. However, it should be noted that the hospital exemption can be used only within the same member state, so any product imported, whether from elsewhere in the EU or a third country, could not use this provision. It is a little difficult to understand why the new hospital exemption was needed at all.

19.3.2.2 Hospital Exemption

Regulation (EC) No 1394/2007 (European Parliament and Council 2007) provides a new provision that has come to be known as the hospital exemption, a point that was hotly debated within the community while drafting this important new regulation. The crux of the argument was that many "hospitals" (apparently even "hospital" was hard to define at the EU level since the legal basis of a hospital varies throughout the EU) already prepare ATMPs for special situations, e.g., skin equivalents for severe burns and corneal epithelium for corneal damage.

Member states did not want these innovative novel treatments to be subject to marketing authorization, since in most cases they were prepared infrequently and with specific patients in mind. It is not clear why Article 5 was not sufficient, but it is important to note that the traceability and pharmacovigilance requirements for ATMPs would not necessarily be enforced by it (based on how the member state applies this provision). The resulting wording says any ATMP

> "which is prepared on a non-routine basis according to specific quality standards, and used within the same Member State in a hospital under the exclusive professional responsibility of a medical practitioner, in order to comply with an individual medical prescription for a custom-made product for an individual patient."

Unlike Article 5 requirements, the manufacturing site must be authorized by the competent authority; "Manufacturing of these products shall be authorised by the competent authority of the Member State." Furthermore:

> "Member States shall ensure that national traceability and pharmacovigilance requirements as well as the specific quality standards referred to in this paragraph are equivalent to those provided for at Community level in respect of advanced therapy medicinal products for which authorisation is required pursuant to Regulation (EC)

No 726/2004 of the European Parliament and of the Council of 31 March 2004 laying down Community procedures for the authorisation and supervision of medicinal products for human and veterinary use and establishing a European Medicines Agency."

Since the hospital exemption is a national exemption, it must be requested in each member state where the product is to be manufactured. Also, it applies only to products made within the same member state, so any company wishing to use this provision would need a manufacturing site in each country where it wished to provide the ATMP. The hospital exemption does not allow for movement between member states and does not allow the company to publicize the product. In effect, the product is exempt from market authorization because it is not on the market.

Although this provision may look attractive to companies, especially those developing orphan medicinal products, it is not intended to replace a marketing authorization because it is only intended to allow *hospitals* to provide rare *custom-made* treatments. Understandably, some companies have shown considerable interest in this route because they see it as a shortcut to the clinic, dodging the requirement for a clinical trials authorization (CTA), although a manufacturing authorization is still required. It must be stressed that a clinical trial cannot be undertaken under hospital exemption since that requires a CTA and consequently the product will be an *investigational medicinal product* (IMP). In principle it should be possible to charge for such treatments, subject to any national laws that prohibit it, although the product cannot be advertised or openly marketed.

Regenerative medicine brings many scientific challenges, and unlike chemical pharmaceuticals, successful efficacy of ATMPs may rely as much on how they are administered (e.g., surgical technique) as their composition. The hospital exemption may allow some companies to get this insight without the cost and effort of formal clinical trials; whether the hospital exemption will be used in this way is yet to be seen. If so, the safety of patients will rely more heavily on ethical committees and other measures in place within member states since the product itself will not be assessed by the competent authority.

19.3.3 REGULATORY GUIDANCE

19.3.3.1 International Convention on Harmonisation

The success of various harmonization activities from the early days of the EEC showed that it was possible to achieve a consensus view among many countries. In parallel with the activities within the EEC, discussions of Europe, Japan, and the U.S. were also under way. At a meeting in Brussels in April 1990, the International Convention on Harmonisation (ICH) was born (ICH 2009). There are now around 70 ICH guidelines covering the main principles of quality, safety, and efficacy (Q, S, and E guidelines) as well as multidisciplinary (M) guidelines. As a result of these guidelines, the fundamental requirements for marketing authorization should now be common to all ICH regions; and since many countries outside ICH follow the same principles or mutually recognizes them, the principles are common to much of the world. However, the "devil is in the detail" and fundamental differences in approaches still exist between regions.

19.3.3.2 EMA/CHMP

The EMA website contains numerous guidelines covering quality, safety, and efficacy, along with product-specific and indication-specific guidelines. Many of these guidelines mention the need for specific products to be dealt with on a case-by-case basis, and for regenerative medicine products this will be doubly true.

The EMA includes a group called the Innovation Task Force (ITF) that serves as an excellent starting point for interactions because it can provide a wealth of procedural advice (EMA 2009a). For early stage interactions, it can be valuable to request a briefing meeting with one or more of the working parties, the obvious one in the context of this chapter being the Cell-based Products Working Party (CPWP). For companies, the main purpose of such meetings is to start the process of educating the assessors about the product or technology, initiate informal discussion on the potential development routes, and begin to understand the regulator concerns. It must be stressed, however, that the views expressed will be the personal views of the experts and not represent the consensus of the CHMP that requires formal scientific advice.

It is both valuable and strategically important to seek assistance from the Scientific Advice Working Party (SAWP) during product development (EMA 2009b), but exactly how often and when will differ for different products. At the very least, it is prudent to get buy-in on pivotal studies that provide the key clinical evidence. It must be stressed, however, that scientific advice should not be sought without considerable thought for the consequences: the advice is endorsed by the CHMP and the applicant is bound by it. Asking the wrong question too early has the potential to cause more harm than good to a development program and not asking the right question early enough can be disastrous.

The downside to centralized scientific advice is that in most cases it follows a 40-day timetable and is undertaken solely through correspondence. With some advice procedures, the timetable is extended to 70 days by the CHMP/SAWP to allow time for an oral hearing limited to one or two aspects identified by the CHMP/SAWP. The kinds of products discussed in this chapter have a higher likelihood of leading to oral hearing due to their relative novelty.

A final comment on EMA scientific advice is that although it is expensive (€75,500 at the time of writing), for ATMP it is reduced 65%, and for small and medium size enterprises (SME) there is a 90% reduction (European Parliament and Council 2007, Article 16.2). The best deal is for orphan drug indications. SMEs get free protocol assistance (Scientific Advice for Orphan Medicinal Products)—a huge benefit for these companies.

19.3.3.3 National Scientific Advice

Not all member states offer scientific advice but the larger agencies do. Although the EMA may appear to be the obvious source of scientific information when the intention is to submit an MAA to the EMA, it can still be useful to request national scientific advice. Because clinical trials are authorized at a national level, it is highly recommended for any novel product manufacturer to enter into early discussion with the national authority to understand its concerns and requirements for a trial. Such meetings can also help identify issues that might be viewed divergently across the EU.

19.3.3.4 Regulatory Advice

Legal and procedural regulatory questions that arise may be addressed by national competent authorities or by the EMA (EMA 2009b). Such advice or response to questions is free.

19.4 REGISTERING MEDICAL DEVICES IN EU/EEA

Medical devices can be marketed after a declaration of conformity is made and verified through a certificate of conformity issued by a notified body—an organization accredited to confirm compliance with the essential requirements of the directives. After certification, such devices must bear the CE mark before they can be sold in the European Community. Member states must accept that devices bearing the CE mark are compliant and not hinder movement or trade. According to the European Commission website, over 2,000 notified bodies have been registered although not all are still active (European Commission 2009b).

A conformity assessment reflects the increasing risks of the various categories and the routes are as follows (MHRA 2008):

Class I—In most cases, Class I medical devices can be self-declared unless they are sterile or have metrology functions, thus usually negating the need to use a notified body. When a manufacturer is satisfied that a product meets all the relevant essential requirements, it must register with the national competent authority, after which the product can bear the CE mark and be placed on the market.

Class IIa—As for Class I, this declaration must be backed up in all cases with a conformity assessment by a notified body. This assessment is achieved in one of three ways: (1) by examination and testing; (2) by audit of the production quality assurance system (ISO 13485:2003, excluding design); or (3) by audit of final inspection and testing (ISO 13485:2003, excluding design and manufacture). After the notified body certifies the product, it can carry the CE mark and be placed on the market.

Class IIb—This can be obtained (1) via the procedure for IIa with the addition of a type examination or (2) through a full quality assurance system audit (ISO 13485:2003).

Class III—This class is similar to IIb but with the additional requirement for a manufacturer to submit a design dossier to the notified body for approval.

Any discussion on requirements is beyond the scope of this chapter.

19.5 CONCLUSION

Regulation of medicinal products poses a challenge to competent authorities to keep up with the advances in science and enact legislation that permits innovation while simultaneously protecting public safety. This is no small challenge, particularly since science seems to move faster than legislators can react. Hopefully this chapter has explained how the EU/EEA attempt to accomplish this, and also shows that for the greater benefit of humankind, harmonization can help to achieve this aim. Harmonization and standardization are possible—the European experiment demonstrates that—and presumably economists can show the financial benefits, but the pundits often forget the greater social benefits when complaining about small battles lost. Would a study reveal that Europeans have access to a larger array of medicinal products than they had earlier? One would hope so.

Does the EU/EEA have a clear regulatory pathway for regenerative medicine products? Yes, and this has been so for several years, although the most recent changes bring welcomed improvements in clarity. Is there a reasonable chance of getting a regenerative medicine product licensed in the EU/EEA? Again, yes. It is true that before 2009 no ATMPs had been centrally licensed, but a large number of recombinant proteins have been; many (e.g., G-CSF) are regenerative medicine products. It is possible to license biosimilar products in the EU/EEA. Like the U.S., the EU has only one centrally authorized cell therapy but unlike the U.S. it did not approve a number of early tissue engineered products as devices. The FDA's thinking has now also moved toward viewing tissue engineering as medicinal product and not a medical device. As a consequence of near-global harmonization of the fundamental principles of medicine regulation, companies may not necessarily target the U.S. before the EU any longer.

ACKNOWLEDGMENT

I would like to thank the editors and Dr. Sharon Longhurst for their helpful comments during the preparation of this chapter.

REFERENCES

Committee for Advanced Therapies. (2009). Press Release: EMEA holds first meeting of Committee for Advanced Therapies (CAT), January 16. http://www.emea.europa.eu/pdfs/human/cat/786609en.pdf

Committee for Human Medicinal Products. (2009a). Gemesis EPAR refusal. http://www.emea.europa.eu/humandocs/Humans/EPAR/gemesis/gemesisR.htm

Committee for Human Medicinal Products. (2009b). ChondroCelect European Public Assessment Report. http://www.emea.europa.eu/humandocs/Humans/EPAR/chondrocelect/chondrocelect.htm

Committee for Human Medicinal Products. (2009c). Cerepro withdrawal EPAR. ttp://www.emea.europa.eu/humandocs/Humans/EPAR/cerepro/cereproW.htm

Committee for Human Medicinal Products. (2009d). Advexin withdrawal EPAR. http://www.emea.europa.eu/humandocs/Humans/EPAR/advexin/advexinW.htm

Committee for Human Medicinal Products. (2009e). Contusugene ladenovec Gendux withdrawal EPAR. http://www.emea.europa.eu/humandocs/umans/EPAR/contusugene-ladenovecgendux/contusugeneladenovecgenduxW.htm

Conference Board. (2009). Total Economy Database. http://www.conference-board.org/economics

European Economic Area. (1994). Agreement on the European Economic Area; Final Act; Joint Declarations; Declarations by Governments of Member States of the Community and EFTA States; Arrangements; Agreed Minutes - Declarations by One or Several of the Contracting Parties of the Agreement on the European Economic Area. *Official Journal of the European Communities*, L1, 3–36.

European Economic Community. (1957). Treaty Establishing the European Economic Community, March 25. http://eurlex.europa.eu/en/treaties/dat/11957E/tif/TRAITES_1957_CEE_1_EN_0001.tif

European Economic Community. (1987). Single European Act 1986. *Official Journal of the European Communities*, L169, 1–28.

European Free Trade Association. (1960). Convention Establishing the European Free Trade Association, January 4. http://www.efta.int/content/legaltexts/eftaconvention/eftaconventiontexts/conventionstockholm

European Medicines Agency. (2009a). Regulatory and procedural guidance. http://www.emea.europa.eu/htms/human/raguidelines/itf.htm

European Medicines Agency. (2009b). Regulatory and Procedural Guidance: Scientific Advice and Protocol Assistance. http://www.emea.europa.eu/htms/human/raguidelines/sa_pa.htm

European Commission. (2003). Commission Directive 2003/63/EC of 25 June 2003 amending Directive 2001/83/EC of the European Parliament and the Council on the Community Code relating to medicinal products for human use (text with EEA relevance). *Official Journal of the European Communities*, L159, 46–94.

European Commission. (2006a). Commission Directive 2006/17/EC of February 8, 2006 implementing Directive 2004/23/EC of the European Parliament and the Council as regards certain technical requirements for the donation, procurement and testing of human tissues and cells (text with EEA relevance). *Official Journal of the European Communities*, L294, 38–40.

European Commission. (2006b). Commission Directive 2006/86/EC of October 24, 2006 implementing Directive 2004/23/EC of the European Parliament and the Council as regards traceability requirements, notification of serious adverse reactions and events and certain technical requirements for the coding, processing, preservation, storage and distribution of human tissues and cells (text with EEA relevance). *Official Journal of the European Communities*, L294, 32–50.

European Commission. (2009a). Commission Directive 2009/120/EC of September 14, 2009 amending Directive 2001/83/EC of the European Parliament and the Council on the Community Code relating to medicinal products for human use as regards advanced therapy medicinal products (text with EEA relevance). *Official Journal of the European Communities*, L242.

European Commission. (1965). List of Notified Bodies. http://ec.europa.eu/enterprise/newapproach/nando/index.cfm?fuseaction=notifiedbody.main

European Council. (January 26, 1965). Council Directive 65/65/EEC 65 on the approximation of provisions laid down by law, regulation or administrative action relating to medicinal products. *Official Journal of the European Communities*, L22, 369–373.

European Council. (1975a). Council Directive 75/318/EEC of May 20, 1975 on the approximation of the laws of Member States relating to analytical, pharmaco-toxicological and clinical standards and protocols in respect of the testing of proprietary medicinal products. *Official Journal of the European Communities*, L147, 1–12.

European Council. (1975b). Second Council Directive of May 20, 1975 on the approximation of provisions laid down by law, regulation or administrative action relating to medicinal products (75/319/EEC). *Official Journal of the European Communities*, L147, 13–22.

European Council. (1983). Council Directive 83/189/EEC laying down a procedure for the provision of information in the field of technical standards and regulations. *Official Journal of the European Communities*, L109, 8–12. March 28.

European Council. (1985). Council Resolution on a new approach to technical harmonization and standards. *Official Journal of the European Communities*, C136, 1–9.

European Council. (1989). Council Directive 89/341/EEC of May 3, 1989, amending Directives 65/65/EEC, 75/318/EEC and 75/319/EEC on the approximation of provisions laid down by law, regulation or administrative action relating to proprietary medicinal products. *Official Journal of the European Communities*, L142, 11–13.

European Council. (1990). Council Directive 90/385/EEC of June 20, 1990, on the approximation of the laws of the Member States relating to active implantable medical devices. *Official Journal of the European Communities*, L189, 17–36.

European Council. (1993a). Council Directive 93/42/EEC of June 14, 1993, concerning medical devices. *Official Journal of the European Communities*, L169, 1–43, June 14.

European Council. (1993b). Council Directive 93/68/EEC of 22 July 1993, amending Directives 87/404/EEC (simple pressure vessels), 88/378/EEC (safety of toys) ... 90/385/EEC (active implantable medicinal devices) .... *Official Journal of the European Communities*, L220, 1–22.

European Council. (1993c). Council Regulation (EEC) No 2309/93 of 22 July 1993, laying down Community procedures for the authorization and supervision of medicinal products for human and veterinary use and establishing a European Agency for the Evaluation of Medicinal Products. *Official Journal of the European Communities*, L214, 1–21.

European Council. (1999). Council Decision of 28 June 1999, laying down the procedures for the exercise of implementing powers conferred on the Commission (1999/468/EC). *Official Journal of the European Communities*, L184, 23–26.

European Council. (2009). Council Directive 76/768/EEC of July 27, 1976, on the approximation of the laws of the Member States relating to cosmetic products. *Official Journal of the European Communities*, L262, 169.

European Parliament and Council. (1998). Directive 98/34/EC of the European Parliament and of the Council of 22 June 1998, laying down a procedure for the provision of information in the field of technical standards and regulations. *Official Journal of the European Communities*, L 204, 37–48.

European Parliament and Council. (2001a). Directive 2001/20/EC of the European Parliament and of the Council of 4 April 2001, on the approximation of the laws, regulations and administrative provisions of the Member States relating to the implementation of good clinical practice in the conduct of clinical trials on medicinal products for human use. *Official Journal of the European Communities*, L121, 24–44.

European Parliament and Council. (2001b). Directive 2001/83/EC of the European Parliament and of the Council of 6 November 2001, on the Community code relating to medicinal products for human use. *Official Journal of the European Communities*, L311, 67–0128.

European Parliament and Council. (2003). Directive 2002/98/EC of the European Parliament and of the Council of 27 January 2003, setting standards of quality and safety for the collection, testing, processing, storage and distribution of human blood and blood components and amending Directive 2001/83/EC. *Official Journal of the European Communities*, L33, 30–40.

European Parliament and Council. (2004a). Directive 2004/23/EC of the European Parliament and of the Council of 31 March 2004, setting standards of quality and safety for the donation, procurement, testing, processing, preservation, storage and distribution of human tissues and cells. *Official Journal of the European Communities*, L102, 48–58.

European Parliament and Council. (2004b). Directive 2004/27/EC of the European Parliament and of the Council of 31 March 2004, amending Directive 2001/83/EC on the Community code relating to medicinal products for human use (text with EEA relevance). *Official Journal of the European Communities*, L136, 34–57.

European Parliament and Council. (2004c). Regulation (EC) No 726/2004 of the European Parliament and of the Council of 31 March 2004, laying down Community procedures for the authorisation and supervision of medicinal products for human and veterinary use and establishing a European Medicines Agency. *Official Journal of the European Communities*, L136, 1–33.

European Parliament and Council. (2007). Regulation (EC) No 1394/2007 of the European Parliament and of the Council of 31 March 2004, on advanced therapy medicinal products and amending Directive 2001/83/EC and Regulation (EC) No 726/2004. *Official Journal of the European Communities*, L324, 121–137.

European Union. (1992). Treaty on European Union. *Official Journal of the European Communities*, C191, 1–112.

Heads of Medicines Agencies. (2007). Fiftieth meeting of the Heads of Medicines Agencies (HMA). Human and veterinary medicines: Highlights and achievements. http://www.hma.eu/uploads/media/hma_highlights_achievements_01.pdf.

International Conference on Harmonisation. (2009). http://www.ich.org/cache/compo/276-254-1.html

MHRA. (2008). Conformity assessment procedures (medical devices regulations). http://www.mhra.gov.uk/home/groups/es-era/documents/publication/con007492.pdf

20 United States Regulatory Reimbursement, Political Environment, and Strategies for Reform

Michael J. Werner

CONTENTS

20.1 Introduction 323
20.2 Background..324
20.3 Federal Funding Policies 326
20.4 Regulatory Issues...328
20.5 Reimbursement Issues 330
20.6 Overall Political Environment for Regenerative Medicine 331
20.7 Conclusion 332
References..332

20.1 INTRODUCTION

There is no doubt that the United States faces tremendous economic and healthcare pressures from the aging of its population and the effects of a range of diseases and disabilities. The numbers are staggering. For example:

- Following the current path, diabetes cases will increase by 52.9% between 2003 and 2023. The cost for treating complications of Type 2 diabetics over a 30-year period is estimated to be $47,240 per patient (Cutler and McClellan 2010).
- As the United States population ages, researchers estimate that the prevalence of Alzheimer's disease will come close to quadrupling over the next 50 years, when 1 person in 45 may be living with the disease (Brookmeyer, Gray, and Kawas 2010).
- Following current trends, heart disease cases will increase by 41% between 2003 and 2023 (Klowden et al. 2007).

Regenerative medicine holds the promise of treating and curing diseases like these and meeting other currently unmet medical needs. In some cases, this promise has already been realized as companies are commercializing regenerative therapies that have successfully treated diseases and disabilities. Other products, while not yet approved for marketing, have shown success in academic laboratories or in clinical trials.

Successes to date represent just the tip of the iceberg. Regenerative medicine technologies represent opportunities to both respond to unmet medical needs and provide better and more cost-effective care to patients. Thus, regenerative medicine could dramatically change the way healthcare is provided in the years ahead. It could also reduce overall health expenditures. Key to reaching these objectives, however, is a favorable regulatory, political, and funding environment. U.S. policies in these areas will play an important role in determining the success of regenerative medicine (see Chapter 2 by Klein and Trounson). This chapter will address the regulatory, reimbursement, and funding challenges confronting regenerative medicine efforts and suggest possible policy options to enable researchers and companies to overcome those and other development issues. While there are many definitions of regenerative medicine, for purposes of this chapter, I define it as:

> A rapidly evolving interdisciplinary field in healthcare that translates fundamental knowledge in biology, chemistry, and physics into materials, devices, systems, and therapeutic strategies which augment, repair, replace or regenerate organs and tissues.

This definition was adopted by the Alliance for Regenerative Medicine, an organization whose mission is to advocate for favorable policies to promote regenerative medicine (www.alliancerm.org).* The use of stem cells from embryos, cord blood, other adult tissues, induced pluripotent stem cells, as well as tissue engineering and related technologies all fall under this definition. Moreover, the term includes uses of stem cells as therapeutics and for drug discovery and development.

20.2 BACKGROUND

Many observers have noted the potential of regenerative medicine to address numerous diseases and disabilities that have plagued patients and their families for decades. Regenerative medicine has been touted at the highest levels of government. The three most recent U.S. Presidents—Clinton, Bush, and Obama—have all described the attributes of stem cells and their potential to ease human suffering. Their administrations all funded this work.†

Dr. Harold Varmus, the former Director of the U.S. National Institutes of Health (NIH) summed up the power of this technology by saying, "The development of cell lines that may produce almost every tissue of the human body is an unprecedented scientific breakthrough. It is not too unrealistic to say that this research has the potential to revolutionize the practice of medicine and improve the quality and length of life." (Varmus 1998).

---

* The author is one of the founders of the organization and currently provides the Alliance with legal representation.
† Although President Bush limited federal funding of research using human embryonic stem cells, he supported other stem cell and regenerative medicine research.

A 2006 report issued by the U.S. Department of Health and Human Services (DHHS) titled *2020: A New Vision: A Future for Regenerative Medicine* recognized the value of regenerative medicine. It noted that:

> This new field holds the realistic promise of regenerating damaged tissues and organs *in vivo (in the living body)* through reparative techniques that stimulate previously irreparable organs into healing themselves. Regenerative medicine also empowers scientists to grow tissues and organs *in vitro (in the laboratory)* and safely implant them when the body is unable to be prompted into healing itself. This revolutionary technology has the potential to develop therapies for previously untreatable diseases and conditions. Examples of diseases that regenerative medicine can cure include diabetes, heart disease, renal failure, osteoporosis and spinal cord injuries. Virtually any disease that results from malfunctioning, damaged, or failing tissues may be potentially cured through regenerative medicine therapies (U.S. Department of Health and Human Services 2006).

In addition to the health implications of treating disease, regenerative medicine holds the promise of economic benefit. Specifically, by treating or perhaps curing chronic diseases, regenerative medicine may significantly reduce healthcare costs. According to the DHHS report, 250,000 patients receive heart valves, at a cost of $27 billion annually; 950,000 people die of heart disease or stroke, at a cost of $351 billion annually; and 17 million patients with diabetes are treated at a cost of $132 billion annually (U.S. Department of Health and Human Services 2006).

Regenerative medicine has the ability to prevent many of these conditions by replacing or repairing malfunctioning tissues. The economic value of these breakthroughs is potentially enormous. A recent analysis based on data from one state (Michigan) found that a 1% reduction in costs for common conditions most likely to be treated by regenerative medicine technologies (Type 1 diabetes, stroke, heart disease, Parkinson's Disease, and spinal cord injuries) indicates benefits in the form of reduced treatment costs of approximately $79.3 million annually (Goodman and Berger 2008). Moreover, as a new, growing technology, regenerative medicine can lead to long-term economic growth as research by academia and non-profit organizations expands and new and existing companies commercialize the technology. States that have enacted regenerative medicine initiatives project significant economic benefits.

For example, a recent Stanford University report analyzing the impact of the efforts of the California Institute for Regenerative Medicine (CIRM) said, "The information available to date suggests that California is already starting to see new economic activity resulting from CIRM and that there is significant potential for additional future economic benefits" (Baker 2008).

A separate analysis of the job climate in Michigan reported that "As small as a 1% increase in biotech employment due to [regenerative medicine] would lead to approximately 443 new jobs with a commensurate payroll increase of $32 million per year; another 354 jobs in induced employment, for a total of approximately 797 jobs; increased payroll of $51 million, with proportional impacts on tax receipts" (Goodman and Berger 2008). In sum, as noted by the DHHS report, "The potential benefits of regenerative medicine in improved healthcare and economic savings are enormous" (U.S. Department of Health and Human Services 2006).

Some regenerative medicine products are already on the market (see Chapter 15 by Buckler, Margolin, and Haecker). For example, Dermagraft is indicated for the treatment of certain diabetic foot ulcers. In addition, several regenerative medicine projects have succeeded in the clinic and are in late-stage clinical trials for conditions such as leukemia in pediatric patients, ischemic heart failure, and critical limb ischemia.

However, federal policies that support regenerative medicine research and commercialization, create a stable regulatory pathway for product approval, and foster a reimbursement environment conducive to private investment, are essential if this technology is going to reach its full potential and achieve widespread health and economic objectives.

A new organization was formed in 2009 to promote these policies. The Alliance for Regenerative Medicine (the Alliance) is a non-profit organization based in Washington, DC. Its mission is to educate key policymakers about the potential for regenerative medicine and advocate for public policies—funding, regulatory, reimbursement, and others—that will facilitate advances in the field. The Alliance is the first advocacy organization exclusively dedicated to representing the regenerative medicine community with a focus on product development. Its membership includes biotechnology and pharmaceutical companies, university-based research organizations, independent research organizations, venture capital firms, patient advocacy organizations, and service organizations that share the goal of advancing cell-based therapies, biomaterials, and tissue engineering products to the marketplace.

20.3 FEDERAL FUNDING POLICIES

The federal government spends hundreds of millions of dollars funding regenerative medicine and tissue engineering research. The NIH is one of the primary sources of funds awarded through several of its institutes. While NIH funding of human embryonic stem cell research has been controversial, the agency has been supporting other regenerative medicine and tissue engineering research efforts for many years and provided hundreds of millions of dollars for regenerative medicine projects. In addition, the Department of Defense funds regenerative medicine projects through the Armed Forces Institute for Regenerative Medicine (AFIRM) program. Through federal, state, industry, and academic sources, AFIRM will provide up to $250 million for regenerative medicine research that focuses on conditions faced by military personnel. While these programs are laudable efforts, more needs to be done.

Specifically, the NIH must fund more regenerative medicine translational research projects in non-profit research institutions or academia and fund collaborations between the entities and commercial enterprises. In addition, federal research efforts should be better coordinated and designed to accomplish research goals developed by funding agencies in consultation with non-profit research institutions, academic researchers, and commercial entities. Moreover, the federal government should take steps to support regenerative medicine companies attempting to commercialize such technology. If regenerative medicine projects are to be developed into therapies and diagnostics to help patients, these companies must receive necessary support.

It should be noted that the U.S. government has launched initiatives designed to support and promote other technologies deemed to be in the national interest. For example, for fiscal year 2010, Congress appropriated $2.2 billion "to increase investments in technologies that use energy more effectively and produce clean, inexpensive energy from domestic sources" (U.S. Congress 2009). In addition, the American Recovery and Reinvestment Act of 2009 appropriated $7.2 billion to "expand broadband access to unserved and underserved communities across the U.S., increase jobs, spur investments in technology and infrastructure, and provide long-term economic benefits" (Broadband Use 2009). Moreover, the White House set up a job creation initiative announced in December, 2009 that among other things would "expand existing programs designed to leverage private investment in energy efficiency and create clean energy manufacturing jobs" (White House 2009).

Given the potential of regenerative medicine to reduce human suffering, reduce healthcare costs, and create jobs, it is appropriate that the government support this industry. The Alliance has advanced a number of ideas regarding government support including:

- Funding researchers who collaborate with commercial entities by providing matching grants for researchers at academic institutions and other non-profit research organizations who are working on projects with industry. Grant amounts would match those made by industry and the funds would be used to support relevant research.
- Funding for research and other efforts designed to facilitate filing of investigational new drug applications (INDs) and investigational device exemptions (IDEs) with the U.S. Food and Drug Administration (FDA). Modeled after a CIRM program, this program would provide funds for preclinical science or regulatory work to be performed in advance of an IND or IDE filing.
- Funding for regenerative medicine companies. The federal government should provide financial assistance to regenerative medicine companies seeking to develop treatments and cures. Such aid in the form of tax credits or grants should focus on the companies' commercialization projects.
- Better coordination of federal regenerative medicine activities. While several institutes of the NIH fund regenerative medicine and tissue engineering projects, mechanisms should be in place to coordinate this research. In addition, policies of other DHHS agencies, including FDA, CMS, the Department of Defense, and the National Institutes of Standards and Technology, will impact developments in regenerative medicine. Thus, a mechanism is needed to coordinate the policies of all relevant federal agencies. Furthermore, this multi-agency group should set national goals for regenerative medicine and track the progress of government and private efforts to meet these objectives.
- Creation of mechanisms for federal agencies to interact with the regenerative medicine community outside the context of specific projects. Current interactions between federal agencies and stakeholders usually focus on a specific research project or potential product. As federal agencies develop more policies, researchers and companies should have the opportunities to

interact with regulators and policy makers to discuss general research and development issues.
- Funding for the FDA to perform regulatory research in regenerative medicine or hire sufficient personnel with expertise in regenerative medicine. The agency should have additional funds to perform research that will facilitate speedy scientific review of regenerative medicine products and it needs sufficient staff with necessary expertise to review these products.

Congress is currently developing legislation incorporating these principles.

Health reform legislation enacted last year contains provisions that would accomplish some of these goals although they are not aimed specifically at regenerative medicine. The legislation created the Cures Accelerator Network (CAN) whose mission is to support medical research and product development for diseases for which there is little economic incentive for product development. The provisions apply to all biomedical research. Under the program, NIH will fund research institutions, life sciences companies, medical centers, and patient advocacy groups for research and product development activities and also encourage communication between life sciences companies and the FDA. The program will be run (and funds dispersed) by the NIH director based on advice of a board comprised of different stakeholders including researchers, patients, private investors, and representatives from the biotechnology and pharmaceutical industries (U.S. Congress 2009).

The legislation also created a new tax credit or grant program to financially assist biotechnology and medical device companies developing products for unmet medical needs (U.S. Congress 2009). As noted, these provisions apply to all biomedical research not simply regenerative medicine. Nonetheless, they demonstrate that policymakers recognize the need to implement new programs to jump-start life sciences research and product development.

20.4 REGULATORY ISSUES

Responsible and predictable FDA regulatory policies are essential to ensure that safe and effective regenerative medicine products reach patients as soon as possible. The regulatory framework must appropriately balance risk and benefit. The FDA currently regulates cell therapy products under its Good Tissue Practices (GTP) rules. Under GTP, cell therapy products are regulated based on their level of risk. The threshold standards in the regulations for all cell products are intended to prevent the use of contaminated tissues and the improper handling and processing of materials and products, as well as to ensure clinical safety and effectiveness for highly processed tissues (21 Code of Federal Regulations [CFR] Part 1271; U.S. FDA 2009a and b).

The standards address issues such as registration of facilities, ensuring proper storage and handling, and donor eligibility and screening. If a stem cell-based product entails more than "minimal manipulation" or cells for other than only *homologous use*, the FDA will also impose its regulations for biologics or devices. Manufacturers will also be required to comply with the FDA's Good Manufacturing Practice (GMP) requirements. Processes that entail more than minimal manipulation are defined as those "that alter the relevant biological characteristics of cells or tissues." Common ways of creating, maintaining, and using cell lines, such as inducing multipotency of adult cells or causing targeted differentiation of an embryonic stem cell line, are considered processes that alter biological characteristics (21 CFR 1271). According to 21 CFR 1271.3, *homologous use* means the repair, reconstruction, replacement, or supplementation of a recipient's cells or tissues with a human cell-, tissue-, cellular-, or tissue-based product (HCT/P) that performs the same basic function or functions in the recipient as in the donor. It is expected that most stem cell products as currently envisioned will be regulated in this manner, that is, manufacturers will have to comply with GTPs and the standards for approval of biologics and devices, including compliance with GMPs (Kessler and Halme 2006).

In addition to these rules, the FDA published several guidance documents designed to inform product developers of its requirements for product approval. Observers have identified a series of regulatory questions related to regenerative medicine products including how to comply with existing FDA GTP and GMP rules and how to ensure product consistency and cell stability (Carpenter et al. 2009).

Other specific questions that need to be resolved by the FDA include: whether regenerative medicine technologies should be regulated like other biologics and devices (and if not, what are the relevant differences?); how to use existing and developing clinical testing and imaging techniques and technologies to determine the safety of regenerative medicine products; what standards should be used to determine the safety and efficacy of new products; and the importance of the proposed clinical indication of the new product on the regulatory pathway (Werner 2009).

Members of the Alliance have identified several ways to improve the regulatory climate for regenerative medicine products. One proposal entails improving communications among FDA and researchers and product developers to better identify the key regulatory issues relating to product development. A process whereby researchers and product developers can educate the FDA about the latest science while agency representatives can educate the field about mechanisms to achieve safety holds enormous promise (Baker 2009). Unfortunately, this can be challenging because most interactions between the agency and product developers are in the context of a specific product or application. Thus, the industry needs improved vehicles for communication to and from the FDA including:

- Developing and instituting a mechanism whereby sponsors (non-profit institutions and companies) can meet with the FDA in a consultative and collaborative manner outside the typical regulatory process to exchange feedback on key development issues. This mechanism could be used to educate the agency, for example, about the novelty of a technology, unique challenges of regenerative medicine, and other matters. This mechanism should exist throughout the development process.
- Development of a mechanism outside the guidance process for the agency to provide information to sponsors about important issues as scientific knowledge evolves.

A dialogue between the regenerative medicine community and the agency is likely to be well received by the agency. FDA officials recently stated that "efforts are underway at FDA to promote communication and exchange of scientific information between the agency and its stakeholders, with the goal of facilitating development of safe and effective stem cell-based products" (Fink 2009).

In addition to an expanded dialogue with product developers, the agency should address key regulatory requirements such as modifying the use of animal models and in vitro potency assays because existing rules are often inapplicable to regenerative medicine products; and modifying the regulatory pathway requirements for cell therapy products to ensure that the requirements distinguish between autologous and allogeneic products. Moreover, since some cell therapy products are delivered via novel devices, important regulatory issues arise including how the products should be regulated (as a biologic, device, or combination product) and how the relevant FDA review divisions will coordinate when reviewing such a product.

The Alliance has also advocated for the FDA (and NIH) and the regenerative medicine community to engage in other collaborative efforts that will lead to commercialization of scientific discoveries including:

- Identifying priorities for regulatory research
- Developing consensus on scientific issues related to regulatory approval

The FDA Commissioner and NIH Director recently annouced the formation of the NIH-FDA Joint Leadership Council to provide a forum for the leadership of both agencies to work together and coordinate efforts. This Council provides an ideal forum for the agencies to work collaboratively with the regenerative medicine community on projects designed to promote translation of scientific discoveries into products for patients.

20.5 REIMBURSEMENT ISSUES

The reimbursement environment will significantly impact the future of regenerative medicine (see Chapter 21 by Meurgey and Wille and Chapter 22 by Faulkner). A market-based reimbursement system that rewards innovation and ensures patient access to new products is critical.

Some regenerative medicine products are already on the market and being paid for by insurers. For example, Dermagraft is currently reimbursed by both Medicare and private insurers. However, with health costs growing, health insurers have developed technology assessment programs designed to review new products to ensure that they pay only for what they deem appropriate treatments. The Blue Cross and Blue Shield Association Technology Evaluation Center uses five criteria to assess whether a technology improves health outcomes such as length of life, quality of life, and functional ability:

1. The technology must have final approval from the appropriate governmental regulatory bodies.
2. The scientific evidence must permit conclusions concerning the effect of the technology on health outcomes.

3. The technology must improve the net health outcome.
4. The technology must be as beneficial as any established alternatives.
5. The improvement must be attainable outside the investigational settings (Blue Cross/Blue Shield Association 2009).

Even if a new product overcomes these hurdles, payers often use a variety of techniques to reduce the use of high-cost products such as formularies, pre-approval requirements, and preferred drug lists. Use of technology assessment programs and other mechanisms could reduce the reimbursements for new products, including regenerative medicine therapeutics.

In addition, increased focus on the comparative effectiveness of medical treatments could pose another challenge for product developers. The American Recovery and Reinvestment Act (ARRA) enacted by Congress in early 2009 provided over $1 billion for government-funded comparative effectiveness research. Funds are divided between the NIH, AHRQ, and DHHS. Subsequently, the Institute of Medicine recently issued its recommended priorities for research topic areas to be addressed by ARRA funds. Some of these topics, including use of biomarkers for treatment of various cancers, are clearly relevant to regenerative medicine. Moreover, a new independent organization was created in health reform legislation that was empowered to perform comparative effectiveness research (U.S. Congress 2009).

Consequently, as regenerative medicine technologies mature and more products approach marketability, the regenerative medicine community needs to lay the groundwork for a favorable reimbursement climate. It should begin educating payers about the scientific and healthcare potential of these technologies and must demonstrate to public and private insurers that regenerative medicine products hold the promise of delivering improved clinical outcomes in a cost-effective manner. Specific reimbursement issues are likely to include (1) determining what evidence of improved care and cost effectiveness payers will seek when making coverage and payment decisions; (2) providing input into such evaluations; and (3) developing strategies for companies to obtain that information prior to product marketing.

Alliance members intend to establish a dialog with Centers for Medicare and Medicaid Services (CMS) and private insurers to educate them about the value of regenerative medicine technologies and ensure communication and input into agency and insurer decision making including cost effectiveness and comparative effectiveness standards. Alliance members will also work to ensure that federal comparative effectiveness research is done appropriately. For example, since the NIH issued proposals for comparative effectiveness research in regenerative medicine, the Alliance will engage the agency to discuss research priorities for grants in this area.

20.6 OVERALL POLITICAL ENVIRONMENT FOR REGENERATIVE MEDICINE

The time is right to pursue policy changes to promote regenerative medicine. After years of political debate over the ethical virtues of human embryonic stem cell research, federal agencies have finally put this dispute in the rearview mirror.

President Obama issued an Executive Order on March 9, 2009 that authorized the NIH to fund research using human embryonic stem cells (hESCs) regardless of the date of derivation and called on the NIH to develop implementation guidelines outlining the funding criteria (White House 2009). The guidelines were published on July 7, 2009. All institutions and researchers seeking to use federal funds for embryonic stem cell research must follow the rules outlined therein. These actions open the door for greater NIH funding of hESC research.

Since then, the NIH has declared dozens of cell lines eligible for federal funding. Thus, it appears that the major political controversy surrounding regenerative medicine has subsided.

While the federal debate about stem cell research revolved exclusively around embryonic stem cells, and therefore did not directly affect other regenerative medicine technologies, the 2009 Executive Order removed any controversy that may have dampened enthusiasm for the entire field. However, it should be noted that regenerative medicine and tissue engineering products are impacted by the other political winds blowing in Washington. For example, debates about implementation of a regulatory pathway for biosimilars and price regulation of pharmaceutical and biotech products affect regenerative medicine companies. An environment that is generally hostile to medical innovation and product commercialization will be problematic for regenerative medicine companies.

20.7 CONCLUSION

Regenerative medicine has enormous potential to improve patient care, reduce health costs, and create economic growth. Almost every day a new breakthrough is announced. In addition, there is a track record of success as several regenerative medicine products are on the market and have demonstrated their effectiveness.

While there has been congressional support for funding of stem cell research over the past several years, the regenerative medicine community needs to seek broader gains from the federal government—Congress, the NIH, FDA, CMS, and other agencies—that will create a policy environment conducive to long-term success. That means funding for basic and translational research along with appropriate regulatory and reimbursement policies. How well the regenerative medicine community engages policymakers in Congress, the FDA, CMS, private insurers, and other agencies, will be key to whether—and how quickly—this technology will lead to patient treatments and cures.

REFERENCES

Baker L. 2008. *CIRM Interim Economic Impact Report*, California Institute for Regenerative Medicine, San Francisco, September 10.

Baker, M. 2009. Melissa Carpenter: making stem cells for many, safely. *Nat Rep Stem Cells*. Published online August 27. doi:10.1038/stemcells.2009.113

Blue Cross Blue Shield Association Technology Evaluation Center, Criteria. http://www.bcbs.com/blueresources/tec/tec-criteria.html, accessed December 24, 2009.

Broadband Use. 2009. www.broadbanduse.gov, accessed December 23, 2009.

Brookmeyer, R., S. M. Gray, and C. Kawas. 2010. Projections of Alzheimer's disease in the United States and the public health impact of delaying disease onset. *Am J Publ Health* 88: 1337–1342. Accessed January 10, 2010. www.silverbook.org

Carpenter, M., J. Frey-Vasconcells, and S. R. Mahendra. 2009. Developing safe therapies from human pluripotent stem cells. *Nat Biotechnol* 27: 606–613.

Cutler, D. M., and M. McClellan. 2010. Is technological change in medicine worth it? *Health Affairs*, 20: 11–29. Accessed January 10, 2010. www.silverbook.org

Fink, D. 2009. FDA Regulation of Stem Cell-Based Products, *Science*, June 26, 1662–1663.

Goodman, A. and S. Berger. 2008. Michigan Stem Cell Economic Study, *Mich Prospect*, September.

Kessler, D. and D. Halme. 2006. *New Engl J Med* 355: 1730–1735.

Klowden, K., R. DeVol, A. Bedroussian et al. 2007. In *An Unhealthy America: The Economic Burden of Chronic Disease,* Kim, K. et al., Eds., Milken Institute. Accessed January 10, 2010. www.silverbook.org

U.S. Congress. 2009. Patient Protection and Affordable Care Act, Public Law 111-148, March 23, 2010.

U.S. Congress. 2009. 2009. *FY 2010 Conference Summary: Energy and Water Appropriations.* Committee on Appropriations, September 30.

U.S. Congress. 2009. American Reinvestment and Recovery Act, P.L. 111-5, February 17.

U.S. Department of Health and Human Services. 2006. *2020: A New Vision: A Future For Regenerative Medicine.*

U.S. FDA. 2009a. www.fda.gov/CBER/tissue/tissue.htm, accessed December 24, 2009.

U.S. FDA. 2009b. 21 CFR Part 1271: Human Cells, Tissues, and Cellular Tissue-Based Products. www.fda.gov/CBER/tissue/tissue.htm, accessed December 24, 2009.

Varmus, H. 1998. Statement by Director of National Institutes of Health, U.S. Department of Health and Human Services, before the Senate Appropriations Committee, Subcommittee on Labor, Health and Human Services, Education and Related Agencies, December 2.

Werner, M. 2009. *World Stem Cell Report*, Genetics Policy Institute, Washington, DC.

White House. 2009. Removing Barriers to Responsible Scientific Research Involving Human Stem Cells.

White House 2009. President Obama Announces Proposals to Accelerate Job Growth and Lay the Foundation for Robust Economic Growth.

Section 6

Reimbursement

21 The Fourth Hurdle
Reimbursement Strategies for Regenerative Medicine in Europe

François M. Meurgey and Micheline Wille

CONTENTS

21.1 Regenerative Medicine in Europe: Current Reimbursement Status 337
21.2 Case Study: TiGenix's ChondroCelect 339
21.3 Various Country Environments and Challenges 342
    21.3.1 Germany 343
    21.3.2 United Kingdom 343
    21.3.3 Netherlands 344
    21.3.4 France 344
    21.3.5 Spain and Italy 344
    21.3.6 Belgium 345
    21.3.7 Other European Countries 345
21.4 Conclusion 347
References 348

21.1 REGENERATIVE MEDICINE IN EUROPE: CURRENT REIMBURSEMENT STATUS

Public spending accounts for 50 to 80% of pharmaceutical expenditures in Western Europe; the balance is made up by private health insurance and out-of-pocket payments by patients (OECD 2008). In practice, this means that obtaining reimbursement from public healthcare systems is imperative for commercial success in the European Union (EU).

Unlike regulation regarding the registration of medicinal products, pricing and reimbursement are not harmonized in Europe and fall within the exclusive jurisdictions of national authorities, provided that the basic transparency requirements described in Directive 89/105/EEC of December 21, 1988 (relating to transparency in the regulation of prices for medicinal products for human use and their inclusion in the scope of national health insurance systems) are met. As a consequence, reimbursement mechanisms by private and public health insurers vary from country to country (Mossalios, Mrazek, and Walley 2004; Espín and Rovira 2007; Vogler et al. 2008).

From a reimbursement perspective, the current situation for regenerative medicine is fragmented, confusing, and differs widely from country to country. This is due to the lack of a clear framework for regulations governing cell therapies and tissue engineering along with haphazard decisions on allowing public funding for specific interventions. The situation is slowly being clarified, but much remains to be settled in order to achieve broad patient access to these new therapies and encourage innovation in the European biotechnology industry.

While regulatory oversight for cell therapy and tissue engineering is now firmly a pan-European prerogative of the European Medicines Agency (EMA) as a result of European Regulation 1394/2007 (for information about the evolution of the regulatory framework in Europe and ATmP regulation, see Chapter 19 by Bravery), decisions on pricing and reimbursement for any healthcare product or service—whether a medical or surgical procedure, a medical device, or a medicinal product—remain under the exclusive jurisdiction of competence of each member state.

This led to so-called postcode prescribing (Torstensson and Pugatch 2009; NHS Postcode Lottery weblog), whereby a patient's access to an innovative treatment may depend exclusively on his or her region of residence within Europe, with regional differences evident even in the same country (Hayhurst, Brown, and Lewis 2003; White 2004; Jones 2004).

This certainly applies to regenerative medicine, with marked contrasts in access between different European countries. For example, German patients have had access to autologous chondrocyte implantation (ACI) for many years, while French patients are deprived of any cartilage repair options by the French healthcare system. None of the widely used surgical techniques for cartilage repair available in the rest of the world (e.g., microfracture, mosaicplasty, and ACI) is recognized by the French authorities, and thus none is reimbursed. The only option for French patients with cartilage lesions is treatment through inclusion in clinical trials, which is rare, or travel to a neighboring country for a surgical procedure in the hope that the local sick fund (Caisse Primaire d'Assurance Maladie) may cover the costs of travel and treatment.

There is no general reimbursement for advanced therapy products in Europe, in part due to the relative novelty of the techniques and lack of robust data from clinical trials. However, several countries have established processes to reimburse novel therapies like ACI, although the stakeholder and decision-making pathways vary significantly between territories. In general, since regenerative medicine interventions are considered experimental, typically they are funded only via clinical trials. A case in point is the treatment of knee cartilage lesions with ACI—a technique reviewed in 2005 by both the National Institute for Health and Clinical Excellence (NICE) in the United Kingdom (Clar et al. 2005) and the French Haute Autorité de Santé. Both assessment reports concluded that the technique was promising but too recent to be adopted broadly. More clinical and fundamental research was required to evaluate long-term outcomes, "improve characterization of chondrocytes and define release criteria for chondrocyte culture," (Haute Autorité de Santé 2005), and "explore factors that influence stem cells to become chondrocytes and to produce high-quality cartilage" (Clar 2005).

This led to the launch of a randomized controlled trial referred to as the Autologous Chondrocyte Transplantation/Implantation versus Existing (ACTIVE) treatments program funded by the Medical Research Council, with 22 hospitals in the United Kingdom and 2 in Norway currently recruiting patients (ACTIVE trial website). Unfortunately, patient recruitment has lagged behind the original timelines, leading NICE in May 2008 to postpone an update of its 2005 Healthcare Technology Assessment (HTA) until 2012. In the meantime, the only option for patients with knee cartilage lesions to gain access to ACI within the National Health Service (NHS) is to enter the ACTIVE trial, where they will be randomized to one of three types of ACI or "standard treatment" (debridement, abrasion, drilling, microfracture, mosaicplasty, or autologous membrane-induced chondrogenesis).

In contrast, countries like Germany and Sweden allow routine use of ACI; Germany has a standard reimbursement code for the procedure as part of its case-specific hospital diagnosis-related groups (DRG), Operations und Prozedurenschlüssel (OPS)—Codes 5-801.k (open knee or meniscus surgery for matrix-assisted ACI) or 5-812.h (arthroscopic knee or meniscus surgery for matrix-assisted ACI). It is worth mentioning that reimbursement levels in both countries appear minimal and cover only the variable cost of cell culture.

In another area of regenerative medicine described as "low-hanging fruit" by Plagnol and colleagues (2009), OrganoGenesis announced in August, 2008 that Apligraf®, its living bilayered cell therapy product, gained reimbursement in Switzerland for patients with non-healing wounds. Switzerland was the only country in Europe to reimburse for this treatment at the time (Wound Source). OrganoGenesis commercializes Apligraf in other European countries only on a compassionate use basis (Kälin 2009).

21.2 CASE STUDY: TIGENIX'S CHONDROCELECT

ChondroCelect® (TiGenix NV, Leuven, Belgium) is the first advanced cell therapy product, according to the European regulatory framework for advanced therapy medicinal products (ATMPs). It consists of autologous cartilage cells expanded ex vivo through a highly controlled and consistent manufacturing process. ChondroCelect is indicated for the repair of single symptomatic cartilage defects of the femoral condyle of the knee (International Cartilage Repair Society grade III or IV) in adults. In order to culture cells with the best potential to generate stable cartilage, it is critical that the cells do not lose their cartilage phenotype during the cell culture process. Maintaining their ability to maximally conserve their preculture phenotype and high quality cartilage forming capacity was the key element for product development.

ChondroCelect is intended for use in autologous cartilage repair only and is administered to patients as part of an ACI procedure; the combination of ChondroCelect (product) and ACI (procedure) is referred to as characterized chondrocyte implantation (CCI). The ACI procedure, initially described by Brittberg et al. (1994), classically consists of two steps: a first arthroscopic step to collect a biopsy sample of healthy cartilage in a non-weight-bearing part of the same knee followed 4 to 6 weeks later by an arthrotomy (open-knee surgery) to implant the expanded cells. Although Brittberg originally used a periosteal flap to cover the lesion, surgeons are now increasingly using collagen I/III membranes.

Conducting clinical trials with advanced cell therapy products raises a series of important challenges, including (1) the fact that the impact of the product on the repair processes can be demonstrated only in the long term, (2) the absence of well-established validated endpoints, and (3) the need of surgical intervention in applying the medicinal products. Measuring the impact of knee joint injuries and their treatment on patient functioning and well-being is an essential part of evaluating benefit. Experience with available patient-reported outcomes instruments is limited. The Knee Injury and Osteoarthritis Outcome Score (KOOS) is currently considered the best validated of all the instruments. In addition, structural assessments are often not standardized, with no consensus on the most appropriate outcome measures.

The efficacy of ChondroCelect was studied in a phase III, multicenter, randomized controlled trial known as the TIGACT01 study. Patient recruitment commenced in March 2002. ChondroCelect was compared to microfracture—a surgical procedure considered the standard of care—in the repair of single symptomatic cartilage lesions of the femoral condyles. Histological examination of the repair biopsy at 12 months showed superior structural repair in the ChondroCelect arm compared to the microfracture arm (as measured by computer-assisted histomorphometry; $P = 0.003$; overall histology score $P = 0.010$; Saris et al. 2008). The KOOS measured continuous improvement in the clinical outcomes in both treatment arms up to 36 months. However, the estimated benefit at 36 months was greater for ChondroCelect-treated patients (Saris et al. 2009).

As described in the HTA by Clar and colleagues (2005), long-term modelling of the cost effectiveness of cartilage repair treatments such as ACI is hampered by the lack of long-term data. In addition, there is relatively limited evidence on the natural history of hyaline cartilage lesions (Messner and Maletius 1996; Prakash and Learmonth 2002; Shelbourne, Jari, and Gray 2003). The main expected benefit of ACI could be characterized as the prevention of osteoarthritis and total knee arthroplasty. When assuming that repair of hyaline cartilage is the predictor of long-term clinical benefit, ACI using ChondroCelect performs relatively well.

A decision analytic model using many of the assumptions described by Clar and colleagues was built for ChondroCelect using official Belgian pharmacoeconomic guidelines (Cleemput et al. 2008) and enriched with the clinical and histology results from TIGACT01 (Gerlier et al. 2009). Over a 40-year horizon, ChondroCelect was found to be cost effective, with an incremental cost effectiveness ratio per quality-adjusted life year well below the classical thresholds used by NICE (£20,000) or recommended by the World Health Organization (gross domestic product per capita).

The case study we will describe with ChondroCelect is pertinent because ChondroCelect is the first cell therapy approved as an ATMP by the Committee for Advanced Therapies and also because it encapsulates many of the challenges other regenerative medicine products are likely to face (i.e., application through a surgical procedure, combination of a medicinal product and a medical device, difficult to measure long-term benefits, etc.).

It is also quite relevant because, as outlined below, the cost structure of a product like ChondroCelect is likely to make it a relatively expensive product. Plagnol and colleagues (2009) cited the issues of "high cost and perception of value for money" as classical limitations of regenerative medicine products:

- The logistics involved in collecting a cartilage biopsy, transporting it in temperature-controlled conditions across Europe, and sending back the cultured cells by courier to the Operating Room in similar conditions are expensive.
- The number and extent of microbiology and quality tests lead to a significant cost per product.
- The treatment for each patient is an individual production batch, and the cell culturing process is very labor-intensive (4 to 5 weeks of culture requiring about 1/6 of a full-time equivalent for a highly skilled laboratory technician).
- The cell harvesting and implantation techniques are delicate, and the manufacturer must provide on-site training for the surgeon and the operating room staff, necessitating deployment of a specialized field support team.
- Conducting a full-scale clinical development program that will satisfy EMA requirements and, after approval, funding post-marketing commitments requires a substantial financial investment.

Because ChondroCelect is administered as part of a double surgical procedure and potentially also requires the use of a medical device (collagen membrane), several different parties, as outlined below, are typically involved in pricing and reimbursement decisions:

- Drug pricing and reimbursement authorities for the cell suspension (now considered a medicinal product, and thus typically managed, at least in theory, like a classical pharmaceutical product).
- Medical device pricing and reimbursement authorities for the collagen membrane, usually a different department of the same institution.
- Hospital management authorities for the surgical procedures, particularly groups in charge of defining the DRG case mix or evaluating new medical and surgical procedures.

This complexity and the fact that each country has a different mix of these three types of evaluation and coverage decisions, make ChondroCelect particularly representative of the potential challenges to obtaining reimbursement for regenerative medicine products in Europe.

Finally, a point has to be made about the challenge of demonstrating cost effectiveness and providing long-term clinical results. All authors who reviewed the state of the tissue engineering and regenerative medicine industry commented on these issues (Lysaght and Hazlehurst 2004; Kemp 2006; Martin, Hawksley, and Turner 2009; Plagnol et al. 2009).

Regenerative medicine typically attempts to intervene in the natural history of a disease and modify the function of a diseased organ through repair or regeneration. These processes rarely reveal immediate clinical results. For example, even though the benefits of ChondroCelect can be measured in terms of structural repair at 1 or 2 years (through systematic biopsies of the repair tissue and histology assessments), it may take at least 3 to 5 years to see a clear differentiation in clinical output (compared to other interventions such as microfracture). Assessment of the ultimate goal of cartilage repair (the delay or prevention of knee osteoarthritis) would take a minimum of 15 to 20 years, which is clearly beyond any clinical or commercial feasibility.

Any evaluation of cost effectiveness must be modelled on assumptions of long-term benefits, but in general, the literature contains few data about the natural history of the disease or the long-term impacts of various interventions (Gerlier et al. 2009). This has been and will continue to be a major issue affecting the broad reimbursement of regenerative medicine products.

21.3 VARIOUS COUNTRY ENVIRONMENTS AND CHALLENGES

After registration through a centralized procedure has been obtained from EMA, any company intent on commercializing a regenerative medicine product in Europe faces an enormous challenge requiring substantial resources and time. This challenge stems from the fact that every EU member state has the sole responsibility for funding drug reimbursement within its jurisdiction, and, above all, because the associated healthcare systems are profoundly different and have adopted strikingly dissimilar ways to encourage innovation and facilitate the uptake of new technology.

Each country naturally expects the applicant to file a Pricing & Reimbursement dossier in the national language, which is a further challenge for an SME, since there are 23 official or working languages in the European Union.

Any attempt to seek pricing and reimbursement in Europe must therefore start with a comprehensive evaluation of the likely procedures in each country and identification of the relevant experts best equipped to assist the efforts of the marketing authorization holder (MAH). It is also essential to secure the support of the relevant scientific societies that play a vital role in evaluating the therapeutic efficacy of the product as experts to the reimbursement authorities and in many cases are the only bodies that can legally apply for coverage of a new medical or surgical procedure. Strategic considerations that are common in the pharmaceutical and medical device industries and have to be taken into account at this stage are described below:

- International reference pricing is a widespread practice among European pricing and reimbursing authorities; the maximum price that can be obtained in a given country is typically the average of prices of several other EU countries. The applicant must therefore pay close attention to the order of geographic introduction of its product (Schoonveld 2001; U.S. Department of Commerce 2004; Brekke, Königbauer, and Straume 2007; Brekke, Grasdal, and Holmås 2008).
- The timing and content of submission to HTA agencies must be weighed carefully, particularly NICE in the United Kingdom and the Commission de Transparence in France, because their opinions are public (and posted on the Internet) and carry substantial weight beyond their own borders. A negative opinion based on a poor submission would haunt the regenerative medicine company for months in many other countries. Submissions have to be planned carefully and sometimes prepared years in advance, particularly in the areas of cost effectiveness and quality-of-life data collection.

Because of the unique characteristics of ChondroCelect, and the various stages of maturity of the ACI technique in various countries, TiGenix had to seek price and reimbursement in a variety of ways in different countries:

- **Germany**—ACI was already covered by a DRG. TiGenix applied for temporary hospital reimbursement status while seeking the creation of a new DRG.
- **Netherlands**—TiGenix sought inclusion on a special list of expensive innovative drugs; inclusion requires demonstration of therapeutic and economic effectiveness.
- **Belgium and France**—TiGenix followed the classical drug pricing and reimbursement pathway, although the coverage of the other parts of product administration (surgical procedure and biological membrane) may require extra efforts and take substantially more time.
- **Spain and Italy**—TiGenix followed a "double-barreled" submission because the (maximum) price is granted at the national level, but access is dependent on inclusion into regional hospital formularies.

21.3.1 Germany

TiGenix submitted and obtained in early 2009 a New Diagnostic and Therapeutic Procedures (Neue Untersuchungs und Behandlungsmethoden or NUB) agreement from the Institut für das Entgeltsystem im Krankenhaus (InEK), allowing ChondroCelect to be reimbursed before the attribution of a new DRG code and tariff. Of 546 NUB applications in 2009, the ChondroCelect application was among only 87 that received the highest "status 1" rating from IneK—a testimony to its innovative nature.

The individual hospitals that submitted NUB applications for ChondroCelect are now negotiating budgets with their local Krankenkassen (sick funds) to cover reimbursement for ChondroCelect—a process that usually takes a few months. The first commercial sales for ChondroCelect were recorded in December 2009, almost a year after the granting of a positive NUB by InEK. In parallel, an application for a new OPS code specifically for ChondroCelect implantation is under review by DIMDI (German Institute of Medical Documentation and Information, the publisher of official medical classifications), but a decision is not likely before early 2011 at the earliest.

21.3.2 United Kingdom

As noted in the introduction, NICE conducted an HTA of ACI in 2005 and concluded that it had "insufficient evidence at present ... that ACI is cost-effective compared with microfracture or mosaicplasty" (Clar et al. 2005). It is therefore unlikely that the NHS will cover ACI until the completion of the NICE review of the procedure that has been postponed until 2012. In the meantime, TiGenix is actively pursuing reimbursement through other means such as "pass-through" coverage with individual primary care trusts or from private healthcare insurance companies.

21.3.3 NETHERLANDS

The Dutch system allows various possibilities for funding of new treatments, such as financing by a recently-created "innovation office" (Innovatieloket), inclusion on a list of expensive medications financed outside the hospital DRG system (Lijst Dure Geneesmiddelen), and reimbursement on a named-patient basis for experimental but unapproved therapies.

After consultation with the Dutch Healthcare Insurance Board (College voor Zorgverzekeringen), TiGenix developed an application to be included on the so-called List for Expensive Drugs and submitted it soon after the European Marketing Authorization was granted for ChondroCelect. Inclusion on this list requires the MAH to demonstrate therapeutic and economic effectiveness within 3 years of the formal inscription, usually through the conduct of a Phase IV study specifically on Dutch patients (a so-called effectiveness study or doelmatigheidsonderzoek). Some large pharmaceutical companies have challenged this requirement and prefer to submit data collected by other means (from other countries or in the course of other Phase IIIB or IV studies). It is worth mentioning that a possibility exists to submit a protocol for a Dutch effectiveness study to the Dutch Organization for Health Research and Development (ZonMW) that may fund up to 50% of the study cost.

21.3.4 FRANCE

The independent body in charge of healthcare evaluation, the Haute Autorité de Santé (HAS), comprises three commissions charged with evaluating (1) new medicinal products (the Commission de Transparence), (2) new medical or surgical procedures (the Commission d'évaluation des actes professionnels), and (3) new medical devices (the Commission nationale d'évaluation des dispositifs médicaux et des technologies de santé), respectively. After HAS evaluation, the MAH must make an application for the new product to the Pricing Commission (Comité Économique des Produits de Santé).

Although ChondroCelect is a drug from a regulatory standpoint, substantial ambiguity remains as to the best pathway to gain full reimbursement, since some of the parties involved recommended a classical drug submission in parallel with a request by the concerned scientific societies for reimbursement of the surgical procedure, while other parts of the administration recommended the creation of a new DRG covering all necessary parts of the implantation procedure (both arthroscopic and arthrotomic procedures, the implantation cells, and possibly even the biological membrane). Although this latter approach would have the great advantage of being all-inclusive, it may take several years to complete. TiGenix is currently in the process of contacting all parties involved in hospital financing to assess the best strategy.

21.3.5 SPAIN AND ITALY

In both countries, decentralization has in effect split drug coverage decisions between a national level (which decides the maximum price allowed for a new drug and grants a national reimbursement in principle) and a regional level (which decides on inclusion into a regional formulary and negotiates the final price level). The initial application for drug pricing and reimbursement is therefore a central one (to the Interministerial Commission on Drug Pricing of the Ministry of Health in Spain and the Agenzia Italiana del Farmaco in Italy) followed by regional applications that include assessments by 17 regional health departments (each with its own evaluation agencies) leading to a coverage recommendation in Spain (Durán, Lara, and van Waveren 2006) and negotiation for inclusion on regional "positive lists" (Prontuario Terapeutico Ospedaliero Regionale) in 21 regional governments and their local health units (Aziende Sanitarie Locali) in Italy (Donatini et al. 2001; Martini et al. 2007). Any regional decision is typically followed by a direct negotiation with the hospital or group of hospitals, with price and discount levels negotiated case by case.

21.3.6 Belgium

Price and reimbursement decisions in Belgium are separated. The first step is a price submission to the Ministry of Economic Affairs (accompanied by a justification of the price structure based on costs), followed by an application for public insurance coverage to the Drug Reimbursement Commission (Commission de Remboursement des Médicaments [CRM] or Commissie Tegemoetkoming Geneesmiddelen [CTG] in bilingual Belgium). Voting members of the CRM/CTG include representatives from the sick funds (Mutuelles or Mutualiteiten), pharmacists, physicians, and academics. The commission evaluates the therapeutic value, cost effectiveness, and budget impact of new products, negotiates price levels and reimbursement conditions with applicants, and makes recommendations concerning reimbursement to the Minister of Health after advice from the Budget Minister.

21.3.7 Other European Countries

Among other European countries, several (Austria, Denmark, and Sweden) do not require national price and reimbursement submissions for hospital-only products as required by other countries (Finland, Norway, and Portugal). A brief assessment of procedures in each country is presented below:

- **Austria**—There appears to be no central pricing process for hospital drugs; prices of drugs, medical devices, and procedures for the hospital sector are negotiated directly between the companies and the hospitals (mainly through drug commissions or centralized purchasers). Inpatient hospital costs are covered by a DRG system called Leistungsorientierte Krankenanstaltenfinanzierung (LKF or performance-oriented hospital funding). The adoption of a new procedure into the LKF catalogue is based on an assessment using the criteria of evidence-based medicine and a cost calculation. The LKF model is updated annually by July 15 for application on January 1 of the following year. The national uniform standard is a task of the Federal Health Agency. The responsible institution of the agency that decides on the LKF model is the Federal Health Commission (Hofmarcher and Rack 2006; Leopold and Habl 2008).

- **Denmark**—Prices of hospital-only and ambulatory drugs used in a hospital setting are not controlled. In Denmark, all hospital treatment including pharmaceuticals is free of charge to patients. Virtually all (95% by value) pharmaceutical sales to hospitals are made via a central purchasing agency AMGROS (owned by the regions) that invites tenders for pharmaceutical contracts. Average discounts of 20 to 25% against list price (pharmacy purchase price) are usually observed (Strandberg-Larsen et al. 2007; Thomsen et al. 2008).
- **Finland**—Even though prices are theoretically free, they are actually controlled through a reimbursement system. Pharmaceutical companies wishing to make their products eligible for reimbursement must apply for approval of a so-called reasonable wholesale price from the Pharmaceuticals Pricing Board (Lääkkeiden hintalautakunta or HILA). Procurement of hospital drugs seems relatively decentralized and expected discounts are not public (Järvelin 2002; Peura et al. 2007).
- **Norway**—The Norwegian Medicines Agency is responsible for setting maximum pharmacy purchase prices. All suppliers of prescription pharmaceuticals must apply for maximum prices, whether or not they seek reimbursement for the products. Pharmaceuticals can only be sold at or below the maximum price level. Hospital-only drugs are subject to the same maximum price rule as primary care drugs—the average of the three lowest market prices in nine EU countries. Procurement of hospital drugs is highly centralized, with most purchasing carried out by tender through the Norwegian Drug Procurement Co-Operative (Legemiddelinnkjøpssamarbeidet) with discounts of approximately 30% of maximum price. A pharmacoeconomic evaluation is mandatory for any drug applying for reimbursement (Johnsen 2006; Frostelid et al. 2007).
- **Portugal**—Pricing and reimbursement for all drugs, including inpatient drugs, are negotiated centrally. The Direcção Geral das Actividades Económicas (DGAE) is responsible for setting the maximum prices of pharmaceuticals (not including over-the-counter and hospital-only medicines). Reimbursement decisions are made by the Instituto Nacional de Farmacia e do Medicamento (INFARMED), which also manages the National Prescribing Formulary and the National Hospital Pharmaceutical Formulary. Reimbursement of hospital-only drugs requires demonstrating cost effectiveness. Hospitals purchase drugs by negotiating directly with suppliers or by a public procurement process (managed by the Administração Central do Sistema de Saúde). In the latter process, offered discounts are important aspects of the decision-making process. In 2007, a maximum price was introduced along with budget controls for new pharmaceuticals in public hospitals (Decree Law 195/2006, October 3, 2006; Pedrosa Vasco and Alves da Silva undated; Pita Barros and de Almeida Simões 2007; Teixeira and Vieira 2008).
- **Sweden**—Hospital-only products are not covered by the national benefits scheme and are thus excluded from any centralized pricing or reimbursement system. Inpatient drugs are purchased at the local level by the 21 county

councils that negotiate significant discounts. Cost effectiveness analysis is performed by local formulary committees (Glenngård et al. 2005; Redman and Köping Höggård 2007; ISPOR 2009).

Finally, no overview of pricing and reimbursement in Europe is complete without mention of the emerging role of private health insurance in a number of countries (Mossalios and Thompson 2004). Approximately 10% of the German population (in particular, individuals with annual gross incomes over a certain threshold, e.g., €47.250 in 2006) are insured by for-profit insurance carriers. Private companies are popular choices for complementary health insurance in France (92% of the French population is covered this way, usually through the so-called mutuelles). Historically, these private companies have mainly covered portions of patient co-payments for drugs or hospital stays, but they are increasingly exploring paying for treatments not covered by the public healthcare insurance, often as a means of attracting new customers.

21.4 CONCLUSION

As discussed, the European pricing and reimbursement landscape is complex and represents a substantial challenge for small and medium-sized enterprises that comprise most companies active in regenerative medicine. In addition to the multi-disciplinary skills and significant effort required to navigate this intricate and disconcerting environment, there are fundamental issues with the classical drug pricing and reimbursement process, as summarized below, that make it particularly ill-suited for the evaluation of regenerative medicine products:

- It typically reviews drug files that include multiple trials, often with several thousand patients, while a regenerative medicine product such as a cell therapy is likely to be approved on the basis of one or two trials with much smaller sample sizes (perhaps a few hundred patients).
- The time horizons are radically different; drug trials rarely exceed a few weeks or months. Most regenerative therapies may require years to demonstrate their full effectiveness.
- As noted, many cell therapy and regenerative medicine products are likely to involve substantial immediate costs (linked to the labor-intensive cell culture process, the burdensome logistics of individual deliveries to treatment centers, and surgical procedures to implant the products), while most of the expected cost savings are likely to be demonstrated in the long term. Demonstrating cost effectiveness with reasonable certainty at launch may therefore present a real challenge.

We hope that this superficial overview of the European reimbursement landscape will not paint a discouraging picture for companies involved in the development of regenerative medicine therapies. However, it should serve as a warning that the "fourth hurdle" (beyond the classical regulatory requirements of safety, efficacy, and quality) is a reality, and careful planning as early as Phase II is an absolute must.

From a public health view, we hope that national authorities will extend helping hands to small enterprises struggling to navigate the maze of different national systems and show flexibility in the application of classical reimbursement rules. European governments that have often invested substantial public funds to encourage biotechnology, and particularly regenerative medicine projects, must realize that no industry can flourish without a viable commercial outlet for its products. Given the economics of the European healthcare market, and specifically its pharmaceutical component, this requires *de facto* allowing *public* coverage of innovative regenerative medicine products, at least on a trial basis. This may force public bodies such as NICE, IQWiG (German Institute for Quality and Efficiency in Healthcare), and the French HAS to adopt specific rules or procedures that would take into account the unique characteristics and delivery modes of these products. If this does not happen, it can be safely forecast that the regenerative medicine industry in Europe will remain an academic pursuit, without any commercial potential or it will remain restricted to the realm of private payers, i.e., a luxury product available only to the rich.

REFERENCES

ACTIVE trial website. www.active-trial.org.uk/Templates/main.dwt (accessed November 23, 2009).

Brekke K. R., A. L. Grasdal, and T. H. Holmås. 2008. Regulation and pricing of pharmaceuticals: reference pricing or price cap. *Eur Econ Rev.*

Brekke K. R., I. Königbauer, and O. R. Straume. 2007. Reference pricing of pharmaceuticals. *J Health Econ.* 26: 613–642.

Brittberg M., A. Lindahl, A. Nilsson et al. 1994. Treatment of deep cartilage defects in the knee with autologous chondrocyte transplantation. *New Engl J Med.* 331: 889–895.

Clar C, E. Cummins, L. McIntyre et al. 2005. Clinical and cost-effectiveness of autologous chondrocyte implantation for cartilage defects in knee joints: systematic review and economic evaluation. *Health Technol Assess* 9: 1–98.

Cleemput I., P. Van Wilder, F. Vrijens et al. 2008. *Guidelines for Pharmacoeconomic Evaluations in Belgium.* KCE reports 78C.

Donatini A., A. Rico, M. G. D'Ambrosio et al. 2001. *Health Care Systems in Transition*, Rome: European Observatory on Health Care Systems and Policies.

Durán A., J. L. Lara, and M. van Waveren. 2006. Spain: health system review. *Health Care Systems in Transition,* Vol. 8. Rome; European Observatory on Health Care Systems and Policies.

Espín J. and R. Rovira. 2007. Analysis of differences and commonalities in pricing and reimbursement systems in Europe. A study funded by DG Enterprise and Industry of the European Commission. Escuela Andaluza de Salud Pública (Andalusian School of Public Health) and University of Barcelona.

Frostelid T., T. Hansen., K. Kim Sveen et al. 2007. Pharmaceutical pricing and reimbursement information: Norway. Gesundheit Österreich GmbH/Geschäftsbereich ÖBIG.

Gerlier L., M. Lamotte, M. Wille et al. 2009. Cost utility of autologous chondrocyte implantation using ChondroCelect in symptomatic knee cartilage defects. Poster presented at ISPOR 12th Annual European Congress, October 24–27, Paris.

Glenngård A. H., F. Hjalte, M. Svensson et al. 2005. Sweden: health system review. *Health Care Systems in Transition,* Vol. 7. Rome: European Observatory on Health Care Systems and Policies.

Haute Autorité de Santé. 2005. Service évaluation en santé publique: Évaluation de la Greffe Chondrocytaire Autologue du genou: Rapport d'Étape.
Hofmarcher M. M. and H. M. Rack. 2006. Austria: health system review. *Health Care Systems in Transition*, Vol. 8. Rome: European Observatory on Health Care Systems and Policies.
Hayhurst K. P., P. Brown, and S. W. Lewis. 2003. Postcode prescribing for schizophrenia. *Br J Psychiatr* 182: 281–283.
ISPOR. 2009. Description of Swedish payment and reimbursement system. www.ispor.org/htaroadmaps/Sweden.asp (accessed November 23, 2009).
Järvelin J. 2002. Finland, Health Care Systems in Transition, Vol. 4 No. 1, *European Observatory on Health Systems and Policies*.
Johnsen J. R. 2006. Norway: health system review. *Health Care Systems in Transition*, Vol. 8. Rome: European Observatory on Health Care Systems and Policies.
Jones R. W. 2004. Dementia, postcode prescribing and NICE. *Age and Ageing* 33: 331–332.
Kälin S. 2009. OrganoGenesis Switzerland, personal communication, November.
Kemp P. 2006. History of regenerative medicine: looking backwards to move forwards. *Regenerative Med* 1: 653–669.
Kraus L. 2004. Medication misadventures: the interaction of international reference pricing and parallel trade in the pharmaceutical industry. *Vanderbilt J Transnat Law* 37.
Leopold C. and C. Habl. 2008. Pharmaceutical pricing and reimbursement information: Austria. Gesundheit Österreich GmbH/Geschäftsbereich ÖBIG.
Lysaght M. J. and A. L. Hazlehurst. 2004. Tissue engineering: the end of the beginning. *Tissue Eng* 10: 309–320.
Lysaght M. J., A. Jaklenec, and E. Deweerd. 2008. Great expectations: private sector activity in tissue engineering, regenerative medicine and stem cell therapeutics. *Tissue Eng* 14: 305–315.
Martin P., R. Hawksley, and A. Turner. 2009. The commercial development of cell therapy: lessons for the future? Part 1: Summary of findings. EPSRC report. www.nottingham.ac.uk/iss (accessed November 23, 2009).
Martini N., P. Folino Gallo, and S. Montilla. 2007. Pharmaceutical pricing and reimbursement information: Italy. Gesundheit Österreich GmbH/Geschäftsbereich ÖBIG.
Messner K. and W. Maletius. 1996. The long-term prognosis for severe damage to weight-bearing cartilage in the knee: a 14-year clinical and radiographic follow-up in 28 young athletes. *Acta Orthop Scand* 67: 165–168.
Mossalios E., M. Mrazek, and T. Walley, Eds. 2004. *Regulating Pharmaceuticals in Europe: Striving for Efficiency, Equity and Quality*. European Observatory on Health Care Systems Series, Open University Press.
Mossalios E. and S. Thomson. 2004. Voluntary health insurance in the European Union. World Health Organization, on behalf of European Observatory on Health Care Systems and Policies.
NHS Postcode Lottery weblog, http://nhspostcodelottery.blogspot.com/ (accessed November 27, 2009).
OECD. 2008. Health Policy Studies, Pharmaceutical Pricing Policies in a Global Market.
Pedrosa-Vasco J. and E. Alves da Silva F. Country Profile: Portugal: Pharmaceutical Pricing and Reimbursement. DGCC, Ministry of Finance, and INFARMED. ec.europa.eu/enterprise/phabiocom/docs/tse/portugal.pdf (accessed November 23, 2009).
Peura S., S. Rajaniemi, and U. Kurkijärvi. 2007. Pharmaceutical pricing and reimbursement information: Finland. Gesundheit Österreich GmbH/Geschäftsbereich ÖBIG.
Pita Barros P. and J. de Almeida Simões. 2007. Portugal: health system review. *Health Care Systems in Transition*, Vol. 9. European Observatory on Health Care Systems and Policies.
Plagnol A. C., E. Rowley, P. Martin et al. 2009. Industry perceptions of barriers to commercialization of regenerative medicine products in the UK. *Regen Med* 4: 549–559.

Prakash D. and D. Learmonth. 2002. Natural progression of osteochondral defect in the femoral condyle. *Knee* 9: 7–10.

Redman T. and M. Köping Höggård. 2007. Pharmaceutical pricing and reimbursement information: Sweden. Gesundheit Österreich GmbH/Geschäftsbereich ÖBIG.

Saris D. B., J. Vanlauwe, J. Victor et al. 2008. Characterized chondrocyte implantation results in better structural repair when treating symptomatic cartilage defects of the knee in a randomized controlled trial versus microfracture. *Am J Sports Med* 36: 235–246.

Saris D. B., J. Vanlauwe, J. Victor et al. 2009. Treatment of symptomatic cartilage defects of the knee: characterized chondrocyte implantation results in better clinical outcome at 36 months in a randomized trial compared to microfracture. *Am J Sports Med* 37: 10S–19S.

Schoonveld E. 2001. Market Segmentation and International Price Referencing. Remarks at the WHO-WTO Workshop on Differential Pricing and Financing of Essential Drugs. www.wto.org/english/tratop_e/trips_e/hosbjor_presentations_e/36schoonveld_e.doc (accessed November 27, 2009).

Shelbourne K. D., S. Jari, and T. Gray. 2003. Outcome of untreated traumatic articular cartilage defects of the knee: a natural history study. *J Bone Joint Surg Am* 85: 8–16.

Strandberg-Larsen M., M. B. Nielsen, S. Vallgårda et al. 2007: Denmark: health system review. *Health Care Systems in Transition,* Vol. 9. European Observatory on Health Care Systems and Policies.

Teixeira I. and I. Vieira. 2008. Pharmaceutical pricing and reimbursement information: Portugal. Gesundheit Österreich GmbH/Geschäftsbereich ÖBIG.

Thomsen E., S. Safiye Er, and P. Fonnesbæk Rasmussen. 2008. Pharmaceutical pricing and reimbursement information: Denmark. Gesundheit Österreich GmbH/Geschäftsbereich ÖBIG.

Torstensson D. and M. Pugatch. 2009. Europe's postcode lottery: the challenge of central authorisation versus national access to medicines. Stockholm Network.

U.S. Department of Commerce. 2004. Pharmaceutical price controls in OECD countries: implications for U.S. consumers, pricing, research and development, and innovation. International Trade Administration, Washington, December.

Vogler S., C. Habl, C. Leopold et al. 2008. Pharmaceutical pricing and reimbursement information: Austria. Gesundheit Österreich GmbH/Geschäftsbereich ÖBIG.

White C. 2004. NICE guidance has failed to end postcode prescribing. *Br Med J* 328: 1277.

Wound Source. www.woundsource.com/news/organogenesis-inc-announces-apligraf-cell-therapy-reimbursed-switzerland (accessed November 23, 2009).

22 Cellular Therapies and Regenerative Medicine
Preparing for Reimbursement in the United States

Eric Faulkner

CONTENTS

22.1 Overview 351
22.2 Influence of Regulatory Pathway on Reimbursement Opportunities 352
22.3 Coding and Payment Considerations 353
      22.3.1 Coding and Payment System: General Considerations 354
22.4 Health Technology Assessment and Coverage 358
      22.4.1 Early Evidence of Payer Positions on Cellular Therapies and Regenerative Medicine 359
22.5 Costs and Economic Considerations 365
22.6 Conclusions 366
References 367

22.1 OVERVIEW

Cellular and regenerative medicine technologies are poised to improve treatment options and health outcomes in a variety of disease areas in the coming years. These innovative technologies not only hold the potential to transform healthcare in a manner similar to the ways in which monoclonal antibodies and other biologics have done so over the past decade. Some may even offer cures or prolonged therapeutic effects in certain chronic diseases.

Despite this promise, U.S. payers, providers, and policy makers currently face the daunting reality of integrating a rising tide of emerging health technologies into a national health system fraught with quality and cost challenges. On the front lines of healthcare scrutiny are emerging technologies including diagnostics and imaging modalities, personalized medicines, complex combination products, biologics, and regenerative medicine which promise increasingly personalized, more efficient, and cost-effective care delivery. The Congressional Budget Office (CBO), Institute of Medicine (IOM), and other influential groups have cited the availability of innovative and breakthrough healthcare technologies and lack of clarity surrounding their use as a significant driver of increased healthcare spending (U.S. Congressional Budget Office 2008; National Academy of Sciences 2008).

Expansion of biologics and targeted therapies is illustrative of the opportunities and challenges associated with acceptance of new technologies. In 2005, approximately 350 biologics were in Phase III trials or undergoing FDA review; over 2,000 others were in early development (Watkins et al. 2007). A recent study of Blue Cross/Blue Shield plans reported that spending on specialty pharma products had risen almost 35% between 2002 and 2003 and these products were estimated to represent 25% of all outpatient pharmacy spending by 2008 (Mullins et al. 2007; Watkins 2007; Stern and Reissman 2006). As the cost to patients of some specialty pharma products approaches or exceeds $10,000 per month, overall affordability and access are key considerations for U.S. healthcare, despite the potential values of such treatments (Fish 2006). While some treatments may markedly improve mortality and quality of life through targeted treatment, others will only offer marginal benefits—because of these trends, U.S. payers are becoming less accepting of costly technologies as focus on cost controls and health reform has increased.

Of all factors influencing the viability and success of cellular therapies and regenerative medicine, securing sufficient reimbursement is one of the most formidable hurdles that that any new health technology must overcome. Because many regenerative medicine therapies are truly novel, payers are likely to approach acceptance with caution and require significant evidence regarding treatment safety, effectiveness, and duration of therapeutic effect until their use becomes more transparent and common. While some early regenerative medicine technologies exhibit similar manufacturing and delivery characteristics compared to conventional biologics and medical devices, others will be extremely complex, costly, and face significant market access risks and hurdles. Because of these and other factors, regenerative medicines require strong clinical and economic value propositions to establish themselves in an increasingly restrictive reimbursement environment.

This chapter considers cellular therapies and regenerative medicine technologies in the context of a rapidly evolving U.S. reimbursement environment.

22.2 INFLUENCE OF REGULATORY PATHWAY ON REIMBURSEMENT OPPORTUNITIES

One of the key considerations in assessing the reimbursement potential of cellular and regenerative medicine therapies is whether the technology will be regulated as a biologic, drug, or medical device. Regulatory designation of regenerative medicine therapies plays a strong role in how public and private payers will view the therapy for reimbursement purposes and can accelerate or diminish market access upon market clearance.

Just as regulatory approval for devices, procedures, and drugs often flows through different channels in many markets, U.S. health technology assessment is frequently conducted by myriad health technology assessment (HTA) review bodies. Reimbursement may involve multiple payment systems, depending on the site of care and process of administration. Regulatory designation of a new regenerative medicine technology as a medical device, procedure, service, or drug may enhance or preclude reimbursement as a drug in some global markets and/or care settings—particularly if regulatory designation of the technology differs from the intended reimbursement applications (i.e., approval as a device may preclude reimbursement as a drug).

For example, Genzyme's autologous chondrocyte therapy (Carticel) was the first cellular product approved by the U.S. Food & Drug Administration (FDA) in 1997. While this therapy requires two procedural steps (arthroscopy and biopsy to obtain the cells followed by surgical preparation of the cartilage defect and attachment of a periosteal patch behind which Carticel is injected), it was developed and is regulated and reimbursed as a *drug* with its own unique "J" code, enabling separate billing for the cells in the outpatient setting (De Bie 2007).

On the other hand, some cellular therapies in development for cardiac indications may involve complex procedural steps, depend on multiple medical devices for delivery, and require inpatient observation periods following cell administration. Under such scenarios, it is unclear whether U.S. payers will consider the cell therapy as part of a bundled procedural payment (e.g., a diagnosis-related group [DRG] or case-rate payment ceiling) or allow separate payment for the cells as a drug, enabling greater pricing latitude.

In addition, fulfillment of regulatory requirements for market clearance is often minimally sufficient to meet payer requirements for reimbursement (Raab and Parr 2006). This may be particularly true for regenerative medicine technologies regulated as medical devices due to differences in clinical trial design expectations and under scenarios in which device approval is based on demonstration of equivalency to a predicate device. Because regenerative medicine technologies are new and relatively few on-market examples exist, understanding the consequences of regulatory designation on reimbursement strategy is a critical first step. As noted above, this may depend on a variety of factors, including procedural complexity, site of administration, and cost of the therapy.

22.3 CODING AND PAYMENT CONSIDERATIONS

Reimbursement in the U.S. consists of three core elements: coding, coverage, and payment. These three elements comprise the "dogma" required to obtain reimbursement and failure to sufficiently achieve any one of these three elements often precludes market access to emerging technologies (see Figure 22.1). Well designed clinical and economic evidence justifying the value of emerging technologies is the common-thread foundation or prerequisite for achieving all three components of successful reimbursement. This section covers coding and payment considerations applicable to regenerative medicine therapies and the following section provides an overview of the current coverage landscape for these therapies.

In the U.S., we do not pay for technologies or procedures. Payers fund the codes that describe the procedures. Inadequate or overly specific coding descriptions may preclude compliant use of specific codes to file payment claims for new technologies. Where appropriate coding is inadequate or unavailable, a manufacturer may be able to use non-specific codes or opt to pursue a new code—a long and resource-intensive process in the U.S.

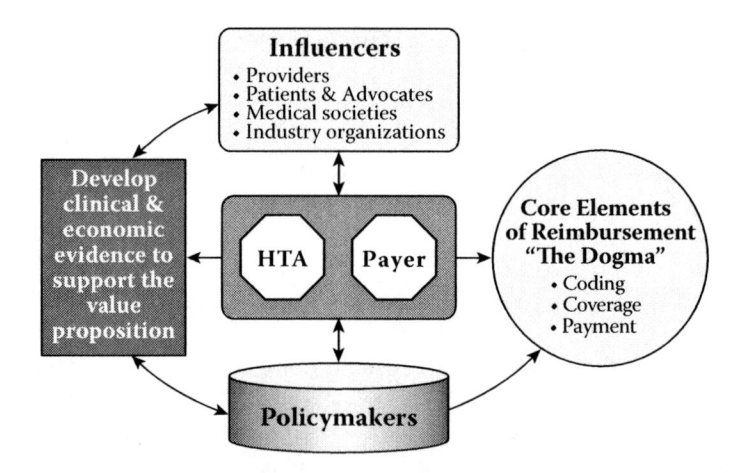

FIGURE 22.1 Core elements of U.S. reimbursement.

Coverage policies represent another core element of reimbursement in the U.S. because they define specific indications and patient populations (including subpopulations) for which coverage is approved. These policies also spell out non-coverage of technologies or applications deemed experimental or investigative, contraindications associated with technology use, and other applicable key coverage restrictions.

Finally, payment systems applicable to inpatient, outpatient, and physician office settings dictate whether an individual technology payment is bundled into an overall procedural payment (as in the case of the Medicare DRG system) or is eligible for separate payment (and therefore often have greater pricing latitude). In scenarios where payment rates associated with specific billing codes are well established, manufacturers must assess whether payment associated with those codes is adequate to cover the cost of providing the technology and the likelihood of achieving a higher payment.

22.3.1 Coding and Payment System: General Considerations

Several coding systems are commonly used in the U.S. to describe healthcare services. Some systems are used across specific sites of care (e.g., inpatient, outpatient, physician office) and others typically are relevant only to single specific sites of care. The most common coding systems applicable to regenerative medicine therapies in the U.S. are:

- International Statistical Classification of Diseases and Related Health Problems (ICD) coding is a universally accepted global system published by the World Health Organization (WHO) and used to classify diseases, signs, symptoms, and other causes of injury or disease. It is used to identify a specific disease or condition for which a patient seeks treatment in all payment settings or systems.

- The Diagnosis-Related Group (DRG) system consists of codes used to describe inpatient procedure groups with similar resource use under the Medicare Hospital Inpatient Prospective Payment System (HIPPS). Codes for specific procedures are severity-adjusted and have hierarchical payment rates based on disease severity and DRG payment levels, and include all costs associated with facility use, services, health technologies required for the procedure, and clinician time. While some commercial payers may use the DRG system, they often refer to payments associated with DRG equivalents as case rates.
- The Current Procedural Terminology (CPT) coding system is maintained by the American Health Association (AHA) and uniformly describes medical, surgical, and diagnostic services used in the inpatient, outpatient, and physician office settings. The system is divided into three categories including those describing established technologies, performance measurement, and emerging technologies. CPT is identified as level 1 of the Healthcare Common Procedure Coding System.
- The Healthcare Common Procedure Coding System (HCPCS) is standardized and maintained by the Centers for Medicare and Medicaid (CMS) for describing specific healthcare procedures or services to facilitate consistent claims processing.

Site of care is critical to reimbursement strategy in the U.S. Each site of care (inpatient, outpatient, physician office, clinical laboratory, ambulatory care facility), particularly in the Medicare system, typically involves specific coding and payment requirements (bundling of payments, limitations on obtaining new codes, and associated payment levels). Although commercial payers have adopted and adapted many of the coding systems used by Medicare, they generally maintain discretion in payment level setting for case rates, determination of items or services included in specific payments, and duration of payment. Regenerative medicine therapies, depending upon their complexity, may also engage multiple payment systems, each with its own attendant set of rules and reimbursement implications.

Differential site of care coding and payment considerations are key reasons why designation as a drug or medical device is so important to complex regenerative medicine technologies. Coding systems, with rare exceptions (e.g., drug eluting stents), typically do not describe individual technologies (especially medical devices), but instead describe the procedure for which the technology will be used (diagnosis or treatment). Except for some physician administered and other drugs eligible for HCPCS codes, virtually all drugs have unique identifiers (National Drug Codes [NDCs]) assigned at the time of FDA market clearance that can be used to bill for their use. Even under HCPCS, most drugs have unique identifying codes.

Because of this difference in coding, drugs generally have greater pricing latitude compared to devices that must often price in a manner that supports physician use at or near a prospectively determined procedural payment amount. In addition, unlike drug costs, medical device costs are more likely to be bundled into payment rates that include facility and other costs, depending upon the site of care (see Table 22.1).

TABLE 22.1
General Medicare Site-of-Care Payment Considerations for Medical Devices and Diagnostics

Site of Care	Medical Devices	Drugs
Hospital inpatient	Payment bundled at applicable DRG payment rate under Hospital Inpatient Prospective Payment System (HIPPS)	Payment bundled under applicable DRG payment rate
Hospital outpatient	Payment typically bundled into appropriate Ambulatory Procedural Classification (APC) payment rate described in Hospital Outpatient Prospective Payment System (HOPPS)	Payment for inexpensive drugs ($65 or less) bundled into appropriate Ambulatory Procedural Classification (APC) payment rate. Payment is separate for drugs greater than $65
Physician	Payment typically bundled into appropriate procedural payment under Medicare Physician Fee Schedule (MPFS)	Payment generally separate

Bundled versus separate payment considerations—In an inpatient setting, the costs of devices and drugs are often bundled into a single prospectively determined payment that covers all average costs required to perform a procedure. While provisions to address outlier costs can sometimes be negotiated separately with commercial payers, in general the price of health technologies used in executing a procedure must be such that they do not require providers accessing these technologies to take a financial loss on the procedure. Treatment in the inpatient setting, particularly under Medicare, offers regenerative medicine developers the least flexibility to pursue value-based pricing. Manufacturers must first assess whether an existing DRG appropriately describes the new procedure. Obtaining a new Medicare DRG or add-on payment for new cell therapy procedures is unlikely and manufacturers must generally assess whether a procedure can be offered in a compliant and cost-effective manner under an existing DRG. Commercial payers may be somewhat more flexible in establishing case rates for new procedures and/or negotiating specific contract terms.

In contrast, drugs are often eligible for separate payments in an outpatient or physician office setting, while medical device payment frequently remains bundled as a component of a procedural payment. Because device costs must often "fit" procedural payment requirements, devices often have more limited pricing latitude compared to drugs. Further, even though cell-based regenerative medicine therapies (e.g., autologous or allogeneic cell therapies) may be developed and regulated as drugs, the processes involved in cellular extraction and purification may be heavily reliant on the use of medical devices (apheresis equipment, cell selection and purification devices, cellular expansion systems) whose costs must also be covered by available reimbursement mechanisms.

Reimbursement of key regenerative medicine therapy steps and components—Depending on whether a regenerative medicine technology is based on autologous or allogeneic cell delivery, a variety of codes may be relevant, including but not limited to the following:

- Cell mobilization to induce the bone marrow to produce stem cells
- Apheresis to extract cells from the circulatory system
- Cell purification and selection to derive a purified dose of cells with certain characteristics (specific cell surface markers thought to improve potency)
- Cell thawing and expansion
- A variety of drug administration and/or procedural codes to support cell delivery (e.g., therapeutic catheterization) that may involve use of existing or novel delivery devices that have not yet achieved reimbursement approval

While many steps involved in regenerative medicine (e.g., apheresis) are adequately described under existing coding systems, other steps such as cell purification and/or selection are not. For example, obtaining appropriate reimbursement for cell selection steps is often a concern for many cell therapy developers because the cost of this step can generally range from $2,000 to $5,000 irrespective of the selection method used.

Due to such hurdles, many cell therapy manufacturers are considering options to embed the cost of cell purification and preparation (expansion, manipulation, potency and viability testing, genetic manipulation) into the cost of the therapy (often considerably easier to do if a therapy is developed as a drug). A similar phenomenon is currently occurring in personalized medicine. Targeted treatment manufacturers are considering options to absorb the costs of associated diagnostic tests into the prices of their drugs (Blaire 2008). Failure to obtain reimbursement for any one step in the regenerative medicine procedure can compromise or preclude adequate reimbursement and/or provider adoption of the entire procedure.

Availability of separate payments in the U.S. generally increases provider incentives to select and use a particular treatment because the provider does not have to compromise margins on existing profitable procedures based on technology selection. Alternatively, incorporation of technologies into a bundled procedural payment can negatively impact technology utilization if adoption results in a significant financial loss to a provider and affordable, effective alternatives are available.

Because of the complexity of U.S. coding and payment systems, it is critical for regenerative medicine manufacturers to assess the reimbursement implications of their procedures early in technology development as addressing certain gaps such as the pursuit of a new code may require significant time and resources or preclude certain development options. A manufacturer should seek to answer key strategic questions:

- What site-of-care requirements and payment systems are relevant?
- Are drug or device component costs bundled or available for separate payment in key scenarios? Will payers limit access to risky procedures (e.g., cardiac or neurologic applications) to inpatient coverage or academic centers of excellence?
- Are payment levels associated with existing codes adequate to cover costs of the treatment?

Failure to proactively address such questions at early stages may result in prolonged timelines, unnecessary expenditures, and, in the worst case, pursuit of development scenarios that are not supported by the existing U.S. reimbursement infrastructure.

22.4 HEALTH TECHNOLOGY ASSESSMENT AND COVERAGE

Health technology assessment (HTA) in the U.S. is a highly decentralized process compared to other global markets. HTA is conducted by a variety of stakeholders including government agencies, commercial payers, and private for-profit and non-profit entities. While these organizations generally employ similar methodologies in evaluating new health technologies, they vary considerably in regard to mission, sophistication, and resource availability. U.S. HTA bodies are also influenced by different decision pressures and may reach markedly different conclusions regarding the sufficiency of evidence to support reimbursement. The following list reflects the main HTA stakeholders relevant to the U.S. marketplace.

- Public Payers
  - Centers for Medicare and Medicaid Services (CMS)
  - U.S. Department of Veterans Affairs (VA) and Department of Defense (DoD) for military health benefits
- Government Agencies
  - Agency for Healthcare Research and Quality (AHRQ)
- Commercial (Private) Payers
  - Top five commercial payers (Aetna, Cigna, Humana, United Healthcare, and WellPoint)
  - Hundreds of smaller regional payers including BlueCross/BlueShield plans
- Private and Non-Profit Organizations
  - BlueCross/BlueShield (BCBS) Technology Evaluation Center (TEC)
  - Hayes, Inc.
  - ECRI

Of these HTA groups, CMS, AHRQ, the leading commercial payers, and the BCBS TEC are viewed as the most transparent and influential for establishing coverage. Unlike nationally representative groups like the National Institute for Health and Clinical Excellence (NICE) and the Institute for Quality and Efficiency in Healthcare (IQWiG) in Germany, some HTA groups such as AHRQ and BCBS TEC report only on the sufficiency of evidence. They do not make specific reimbursement recommendations; they are generally left to the individual payer organizations. HTA reports from Hayes Inc. and ECRI are also frequently used by payers on a subscription basis (85% of U.S. commercial payers use Hayes reports to inform coverage determinations for new technologies) and reports from these organizations contain recommendations for reimbursement (including suggested limitations), though HTA reports developed by these organizations have historically not been available for purchase or review by technology manufacturers (Faulkner 2009).

Commercial payers range broadly in terms of the scopes of their HTA approaches, from complex groups of medical and pharmacy directors and support staff across the country responsible for assessment in the largest commercial plans to a single medical or pharmacy director with limited or no support staff in the smallest commercial plans. Thus, the various commercial payer organizations differ significantly in terms of the volume of HTAs conducted annually to support coverage policies and sources of information and approaches used to inform coverage policy development. Smaller plans may leverage publicly available HTAs and systematic reviews of evidence to inform coverage determinations, given more limited resources available to such payers.

22.4.1 Early Evidence of Payer Positions on Cellular Therapies and Regenerative Medicine

Although few cellular therapies and regenerative medicine technologies have entered the U.S. market, payer coverage positions are illustrative of early payer perspectives on these technologies. Prior to the current and expanding emphasis on regenerative medicine, most cellular therapy in the U.S. involved autologous and allogeneic bone marrow transplant (BMT) procedures for oncology indications that have well established reimbursement mechanisms and payment rates. While evaluation of BMT can be informative related to reimbursement of some procedural steps associated with regenerative medicine (cell mobilization, apheresis), consideration of the more limited number of HTAs and coverage policies applicable to the types of therapies due to enter the market in the next 2 to 5 years offers a better view of the U.S. payer landscape for these therapies.

To date, aside from evaluation of BMT, HTAs of regenerative medicine therapies conducted by AHRQ and BCBS TEC have been very limited. This is not surprising considering that most commercially targeted regenerative medicine therapies remain in clinical development at present. In September 2008, the BCBS TEC conducted an assessment of autologous progenitor cell therapy for treatment of myocardial ischemia (BlueCross/BlueShield Technology Evaluation Center 2008). This assessment concluded that while improvement in left ventricular ejection fraction (LVEF) occurred in several studies, improvement was minimal and no evidence of "harder outcomes" (improvements in mortality, morbidity, and quality of life) was reported. In addition, the BCBS TEC noted that (1) study designs and patient sample sizes were insufficient to enable meaningful conclusions regarding effectiveness, (2) the mechanisms of action for cellular therapies have not been sufficiently established, and (3) study protocols were not sufficiently standardized to enable comparison of results across available studies.

To date, CMS positions on regenerative medicine technologies also remain limited. The Medicare Coverage Analysis Group (CAG) rendered National Coverage Determinations (NCDs) on only three applications of autologous cell therapies: for treatment of amyloidosis (CAG-00050R), following high dose chemotherapy in multiple myeloma (CAG-00011R), and for treatment of chronic non-healing wounds (CAG-00190R2). In the case of autologous stem cell treatment for amyloidosis and multiple myeloma, coverage is allowed but restricted to patients who meet one or more inclusion criteria.

CMS rendered a negative NCD on use of autologous therapies for treatment of chronic non-healing wounds, citing inadequate evidence that the therapy improves patient health outcomes and is medically necessary. These existing policies suggest that CMS may support regenerative medicine technologies that adequately improve patient-centered health outcomes (mortality, morbidity, and quality of life), target key areas of unmet medical need, and are supported by adequately designed clinical trials that include sufficient follow-up to enable characterization of longer term treatment effects.

Alternatively, various U.S. commercial payers have issued policies on regenerative medicine technologies. A majority of these are non-coverage policies for therapies that have not yet entered the U.S. market (autologous or allogeneic cell therapies for cardiovascular applications). Issuance of non-coverage policies prior to availability of technology is a rare occurrence since payers must expend significant resources to maintain written policies during recurrent review processes. In the case of regenerative medicine therapies, two likely reasons drive the development of policies prior to introduction of commercial therapies: (1) since some cell therapy applications do not require FDA approval, payers may be concerned about physician billing of investigational cell therapy applications via existing payment codes and (2) payers will require robust, clinically meaningful evidence of safety and impact on patient outcomes and have implemented these policies in advance of claims based upon insufficient evidence. Table 22.2 illustrates how recent U.S. commercial payer coverage policies have addressed regenerative medicine technologies.

With the exception of autologous chondrocyte implantation (ACI) and some limited wound healing applications, commercial payer policies on regenerative medicine technologies have generally classified regenerative medicine therapies as investigational or experimental and thus non-covered. This is not to say that payers will not cover emerging regenerative medicine technologies upon generation of sufficient evidence to clarify the clinical value of treatments, only that at present sufficient evidence for many therapies does not exist to support regenerative medicine therapies as medically necessary. Further, such "pre-emptive" non-coverage policies also suggest that evidence hurdles for regenerative medicine technologies are likely to be substantial. These general conclusions are consistent with the majority of reimbursement recommendations for regenerative medicine technologies recently issued by NICE (UK) and IQWiG (Germany) for ACI and use of autologous cell therapies as an adjunct to high-dose chemotherapies.

Even in the case of therapies that have been on the market for some time (ACI and skin substitutes), payers still have significant concerns about the comparative effectiveness of available regenerative medicine therapies versus existing and in some cases cheaper treatment alternatives. This is particularly true of those strongly dependent on difficult-to-interpret (e.g., patient-reported) outcomes for applications such as ACI and surrogate outcomes for which "hard" clinical endpoints are unavailable. Because of this, second generation regenerative medicine technologies may face higher hurdles to market, particularly if premium pricing is expected, compared to forerunner predicates at the vanguard of regenerative medicine.

TABLE 22.2
U.S. Commercial Payer Coverage Policies on Regenerative Medicine Technologies

Application	Coverage Determination	Coverage Restrictions	Key Supporting Rationale
Autologous chondrocyte therapy [a,b,c]	MN/C	Limited to treatment of femoral condyle only (most policies) Patients must meet all of 5-11 criteria (varies by payer) Limited to patients with certain physical symptoms (e.g., age, BMI) Use restricted where osteoarthritis is primary cause	Prospective longitudinal study of 154 patients with 4-year median follow-up provided favorable results Five-year comparison to microfracture reported favorable results Uncertainty remains about effectiveness including interpretation of functional and quality-of-life data
Cardiovascular applications [a,e,f]	I/E; non-covered	—	Existing studies, systematic reviews, and meta-analyses reported minimally significant improvement in anatomic functions Some studies reported cardiac arrhythmias Outcomes of most studies were surrogate; patient-centered outcomes including morbidity and mortality needed Many studies are of limited duration to characterize long-term treatment outcomes
Autoimmune diseases[g] including multiple sclerosis, rheumatoid arthritis, systemic sclerosis (also known as scleroderma), systemic lupus erythematosus, and juvenile idiopathic arthritis	I/E; non-covered	—	Limited evidence and poor study designs (heterogeneous patient selection, variability of conditioning regimens) Lack of appropriate randomized, controlled trials with sufficient follow-up Higher transplant-related mortality and morbidity compared to non-autoimmune transplants

continued

TABLE 22.2 (continued)
U.S. Commercial Payer Coverage Policies on Regenerative Medicine Technologies

Application	Coverage Determination	Coverage Restrictions	Key Supporting Rationale
Wound healing and tissue grafting[h]	MN/C (Alloderm, Apligraf, Dermagraft, OrCel, Regranex, Transcyte) I/E (Autogel, SafeBlood, Vitagel)	Allogeneic growth factors limited to FDA-approved indication; all other uses are I/E Autologous products evaluated under this policy considered I/E One xenographic skin substitute MN; all others considered I/E	Covered products usually supported by randomized, controlled trials, multiple observational studies, or case series Key outcomes: more rapid healing of skin damage and/or pressure ulcers Autologous products not supported by peer-reviewed publications at time of review
Stem cell transplantation for epithelial ovarian cancer[i]	I/E; non-covered	—	No significant improvement in time to progression or overall survival compared to high-dose chemotherapy No comparative group or low response rate (some clinical studies) High treatment-related morbidity with hematopoietic stem cell therapy

MN/C = medically necessary under certain circumstances.
I/E = investigational or experimental.

[a] Autologous chondrocyte transplantation. Cigna Medical Coverage Policy 0105. 6/15/2009. http://www.cigna.com/customer_care/healthcare_professional/coverage_positions/medical/mm_0105_coveragepositioncriteria_autologous_chondrocyte_transplantation.pdf

[b] Autologous chondrocyte transplantation. Blue Cross/Blue Shield of Rhode Island Medical Coverage Policy. 7/7/2009. https://www.bcbsri.com/BCBSRIWeb/plansandservices/services/medical_policies/AutologousChondrocyteTransplantation.jsp

[c] Autologous chondrocyte implantation. Aetna Clinical Policy Bulletin 0247. 05/08/2009. http://www.aetna.com/cpb/medical/data/200_299/0247.html

[d] Autologous skeletal myoblast transplant/autologous cell therapy for damaged myocardium. Cigna Medical Coverage Policy 0287. 2/15/2009. http://www.cigna.com/customer_care/healthcare_professional/coverage_positions/medical/mm_0287_coveragepositioncriteria_autologous_myoblast_damaged_myocardium.pdf

[e] Autologous cell therapy for the treatment of damaged myocardium. Blue Cross/Blue Shield of North Carolina Corporate Medical Policy 1042. 10/01/2009. http://www.bcbsnc.com/assets/services/public/pdfs/medicalpolicy/autologous_cell_therapy_for_the_treatment_of_damaged_myocardium.pdf

continued

TABLE 22.2 (continued)
U.S. Commercial Payer Coverage Policies on Regenerative Medicine Technologies

f Autologous cell therapy for the treatment of damaged myocardium. Anthem Medical Policy TRANS.00022. 4/22/2009. http://www.anthem.com/medicalpolicies/policies/mp_pw_a053829.htm
g Stem cell transplant for autoimmune diseases. Cigna Medical Coverage Policy 0357. 6/15/2009. http://www.cigna.com/customer_care/healthcare_professional/coverage_positions/medical/mm_0357_coveragepositioncriteria_stem_cell_transplant_autoimmune_diseases.pdf
h Autologous, allogeneic, xenographic, synthetic, and composite products for wound healing and soft tissue grafting. Anthem Medical Policy SURG00011. 10/21/2009. http://www.anthem.com/medicalpolicies/policies/mp_pw_a053309.htm
i Stem cell transplant for epithelial ovarian cancer. Cigna Medical Coverage Policy 0321. 4/15/2009. http://www.cigna.com.pr/customer_care/healthcare_professional/coverage_positions/medical/mm_0321_coveragepositioncriteria_stem_cell_transplant_ovarian_cancer.pdf

Alternatively, payers may accept a lower threshold of evidence for rare, orphan, or ultra-orphan indications with high unmet need and/or lack of available treatment alternatives. One example is the current Cigna policy that covers allogeneic hematopoietic stem cell transplantation for several inherited metabolic disorders including Gaucher's disease, Hurler syndrome, Maroteaux-Lamy syndrome, Krabbe's disease, and other excessively rare indications (Cigna Medical Coverage Policy 2009). Another example is pancreatic islet cell therapy following pancreatectomy, which has been accepted by Aetna in the U.S. and NICE in the UK (Aetna Clinical Policy Bulletin 2009a; National Institute for Health and Clinical Excellence 2008). Although payers noted several flaws in study designs supporting pancreatic islet cell therapy following pancreatectomy, studies demonstrated transparent and long-term outcomes (24 to 85% of patients remained insulin-independent at 6- and 18-month follow-ups and 75% of those responders remained insulin-independent at 5 years) that addressed relatively rare subsets of diabetic patients and addressed a disease area (diabetes) with exceedingly high clinical and economic burdens.

In addition, a minority of commercial payers will not cover certain steps involved in processing stem cells (ex vivo expansion, expansion of specific T cell lines, genetic modification; Aetna Clinical Policy Bulletin 2009b). In particular, genetic modification of stem cells remains the most broadly non-covered cell processing step due to uncertainties regarding the "behavior," safety, and effectiveness of such cells in vivo. As previously discussed, inability to obtain payment for key procedural steps associated with regenerative medicines may jeopardize reimbursement for an entire procedure. In the case of steps such as cellular expansion that have long been applied to biologic products, it is likely that regenerative medicine manufacturers will have to integrate such steps into the prices of cell-based drugs or procedures if separate payment for processing steps is unavailable. On the other hand, demonstration of the value of genetically modified cells is likely to require characterization and longer term value demonstration in clinical trials to gain payer acceptance.

TABLE 22.3
Overview of General U.S.Payer Clinical Evidence Requirements for Cell-Based Regenerative Medicine Therapies

- Studies should be well designed (RCTs preferable) and include homogeneous patient populations or subgroups; study protocols should be standardized to enable comparison to existing regenerative medicine studies for target indication.
- Well designed prospective observational studies may be accepted in U.S. if they include sufficient sample size, focus on payer-relevant endpoints, and enable robust characterization of effectiveness and safety (particularly for high-risk applications like cardiovascular cell therapies).
- Outcomes should be patient-centered; studies should focus on "hard" endpoints (mortality, morbidity, quality of life); reimbursement based on surrogate health outcomes, particularly for high-morbidity or fatal diseases highly uncertain.
- Surrogate endpoints (left ventricular ejection fraction, changes in biomarker levels, walking time) may be insufficient to secure coverage alone or in combination with other surrogate endpoints.
- PRO and HRQoL endpoints should be validated, easily interpretable by inclusion of instruments such as Standard Form 36 (SF-36) and clearly characterize *functional* improvements.
- Where established comparators exist (e.g., microfracture for knee cartilage repair), study designs should enable direct or indirect comparison to standard-of-care treatments
- Because many regenerative medicine technologies are novel and poorly understood compared to well established therapies (e.g., monoclonal antibody treatments), payers require longer term data to ensure that treatments are safe, effective, and cost-effective. Post-market registries may be required to support reimbursement of some regenerative medicine therapies.

The key requirements highlighted in Table 22.3 are remarkably consistent across multiple U.S. and non-U.S. sources of HTA coverage policies on cell-based regenerative medicine therapies to date and reflect general considerations for improving reimbursement potential and favorable pricing. While actual payer evidence requirements must be considered on a case-by-case basis and will be influenced by a variety of factors including unmet clinical need associated with a particular indication, severity of patient disease, and availability and cost of reasonable treatment alternatives, these requirements appear to be uniformly desired by payers across all scenarios.

Given the novelty of these treatments and their potential as means to extend therapeutic effects or possibly cure disease, consideration of unconventional alternatives to secure reimbursement may be warranted. Leading U.S. payers like CMS *may* be willing to consider clinical trial subsidies for regenerative medicine technologies that meet requirements under the current Medicare Clinical Trial Policy or conditional coverage of certain high-impact regenerative medicine under Coverage with Evidence Development National Coverage Determination policies to better characterize promising applications.

Likewise, leading commercial payers *may* consider pilot demonstration programs or outcomes-based risk-sharing agreements to enable value characterization of the most promising regenerative medicine technologies, provided that the indication is sufficiently "on the radar." Some leading payers also have limited clinical trial subsidy programs for indications exhibiting high unmet needs. Although these unconventional opportunities may offer alternative means to support product development and market access, manufacturers are cautioned that pursuits of these channels are frequently unsuccessful, impose high risk, and are currently resource intensive compared to traditional value demonstration approaches.

22.5 COSTS AND ECONOMIC CONSIDERATIONS

Because many, but not all, regenerative medicine technologies will likely require pricing similar to or greater than conventional biologic therapies, communication of the economic value of these therapies is essential. This is particularly true for crowded indications (e.g., biologics for many oncology indications, rheumatoid arthritis, psoriasis, and Crohn's disease) where a comparable clinical and/or economic outcome versus existing therapy is likely a minimal requirement for reimbursement in the evolving U.S. healthcare landscape. Payers will look at both the cost of the cells and also consider the cost implications of the entire procedure. In the most complex cases, procedural costs may rival complex cardiovascular surgical procedures or BMT; such high-cost scenarios will likely require clinically meaningful evidence of superiority versus standard-of-care alternatives to support reimbursement.

In scenarios where premium pricing compared to standard-of-care alternatives is desired or even required due to the additional development costs and complexity associated with cell-based therapies, it is likely that outcomes must be clinically superior to existing alternatives to secure broad coverage in the U.S. Payer uncertainty regarding longer term safety and effectiveness may also push evidentiary requirements toward superior outcomes, especially in high-risk and high-cost applications such as cardiac or neurological indications.

Unlike markets that require cost-effectiveness data as an *explicit* requirement for HTA (United Kingdom, Australia, Canada), U.S. HTA focuses predominantly on assessment of clinical benefit. Coverage policies are generally based on assessment of clinical evidence alone to establish medical necessity and relevant reimbursement restrictions. For example, the Medicare program is the largest U.S. payer, representing approximately 45 million covered lives, and has statutory prohibitions against evaluation of cost effectiveness as a routine component of HTA (Neumann and Greenberg 2009; Gold et al. 2007; Carpenter 2005). This is not to say that public and private payers do not consider volume and cost trade-offs in a general sense because high volume and/or cost technologies are frequently subject to more intense scrutiny. Even at the price negotiation and contracting stage, the vast majority of U.S. payers do not consider cost-effectiveness data (cost per quality adjusted life year) in the same manner as counterparts in markets outside the U.S., and have a difficult time integrating such ratios into their decision making. Even payers that will consider cost-effectiveness data, for example, WellPoint, the largest U.S. commercial payer representing approximately 35 million covered lives, does so in an optional, informal manner and places the burden of validating cost-effectiveness claims on the manufacturer (WellPoint 2008).

However, budget impact and/or cost offset models are considered by U.S. payers and frequently employed during price and contract negotiations. Commercial and public payers carefully consider the financial impacts of coverage decisions including overall budget impact on a health plan and the cost offset potential associated with granting access to a new technology. While U.S. commercial payers are often critical of manufacturer-developed economic models, many will consider such models if they are simple in design, enable the payer to enter its own beneficiary statistics in a plug-and-play manner, and incorporate comparative costs of standard-of-care treatments.

22.6 CONCLUSIONS

Cellular and regenerative medicine therapies show great promise and may offer innovative options to improve the quality and cost of care, including potential to cure disease in some cases as clinical research in this area progresses and initial therapies are tested in real-world medical practice. In the U.S. and other key global markets where regenerative medicine approaches will be commercialized, navigation of increasingly restrictive reimbursement channels and identifying creative evidence development, pricing, and market access strategies will be paramount to the success of these emerging technologies in an era of global health reform and cost containment.

One of the factors most essential to ensuring adequate market access for regenerative medicine technologies is appropriate alignment of development and launch plans with HTA agency and payer requirements. In early development phases, this involves appropriate due diligence review of the market-specific and indication-specific reimbursement landscape and scenario planning to assess coding, coverage, and payment options (in the context of existing therapies). The need for early reimbursement due diligence cannot be overstated for these therapies because the requirements are significantly more complex than those covering many stand-alone medical device and drug development scenarios.

In tandem with secondary research on coding and payment, as well as how decision makers have handled recent proxy technologies and identification of gaps in product value propositions, regenerative medicine therapy developers should consider (1) early payer engagement to review planned study protocols and identify areas of refinement to ensure that payer clinical and economic data preferences are adequately addressed and (2) limited market research to identify avoidable study design risks, validate reimbursement options, and understand payer-driven coverage hurdles.

Opportunities for early payer engagement have expanded in recent years. Groups like NICE and the Canadian Agency for Drugs and Technologies in Health (CADTH) have opened their processes to such initial iterative scientific discussions prior to conduct of pivotal trials. In the U.S., the CMS Coverage Analysis Group has a long established history of such early engagement meetings (similar to early meetings with the FDA). In contrast, commercial payers are less likely to directly engage manufacturers via in-person meetings, and interview-based qualitative research may be necessary to garner similar information from them.

Making a strong clinical value case for regenerative medicine technologies will become particularly important due to the increasingly cost-conscious and restrictive reimbursement environment in the U.S. Based on early HTA and reimbursement decisions, regenerative medicine technologies are squarely "on the payer radar" and will face significant scrutiny, particularly in their most costly iterations. Development of value communication tools such as an evidence-based Academy of Managed Care Pharmacy (AMCP) dossier (the closest equivalent to a reimbursement submission document recognized in the U.S.) and supporting economic model will enable consistent value communication to managed care decision makers and ensure that all elements of the treatment value proposition are clearly communicated (versus leaving individual HTA agencies to glean all aspects of treatment value from their own internal assessments). More broadly, because regenerative medicine technologies are truly novel, substantial efforts to educate payers, physicians, and patients will likely be required to support broad adoption and uptake in the U.S. and other major markets.

Regenerative medicine technologies represent one of the most promising breakthrough innovations evident across the technology landscape in this century. However, the voyage of these therapies to market is anticipated to be complex and require substantial planning en route. At present these technologies remain in nascent stages and at times appear to be precariously positioned between the multi-headed Scylla of reimbursement decision makers and the Charybdis of evolving scientific discovery and clinical research.* As the vanguard of emerging regenerative medicine technologies enters the U.S. and other markets over the next 2 to 5 years, reimbursement issues associated with these technologies will become clearer and more consistently navigable. In the interim, careful planning from a reimbursement perspective will be essential to successfully chart a safe and viable course as these technologies traverse the ever shifting seas toward commercialization and market access.

REFERENCES

Aetna Clinical Policy Bulletin 0601. (2009a). Pancreas transplantation alone (PTA) and islet cell transplantation. http://www.aetna.com/cpb/medical/data/600_699/0601.html

Aetna Clinical Policy Bulletin 0638. (2009b). Donor lymphocyte infusion. http://www.aetna.com/cpb/medical/data/600_699/0638.html

Blair, E. D. (2008). Assessing the value-adding impact of diagnostic-type tests on drug development and marketing. *Mol. Diagn. Therapy* 12: 331–337.

BlueCross/BlueShield Technology Evaluation Center. (2008). Autologous progenitor cell therapy for the treatment of ischemic heart disease. http://www.bcbs.blueresources/tec/vols/23/23_04.pdf

Carpenter, D. (2005). What politics may be telling us about cost-effectiveness analysis. *Health Affairs* 24: 1175–1185.

Cigna Medical Coverage Policy 0386. (2009). Stem-cell transplant for inherited metabolic disorders. http://www.cigna.com/customer_care/healthcare_professional/coverage_positions/Medical/mm_0386_coveragepositioncriteria_stem_cell_for_inherited_metabolic.pdf

De Bie, C. (2007). Genzyme: 15 years of cell and gene therapy research. *Regen. Med.*, 2: 95–97.

Faulkner, E. (2009). Clinical utility or impossibility? Addressing the molecular diagnostics health technology assessment and reimbursement conundrum. *J. Manag. Care Med.* 12: 42–55.

Fish, L. (2006). The case for cost sharing for biologic therapies. *Am. J. Manag. Care* 12: 159–161.

---

* In Homer's Odyssey (H. Rieu, Ed. New York: Penguin, 2003), Scylla and Charybdis are two monsters from ancient Greek mythology that represent impossible obstacles. Crews and ships faced destruction by passing between them or sailing too close to either one of them.

Gold, M. R., S. Sofaer, and T. Sigelberg. (2007). Medicare and cost-effectiveness analysis: time to ask the taxpayers. *Health Affairs* 26: 1399–1406.

Mullins, C. D., D. C. Lavallee, F. G. Pradel et al. (2006). Health plans' strategies for managing outpatient speciality pharmaceuticals. *Health Affairs* 25: 1332–1349.

National Academy of Sciences. (2008). Knowing what works in healthcare: a roadmap for the nation. Washington, DC: Institute of Medicine.

National Institute for Health and Clinical Excellence. (2008). Guidance IPG274. Autologous pancreatic islet cell transplantation for improved glycaemic control after pancreatectomy. http://www.nice.ortg.uk/Guidance/index.jsp?action=byID&o=11920

Neumann, P. J. and D. Greenberg. (2009). Is the United States ready for QALYs? *Health Affairs* 28: 1366–1371.

Raab, G. G. and D. H. Parr. (2006). From medical invention to clinical practice: the reimbursement challenge facing new device procedures and technology. I: issues in medical device assessment. *J. Am. Coll. Radiol.* 3: 694–702.

Stern, D. and D. Reissman. (2006). Speciality pharmacy cost management strategies of private healthcare payers. *J. Manag. Care Pharm.* 12: 736–744.

U.S. Congressional Budget Office. (2008). Technological change and the growth of healthcare spending.

Watkins, J. B., S. R. Choudhury, E. Wong, and S. D. Sullivan. (2007). Managing biotechnology in a network-model health Plan: a U.S. private payer perspective. *Health Affairs* 25:1347–1352.

WellPoint. (2008). Health technology assessment guidelines. Drug submission guidelines for re-evaluation of products, indications and formulation. https://www.wellpointnextrx.com/shared/noapplication/fl/s0/t0/pw_ad080614.pdf

23 Adoption and Evaluation of Regenerative Medicine and the National Health Service

Margaret Parton

CONTENTS

23.1 Delivering the Promise of Regenerative Medicine to United Kingdom Patients 369
    23.1.1 Key Questions Requiring Answers 370
23.2 Focus on United Kingdom 371
23.3 The National Health Service as Innovator 371
23.4 Delivering Solutions to Patients 372
    23.4.1 Horizon Scanning 372
    23.4.2 Evaluation 373
        23.4.2.1 Role of National Institute for Health and Clinical Excellence (NICE) 373
    23.4.3 Remuneration 374
    23.4.4 Infrastructure 375
23.5 Next Steps 375
References 376

23.1 DELIVERING THE PROMISE OF REGENERATIVE MEDICINE TO UNITED KINGDOM PATIENTS

The history of stem cell research had a benign, embryonic beginning in the mid-1800s with the discovery that some cells could generate other cells. In the early 1900s the first real stem cells were discovered when it was found that some cells generate blood cells. However in the past 20 years there has been a much higher level of interest from the media and public as more and more research has emerged demonstrating the significance and potential impact of this technology.

An endless stream of headlines promising relief from diabetes, Alzheimer's disease, blindness, heart failure, and many other diseases has created enormous expectations among patients and the public as a whole (Table 23.1). A strong view that this technology is the "golden bullet" that will alleviate all the major health concerns in the developed world has been created.

TABLE 23.1
Regenerative Medicine Publicity

Stem Cell Therapy Controls Diabetes in Mice

New York Times, **February 21, 2008 (by Andrew Pollack)**
Scientists reported on Wednesday that they were able to control diabetes in mice by harnessing human embryonic stem cells. The work raised the prospect that the embryonic cells might one day be used to provide insulin-producing replacement cells to treat the disease in people.

Science Daily, **July 22, 2009**
UC Irvine scientists have shown for the first time that neural stem cells can rescue memory in mice with advanced Alzheimer's disease, raising hopes of a potential treatment for the leading cause of elderly dementia that afflicts 5.3 million people n the U.S.

Every day more appears in the global media about yet another research program-demonstrating the potential of this technology. How does this media (and scientific) hype equate with reality? What are the likely time scales to move this research into a treatment regime for patients? Some stem cell-based products are already available commercially but at the moment they tend to focus on skin and cartilage and have not been embedded as standards of care in many health care systems as there are no clear routes to market for these transformational technologies. In addition, many unregulated treatments becoming available globally claim efficacy for everything from wrinkles to Alzheimer's disease.

The regulatory processes for these products and/or procedures are still in the early stages of development and much of the effort of the regenerative medicine community as a whole has been in managing this critical area.

There is currently a lack of clarity about the way these technologies will be brought to market. Funding for commercialization has fluctuated widely over the past few years as the complexities of potential business models become apparent.

23.1.1 KEY QUESTIONS REQUIRING ANSWERS

Will regenerative medicine finally be regulated as a product, a procedure, or a combination? Will allogeneic solutions emerge to drive stronger business models? Can an autologous product really be classed as a product?

In the current economic climate, raising venture capital is proving ever more challenging and the ability to answer these questions and present a coherent business model with a clearly defined market is critical (see Chapter 5 by Prescott).

In the UK, only 20% of the stem cell companies in existence in 2004 are still active. A similar rate of attrition is evident across the rest of Europe and the U.S. It has become apparent that moving from a pure research-based activity to a product in clinical trial is a long process requiring high levels of funding. This has created a resistance among investors who now view this sector as very high risk, particularly because no successful business model has really emerged so far and the route to market is still unclear.

The discussion on the difficulties of establishing a suitable regulatory framework for the wide range of approaches based on regenerative medicine and tissue engineered products is covered elsewhere in this book (see Chapter 19 by Bravery and Chapter 20 by Werner). Needless to say, the final decisions on the actual format will significantly impact on how this technology is taken up and reimbursed in the UK.

How can the regenerative medicine community move from research based clinical activity to delivering the very real promise of this technology as a solution to the wide range of diseases?

23.2 FOCUS ON UNITED KINGDOM

Regenerative medicine research in the UK has benefited enormously from a supportive, well constructed, but strictly controlled regulatory environment. This, coupled by a substantial commitment to funding (£200 million), has been provided by the Medical Research Council (MRC), Biotechnology and Biological Sciences Research Council (BBSRC), Engineering and Physical Sciences Research Council (EPSRC), and Technology Strategy Board (TSB) since 2000, and this has enabled UK researchers to make major advances in a wide range of disease areas. However, the route by which these technologies will become available to UK patients is unclear and other health economies across Europe are similarly unprepared for the likely wave of products and procedures emerging from clinical trials over the next few years.

23.3 THE NATIONAL HEALTH SERVICE AS INNOVATOR

The UK National Health Service (NHS) has long been identified as a late and slow adopter of innovation. This has been attributed to internal issues, such as NHS finance and organization and external factors such as research and industrial policies. The financial issues have been addressed through a significant boost to the share of GDP devoted to health in the wake of the 2002 Wanless Report (Wanless 2004). The NHS Institute for Innovation and Improvement (NHSI) has a specific remit to encourage innovative thinking across the NHS. The NHS Technology Adoption Centre (NTAC) works to accelerate the adoption of proven technologies by NHS organizations. The Cooksey Report (2006) assessed the health research landscape in the UK from basic research to technology adoption and identified two particular weak points:

- Failure to translate basic research into potentially useful products
- Failure to achieve the potential economic, health, and social benefits from technologies after they are developed

The difficulties faced in adopting new technologies within the NHS and ways of overcoming them have been the subjects of much recent research (Greenhalgh et al. 2004) but much of this has been focused on the cultures that exist within the NHS and much less on the "systematic approach to adoption" that David Cooksey highlights in his 2006 review.

Dealing with these issues was thought to be important to the continued improvement of population health and to the growth of the UK economy; subsequent policy responses have focused on the joint health and economic benefits. The recent NHS focus on quality, innovation, productivity, and prevention (QIPP) launched by David Nicholson, its chief executive, in August 2009 highlights the need to harness new healthcare technologies and innovation to improve outcomes in terms of quality of patient care and NHS productivity (Nicholson 2009).

23.4 DELIVERING SOLUTIONS TO PATIENTS

What steps should the healthcare economies such as the NHS be taking to ensure rapid and efficient uptake of regenerative medicine solutions to deliver both an improvement in patient outcomes and a more effective solution to many long-term conditions?

There are currently large numbers of clinical trials underway across the world covering a wide range of diseases. These are based on a range of research clinical protocols, time lines, and final outcomes and are often initially in very small groups of patients. The overall impact is to create a view of a "cottage industry" working in isolation to other established healthcare solutions.

The difficulty of establishing suitable clinical endpoints to support claims of clinical efficacy makes it hard to establish whether an intervention can be viewed as a cure or simply an alternative treatment to those currently available. It is currently very difficult to establish which products or procedures will become available to patients within the European Union (EU). However, in order to introduce a product or procedure that has already been through the regulatory process as a treatment option, the NHS will need to:

- Employ horizon scanning for technologies emerging from clinical trials and commencing the regulatory process
- Evaluate both clinical efficacy of a product and its likely impact on patients and the healthcare system. Cost effectiveness will always play a significant role in the introduction of new products and procedures and as the challenge for the healthcare economies becomes ever greater, then some sort of economic impact is likely to be introduced by most countries.
- Remuneration nominating the decision makers: who decides/who pays?
- Infrastructure: developing a framework for delivery of the technology.

23.4.1 HORIZON SCANNING

Based in the Unit of Public Health, Epidemiology, and Biostatistics at Birmingham University, the National Horizon Scanning Centre (NHSC, http://www.haps.bham.ac.uk/publichealth/horizon/topic.shtml) aims to provide advanced notice to the Department of Health and national policy makers including the National Institute for Health and Clinical Excellence (NICE) of selected new and emerging health technologies that may require evaluation, consideration of clinical and cost impacts, or modification of clinical guidance before launch to the NHS. The scope of horizon scanning activity includes pharmaceuticals, medical devices, diagnostic tests and procedures, therapeutic interventions, rehabilitation and therapy, public health, and health promotion interventions. The NHSC plays reactive and proactive roles by accepting input from industry and healthcare professionals and also actively developing relationships with technology providers to broaden the understanding of technologies likely to have a major impact on the provision of healthcare within the UK.

23.4.2 EVALUATION

23.4.2.1 Role of National Institute for Health and Clinical Excellence (NICE)

Launched by the Department of Health in 1999, NICE is an independent organization responsible for providing national guidance on the promotion of good-health and the prevention and treatment of ill health. The initial objectives for NICE were set in November 1999 to improve standards of patient care and reduce inequities in access to innovative treatment. NICE has established processes:

- To identify new treatments and products likely to have significant impacts on the NHS or for other reasons would benefit from the issue of national guidance at an early stage
- To enable evidence of clinical and cost effectiveness to be brought together to inform a judgment on the value of the treatment relative to alternative uses of resources in the NHS
- To issue guidance on whether a treatment can be recommended for routine use by the NHS (and if so, under what conditions or for which groups of patients) together with a summary of the evidence on which the recommendation is based
- To avoid any significant delays to those sponsoring innovations in meeting any national or international regulatory requirements or bringing innovations to market in the UK.

In November 2008, the MHRA, the UK Medicines and Healthcare Products Regulatory Agency, convened a Medical Device Technology Forum made up of experts in the field and those responsible for developing the regulatory framework. The forum noted the perception that for newly CE marked devices and newly licensed medicines, the NHS would not take up a product without an assessment from NICE, but NICE would not assess the product until there was sufficient take up and this would create a major barrier to uptake.

It was suggested that discussions should be opened with NICE regarding the potential for earlier engagement and assessment of products before they were licensed or CE marked. This would assist in early identification of financially unviable products and enable earlier small-scale clinical evaluations that are often delayed due to the lack of a NICE ruling (MHRA 2008).

Two key policy initiatives that may influence the way that NICE develops guidance for innovative medicines and devices have recently been announced and both may exert significant impacts on regenerative medicine. The Department of Health published the *Next Stage Review of the NHS: High Quality Care for All* (Department of Health 2008) overseen by Lord Darzi of Denham, the then Health Minister. This comprehensive review focused strongly on the need to develop more innovative healthcare solutions and identified the need to move these through into widespread use as soon as possible. In the report, Lord Darzi states: "Ensuring that clinically and cost effective innovation in [medicines and] medical technologies is adopted. For new medical technologies, we will simplify the pathway by which they pass from development into wider use, and develop ways to benchmark and monitor uptake."

NICE is now constructing a single evaluation pathway for devices and diagnostics based on the creation of a new Medical Technology Advisory Committee (MTAC). It is expected that the first technologies will move through the new pathway and processes by April 2010 (NICE 2009b).

The new process recognizes the variability in both quality and quantity of evidence from the devices and diagnostics sector when compared to that available for pharmaceuticals. A new range of appraisal processes are being developed in recognition that much evidence for non-pharma healthcare technologies is often developed post-market. It is also clear, however, that NICE would have to create new methods to manage transformational technologies such as regenerative medicine products and procedures.

In addition the recently launched Office of Life Sciences (OLS) created by the UK government in June 2009 published its first Blue Print in October 2009 (OLS 2009). The OLS is a cross-governmental body formed at the request of then Prime Minister Gordon Brown to transform the UK environment for life sciences companies and ensure faster patient access to cutting-edge medicines and technologies. The OLS forms part of the government's active industrial policy of engagement with the UK healthcare industry.

As part of that Blue Print, there was an agreement that the government, working with NICE, would introduce an Innovation Pass to make selected innovative medicines available on the NHS for a limited period. The Pass will be piloted in 2010/2011, with a budget of £25 million. At the end of the pilot year, the government was to evaluate the results before setting budgets for the remaining 2 years of the initiative.

In addition, the Blue Print announced that the Technology Strategy Board (TSB) would launch an £18 million RegenMed investment program to support commercial R&D with additional funding from the Medical Research Council, the Engineering and Physical Sciences Research Council, and the Biotechnology and Biological Sciences Research Council. The TSB has also committed to improve its expertise in the life sciences. The Blue Print commitments could potentially provide ideal opportunities to consider a similar Innovation Pass for regenerative medicine-based interventions.

23.4.3 REMUNERATION

The NHS has a complex structure for funding its healthcare provision. The Primary Care Trusts (PCTs) hold the budgets for their local health economies. Each of the 152 PCTs are based across England and each takes a dual role by (1) providing both the care out in the community via their general practice-based physicians, district nurses, dentists, and others, and (2) procuring services from a range of providers both NHS hospitals and private providers on behalf of their healthcare economy.

If NICE issues positive Guidance from a Health Technology Appraisal on a therapeutic, PCTs have, by law, to make funding available to provide the treatment. Currently the uptake of NICE guidance for other technologies and procedures is not mandatory.

The price that a PCT will pay for a particular procedure and the scope of that procedure are currently set by the Payment by Results (PbR) system that develops the Healthcare Resource Group (HRG) codes—standard groupings of clinically similar treatments that use comparable levels of healthcare resources and cover more than 1,400 procedures. The costs are developed using reference costs for various procedures from a wide range of acute trusts. A median price for a procedure that may also "bundle in" in a diagnostic process is issued.

The PbR system has been heavily criticized for actually blocking the uptake of innovation as it is inflexible and the HRG codes are developed using historical information. This leads to hospital trusts needing to make arrangements for "pass through" funding or local tariffs to introduce procedures based on the availability of new technology. Even when NICE issues guidance on the uptake of new devices this does not lead to an immediate change in the HRG code, which can lead provider organizations such as the Acute Hospitals out of pocket if they seek to deliver better patient outcomes using new technology.

However it is likely that the commissioning of regenerative medicine based solutions will initially lie with the Specialist Commissioning Teams who will make a judgement on the remuneration based on any NICE guidelines issued. Enabling the Specialist Commissioners to understand both the complexity but also the benefits of the emerging technology will be critical to accelerating its uptake within the NHS. Currently the Commissioners do not have regenerative medicine on their agenda and the regenerative medicine community needs to address this with the same urgency as the regulatory processes and the evaluation paradigms.

23.4.4 INFRASTRUCTURE

Current research and early trials are carried out by major tertiary hospital trusts across the UK. If the procedures are to move into the mainstream, the NHS will have to consider the overall infrastructure and support that will be required to enable the rapid uptake of the technology. Some key questions that will need to be asked within the Department of Health and the NHS include:

1. What resources will need to be available for training clinicians and specialist staff?
2. What specialist facilities are likely to be required?
3. Are the procedures likely to be "mainstreamed" into requiring reduced skill levels? And how long will that take?

23.5 NEXT STEPS

The regenerative medicine community and the NHS will need to collaborate on developing a core program in order to manage this transformational technology through the above stages to guarantee a rapid and effective uptake within the NHS and deliver the promised benefits to UK patients.

REFERENCES

Cooksey, D. 2006. *A review of UK health research funding*. HM Treasury, December. http://www.hm-treasury.gov.uk/d/pbr06_cooksey_final_report_636.pdf (accessed June 2009).

Greenhalgh, T, G. Robert, P. Bate et al. 2004. How to spread good ideas: a systematic review of the literature on diffusion, dissemination and sustainability of innovations in health service delivery and organisation. Southampton: National Coordinating Centre for NHS Service Delivery and Organisation R&D. http://www.sdo.nihr.ac.uk/files/project/38-final-report.pdf (accessed May 2009).

High Quality Care for All. Next Stage Review Final Report. NHS. www.dh.gov.uk/prod_consum_dh/groups/dh_digitalassets/@dh/@en/documents/digitalasset/dh_085826.pdf

MHRA. 2008. Medical Device Technology Forum, November. http://www.mhra.gov.uk/home/groups/clin/documents/websiteresources/con035987.pdf

NICE. 2009. Faster Access to Modern Treatment. www.nice.org.uk/aboutnice/whatwedo/niceandthenhs/faster_access_to_modern_treatment.jsp

NICE. 2009. Single Evaluation Pathway. www.nice.org.uk/media/FCC/17/2009_068_-Medical_Technologies_Programme_and_MTAC_APP.pdf

Nicholson, D. *Implementing the next stage review visions: the quality and productivity challenge*. http://www.dh.gov.uk/en/News/Recentstories/DH_101712

Wanless, D. 2004. *Securing Good Health for the Whole Population: Final Report*. HM Treasury, February. http://www.dh.gov.uk/en/Publicationsandstatistics/Publications/PublicationsPolicyAndGuidance/DH_4074426 (accessed May 2009).

Section 7

Insurance and Risk Management

24 Role of Insurance
If You Build It, Will They Insure It?

Matthew Clark

CONTENTS

24.1 Insurance for Regenerative Medicine 379
24.2 If You Build It, Will They Insure It? 380
24.3 Common Traits 381
24.4 Confronting the Challenges 382
24.5 Critical Elements of Insurance Portfolios 383
    24.5.1 Security and Expertise 384
    24.5.2 Research and Development Operations 384
    24.5.3 Changes in Environmental Conditions 384
    24.5.4 Unrecoverable Expenses 384
    24.5.5 Directors and Officers (Management) Liability Coverage 385
    24.5.6 Supplier Dependencies 385
    24.5.7 Machinery Breakdown and Transit Risks 386
    24.5.8 Clinical Trials 386
    24.5.9 Terrorism and Attacks by Activists 386
    24.5.10 Intellectual Property Protection 386
    24.5.11 Self-Insurance Option 387
24.6 Roles of Risk Management and Business Continuity Management 387
    24.6.1 Timelines 389
    24.6.2 Budgeting for Disaster 389
    24.6.3 Collaboration with Insurers 389
24.7 The Future 389

24.1 INSURANCE FOR REGENERATIVE MEDICINE

Despite the media hype that often accompanies some of the more significant advances in medicine and healthcare technology, there seems little doubt that collaboration among scientists of different disciplines is now giving rise to truly astounding developments within the realm of life sciences. Gene therapy, stem cell transplantation, and tissue engineering are examples of the novel, often game-changing approaches brought to bear on hitherto intractable healthcare challenges. However, as with any emerging field of technology, it is important to identify, quantify, and control the new risks that inevitably shadow such innovations.

These technological advances often originate from small or medium-sized science-based businesses. While the dynamism and innovation of these organizations are founded on the strength of their intellectual properties and the qualities of their management teams, to achieve commercial success these enterprises must be supported by a robust platform of sector-specialist service providers. It is no accident that many of the world's most successful science and technology clusters are characterized by an abundance of investors, lawyers, intellectual property (IP) attorneys, accountants, insurance brokers, and risk managers whose expertise aids the process of commercialization.

Insofar as insurance is concerned, there already exists an established, thriving, global niche market populated by insurance companies, brokers, and related professionals offering a comprehensive array of specialist products and services designed solely to meet the often very specific needs of life science companies and healthcare practitioners. While this is good news for new entrants into these sectors, it does not necessarily follow that *all* new products, techniques, and processes—particularly those involved in regenerative medicine—will be readily insurable. We will now examine why this is so and what the implications are for commercialization in this field.

24.2 IF YOU BUILD IT, WILL THEY INSURE IT?

To examine this question it is first necessary to appreciate some of the fundamental elements of insurance. At its core, insurance is a mechanism for risk transfer—a promise to pay if things go wrong. If a product, service, or process gives rise to costly liabilities, most organizations want their insurance companies to protect them. In turn, the insurer expects full disclosure by its policyholders of all material facts that will influence the insurer's decision whether to insure and at what price. It is therefore crucial that an organization imparts reasonable information about its operations and activities to its insurer.

Insurance is also based upon a statistical analysis of the likelihood (frequency) of an event such as a fire, flood, or explosion, and the likely cost (severity) of the resulting harm. To illustrate this point, insurers have offered household insurance policies for many years and accumulated valuable statistical data concerning frequency and severity of loss along the way. An insurer refers to this data when considering whether to offer coverage to a new applicant and at what price. The same holds true of insurers that offer coverage to life science and healthcare organizations; the availability of coverage is largely dependent upon loss experience.

What happens if the product or process for which insurance is sought is so completely new and fundamentally ground breaking that no experience exists upon which the insurer can base an underwriting judgment? The risk exposures associated with the new technology may be poorly understood or entirely unknown. Can an insurer offer coverage? If so, what policy terms, conditions, and exclusions should apply? What premium is to be charged? A clear message is that anyone developing new technology, whether in the area of regenerative medicine or elsewhere, and those whose interests are vested in such organizations should consider the availability of insurance at the formative, conceptual stage, rather than regard it as an automatic right in return for premium paid. That may strike many as a bold observation, but most healthcare products available in the world today probably owe their continued existence as much to the global insurance industry as to those who discovered or invented them.

From wheelchairs to scalpels, from bedpans to cancer drugs, to be commercially viable, a product generally must be *insurable*. Few enterprises would sink resources into developing products for which there is no market. Likewise, why develop a product no company is willing to insure? It is important to acknowledge that there are exceptions to this general rule (for example, military applications or certain elements of social healthcare systems in which commercialization and insurance are not primary concerns). However, in the context of regenerative medicine within the private, non-governmental theater, let us assume that this essential premise holds true. The natural conclusion is that a product's commercial success relies, at least in part, upon its *insurability*.

It is possible that some established healthcare companies may assume that their existing insurance policies automatically protect them for the development of new products or services in regenerative medicine. They would likely discover that their insurance companies would, when consulted, regard new forays into this sector as a material change in the risk profile of the policyholder that should be declared. At the same time, it is unwise to assume that all life science insurance companies will be familiar with regenerative medicine; they might lack a fundamental understanding of the characteristics of regenerative medicine and the potential risks posed.

24.3 COMMON TRAITS

If we examine some of the common traits shared by regenerative medicine organizations, it may be possible to identify those elements of their nature which set them apart from other, more conventional operators in the life science space. This list is not exhaustive and may not necessarily apply to all forms of regenerative medicine, but it can help us to address issues surrounding insurability and commercialization and will inform a broader examination of how insurance for such organizations should be structured.

Enterprises operating in the sector will tend to develop products and work with techniques that push back the boundaries of current technologies, entering new areas of healthcare technology such as gene therapy, stem cell transplantation, tissue engineering, and the like. This inevitably makes risk assessment difficult; insurers have little in the way of loss history for insurers to go on, given how recent much of the technology is.

- Regenerative medicine draws upon a knowledge base that is truly interdisciplinary, involving biology, biochemistry, physics, engineering, and materials science to name a few. This characteristic yields a complex array of risks relevant to each scientific discipline—exposures that have hitherto not been considered collectively are now pushed together by a single organization.
- Existing regulatory frameworks may not be adequate to deal with the rapid advances in certain areas such as the use of embryonic stem cells.

Alternatively, a "bedding-in" period for new regulations will create uncertainty about their untested scopes and applications and may expose practitioners to unknown and unquantifiable personal liabilities.
- Important ethical issues are brought into sharp focus by some regenerative medicine techniques and applications including controversy over the use of embryonic stem cells, patient consent considerations, confidentiality for blood and tissue donors, therapeutic versus cosmetic application issues, synthetic biology, and the use of animals in research operations.
- Unusually lengthy periods of research and development may exert impacts upon the availability of funding and the overall attractiveness of investments in the sector.
- Following on from the above point, there may be very significant reliance on structured funding over several rounds before product is shipped and revenue earned.
- The development of new protocols for product and device trials may present new challenges to contract research organizations and clinical investigators who must test entirely new therapies and devices, for example, in the area of in vivo tissue implantation and transplantation.
- Prodigious levels of research and development property will be produced, much of which will be difficult and time consuming to recreate if lost or damaged; such property may also be susceptible to changes in controlled environmental conditions.
- Product or component manufacturing may ultimately be outsourced, with consequent supply-chain dependencies that create additional risk exposures.
- Regenerative medicine may create new challenges in the formation and management of intellectual property portfolios.

Some products or applications may not have mass market appeal, but will rather benefit a relatively small number of patients suffering from relatively rare conditions. This may create a concentration of product liability risk around a small patient base, with insurance cost implications.

As with other forms of emerging technology, public awareness and consumer "buy-in" are intrinsic to the successful take up of this new science. This creates a natural imperative for those engaged in this field to coalesce around a coherent social message which is clearly communicated to the public; this is both a challenge and an opportunity.

24.4 CONFRONTING THE CHALLENGES

Having highlighted some of the challenges of regenerative medicine from an insurance perspective, let us consider some essential practical measures that may support development and commercialization in this sector.

Given the nature of developments in regenerative medicine and the nature of its target market, insurers may be tempted to restrict their exposure to these novel products by offering more restrictive coverage or lower policy limits. However, insurance professionals, by their nature, are risk takers and tend to focus on reasons *to insure* rather than looking for excuses not to. In dealing with insurers, it is therefore important that an applicant highlight the positive aspects associated with a new technology or application and acknowledge the potential risks. Insurance should be considered an integral part of business—a benefit rather than a cost. Because an insurer shares in an organization's risk exposure, its position is tantamount to a partnership in the company. The organization should therefore be prepared to share with its insurer whatever scientific data it has relating to its new product or technology.

Insurance companies often employ scientists to help them understand the risks they underwrite. It is advantageous for the organization to identify those individuals and liaise with them in establishing appropriate insurance.

- Due to the emergence of new ethical and regulatory considerations surrounding regenerative medicine, the already onerous personal liabilities of directors and managers of new enterprises in this sector will be exacerbated. Directors and officers liability insurance, already an important component of organizational protection, therefore becomes even more critical.
- When designing organizational policies around health, safety, and security it is important that those looking to commercialize intellectual property in this sector strive for implementation based upon current best practices, rather than introducing programs that pay lip service to basic legal imperatives. This includes documenting compliance with standards and sharing this information with their insurers.

As implied above, it may also benefit a company to pool as much data as possible about the known occupational and environmental risks associated with the materials and components involved in its new technology and demonstrate a willingness to share this with its insurer.

It is also prudent for a company to maintain accurate records to ensure full traceability of all raw materials and components. This can be used as evidence that the organization is capable of handling a product recall. Insurers expect their policy holders to maintain rights of recourse against their suppliers and contractors so as to ensure that suppliers and contractors are contractually held to account for any negligence.

- Appropriate contingency planning can be introduced to respond to serious interruptions of research and development operations, such as a freezer failure resulting in spoilage of refrigerated research materials. Similarly, insurers tend to reward investment in duplication and off-site back-up of critical research data with pricing discounts and broader policy coverage.

24.5 CRITICAL ELEMENTS OF INSURANCE PORTFOLIOS

Having considered some of the main challenges to the successful development of the regenerative medicine sector from an insurance perspective, we can now identify some of the key elements that should form part of the insurance armory of any organization looking to commercialize intellectual property in this space.

24.5.1 Security and Expertise

Life science organizations should procure coverage from insurance companies that have expertise and experience in life sciences. It is also important to monitor the financial strength of these insurers in order to continually reaffirm their claims-paying ability. Equally essential is the appointment of a sector-specialist insurance broker who will represent a company in its dealings with its insurers. It is reasonable to demand that a broker demonstrates capability and experience prior to appointment.

24.5.2 Research and Development Operations

Tailored insurance solutions can be established to protect critical research and development projects that form the cores of most life science organizations. The coverage should be configured to insure the specific risks associated with individual research activities. This can include a combination of the following:

- Research and development materials: cultures, cell lines and resulting products, samples, clinical trial samples, and product prototypes represent significant creation costs. Valuations for insurance purposes must take account of the process costs and other overheads involved in the production of such property, as well as the cost of the raw materials.
- Repeat cost expenditures for research and development: unforeseen interruptions to research activities may result in the need to repeat entire research projects. Insurance can help in mitigating the unbudgeted additional costs associated with repeating such work, enabling the organization to re-perform critical research while continuing to meet ongoing fixed overhead costs. The insurance coverage may be configured to include milestone payments and other contractual obligations that may be prejudiced as a result of an interruption.
- Additional expenditures relating to animals: the value of research animals destroyed in an unforeseen event may extend beyond their stock replenishment cost. Insurance may be a useful means of meeting the cost of repeating research (such as in the area of transgenic colonies) as well as the expense of replacing colonies.

24.5.3 Changes in Environmental Conditions

If development processes involve the use of sterile facilities, clean rooms, or temperature controlled environments, insurance arrangements must take account of the lengthy timescales that may be involved in the recommissioning and recertification of such facilities following damage or contamination. Insurance can also extend to protect against the resulting condemnation of stock and research property as well as the costs of debris removal incurred in such circumstances.

24.5.4 Unrecoverable Expenses

If an adverse event leads to the cancellation or postponement of a product launch, clinical trial, or other planned undertaking, it may not be possible to recover all fees and other costs incurred by the organization prior to the event. These committed costs can be very significant and insurance can be a useful means of gaining reimbursement.

24.5.5 Directors and Officers (Management) Liability Coverage

The importance of adequate levels of liability insurance for an organization's management team and non-executive decision makers cannot be overstated. Regulatory and legislative authorities have shown a discernable tendency to place ever higher levels of responsibility on the shoulders of corporate board members. Investigations by regulatory authorities, class actions by shareholders or frustrated investors, and even criminal prosecutions are now common.

In today's litigious climate, it is advisable that those holding senior positions in any organization (regardless of whether it is a commercial enterprise or not-for-profit) first confirm that this form of insurance is in place. This is especially true of those who do business in North America or who may have assets there. In this case, the insurance policy should be configured to provide coverage for American and Canadian court actions and extradition proceedings. Directors and officers liability coverage provides critical protection against potential personal liabilities associated with disputes arising in the following areas:

- Mergers, acquisitions, and disposals
- Investment management
- Competitors and intellectual property disputes
- Employment practices
- Regulatory compliance
- Contract disputes (licensing, collaboration agreements, etc.)
- Health and safety
- Insolvency
- Privacy and confidentiality

24.5.6 Supplier Dependencies

As previously noted, those engaged in commercializing disruptive technologies will often outsource core activities around materials resourcing, product and/or component manufacture, and data handling. Aside from the challenges that this creates in the areas of quality control, compliance, intellectual property ownership, and data security, these relationships also create de facto supplier dependencies. Adverse events that occur at the premises of key supply chain partners can impact the ability of an organization to continue to function as a going concern. Insurance has a role to play in helping avoid these potentially ruinous loss scenarios.

Consequential losses flowing from these events are generally insurable, but attention must be paid to the loss controls in place at the premises of supply chain partners and the availability and speed with which alternative service or product suppliers can be engaged. Insofar as data warehousing and outsourcing are concerned, a completely different set of concerns arise surrounding loss, damage, or theft of data via hacker attack, computer virus, or network failure. Breaches in data security may also result in allegations of privacy violations or defamation. Comprehensive insurance solutions are available and can be configured to pay for the recompilation of lost or damaged data and provide indemnification against legal costs and compensation awards associated with breach of privacy and defamation allegations.

24.5.7 Machinery Breakdown and Transit Risks

The new advances being made in regenerative medicine may depend upon exotic or bespoke machinery or equipment used as part of the research and manufacturing processes. The failure of such items may involve lengthy replacement timescales and consequential loss of revenue or extra expense. Insurance is often the preferred means of meeting these unforeseen costs.

24.5.8 Clinical Trials

In its simplest form, clinical trial liability coverage protects against the cost of settlement or liability if a clinical research subject suffers bodily injury arising from participation in a trial. Typically, trials involve a drug, vaccine, or medical device. While there is a well-established insurance market for this type of protection, it must be borne in mind that contract research organizations and others involved in the process may be trialing an entirely new form of therapy or device. The insurer will therefore need to gain an understanding of the nature of the product, procedure, or device that is to be trialed, carefully considering issues as diverse as trial funding, CRO reputation, claim potential of the subject group, clinical protocols, ethical considerations, and patient documentation and consent issues.

24.5.9 Terrorism and Attacks by Activists

Standard insurance policies often limit or exclude coverage for losses resulting from acts of terrorism; insurance companies prefer to leave central government to provide protection in the event of loss or damage. As a result, some western governments have developed special funds that act as reinsurance pools that enable insurance companies to offer terrorism insurance in return for a premium payment based upon predefined rates. Given that the legal definition given to "acts of terrorism" might be insufficient to encompass the actions of animal rights protectors, insurance companies operating within the life science sector customarily offer terrorism coverage that specifically includes loss or damage arising from animal rights activists. This insurance protection is a prerequisite for all life science firms.

24.5.10 Intellectual Property Protection

Aside from human resources, an organization's most precious asset is arguably its proprietary technologies. Most businesses do not hesitate to insure their tangible assets and, in the same way, it is advisable to insure the organization's intangible assets: intellectual property comprising patents, copyrights, trademarks, etc. Any infringement of, or attempt to pass off, any patents, trademarks, designs, formulas, or copyrights by a third party could have a devastating impact on the organization's finances and reputation. Intellectual Property Litigation Insurance can help mitigate the costs of self-funding a legal action against an infringer of intellectual property. Merely having the insurance serves as a powerful deterrent to a potential infringer. The insurance policy can also be configured to provide valuable protection in the event that the organization needs to defend against an allegation that its products, services, or know-how infringe a third party's intellectual property. However, the availability of this type of policy is somewhat restricted owing to the relatively small number of insurance companies offering the protection. Consequently, insurance costs can be significant.

24.5.11 SELF-INSURANCE OPTION

Larger organizations, particularly global pharmaceutical companies, often devise alternative means of handling their operational risks. Those having significant financial resources may opt to self-insure some critical exposures rather than pay costly insurance premiums. While this means that they must absorb big losses, it also encourages a culture of proactive risk management, which can result in considerable cost savings over the long term. However, most life science organizations will lack the financial muscle to handle significant levels of risk in house and will require insurance from the open market.

24.6 ROLES OF RISK MANAGEMENT AND BUSINESS CONTINUITY MANAGEMENT

It is natural for management to be concerned with the day-to-day practicalities in developing product pipelines and getting finished products or services to market. Attaining regulatory compliance, validation of manufacturing facilities, driving research and development activity, ensuring that the organization is fully funded, establishing the best management team, and talking to customers and collaboration partners are core management functions vital to the success of an enterprise. It is easy to lose sight of the myriad risk exposures that can irreversibly damage an organization.

How does a company prepare for the unexpected and the unforeseen—unanticipated outcomes that can damage or even destroy a business? It is increasingly clear that the detailed examination and control of operational risk—which we shall refer to as risk management—should be seen as a fundamental function of management rather than as an option. The practice of Risk Management is far too broad a subject to adequately explore here, but organizations that do not take the time to properly assess their risk exposures will, at best, experience difficulty in obtaining insurance and may even fail to secure the coverage they need at any price. It is incumbent upon the management of modern life science companies to take a holistic view of their business risks. Insurance is a critical element, but not a panacea. Insurance in the life science sector is increasingly regarded as a precious commodity, something to be earned, to prove oneself worthy of—rather than an automatic entitlement.

One particularly important by-product of risk management is business continuity management (BCM), also known as disaster recovery planning. BCM is a business management function that ensures the continued availability of core resources and services. It ensures that an organization can respond to an interruption of its fundamental operational activities and recover from the consequences of the interruption. The process is founded upon the principle that risk is inherent in any organization and strategies are required to shield it from the adverse impacts of negative events. The following are examples of events that should be considered by any regenerative medicine organization.

- An electrical outage in the lab that causes spoilage of critical cell lines or other cultures, putting the R&D effort back months, even years—and perhaps jeopardizing a milestone payment from a key investor
- A small electrical fire or flood that contaminates clean rooms
- The theft of vital laboratory notebooks or other R&D records by an employee or intruder
- A malicious hacker attacking or destroying the database
- A valuable researcher becoming injured or too ill to work
- A key supplier cutting off vital raw materials after themselves suffering a devastating fire
- The damage to corporate reputation, or loss of investor attraction, following an adverse event
- Loss of committed expenditure and other resources in the event of a new product launch delay, recall, or withdrawal
- Unforeseen contamination clean-up costs impacting the balance sheet
- A disparaging remark in an internal email bringing a defamation suit from a competitor
- A legal challenge that a product infringes another firm's intellectual property
- Refusal of a disgruntled customer to pay, alleging that an overzealous salesman misrepresented the quality or capability of a product or service
- A large customer, making up the bulk of the firm's revenue, suffers a devastating flood and cancels all orders
- A member of staff becomes the target of animal rights activists
- Political interference or a change in government policy limiting access to materials vital to a company's R&D efforts

It is possible to categorize primary operational risk elements to enable separate consideration of each risk classification. It is important to remember that not all risks can be insured. The idea is not to create a list of insurable events, but rather to schedule and prioritize risk elements that have the potential to adversely interrupt the business and plan how to respond to each risk in turn. Many life science sector insurance companies (and some brokers) retain in-house expertise in risk management and BCM and offer their services to policyholders. It is in the insurers' interests to retain such experts because well managed insurance portfolios based on managed risks will generally yield fewer claims. Some additional considerations for small and early-stage organizations are listed below.

24.6.1 Timelines

Start-up or early-stage organizations sometimes make the mistake of thinking that size matters and that they are too small to consider BCM. However, no organization is too small for BCM. Planning for the unexpected is always a prerequisite for a successful enterprise and it is easier to instill a culture of risk management from the outset instead of attempting to implement it several years later.

24.6.2 Budgeting for Disaster

Effective risk management costs money and it is necessary for an organization to budget accordingly. Although maintaining cash reserves may appear an unsophisticated approach, cash assets can be effective ointments for a bruised balance sheet.

24.6.3 Collaboration with Insurers

Having gone to the time and expense of creating a bespoke BCM plan, communicating it to staff, and testing the various elements with regular drills, a company should share its BCM strategy with its insurers and keep them informed of updates and revisions. This helps control the cost of insurance and enhances long-term insurability.

24.7 THE FUTURE

We have identified significant challenges for the introduction of new products and services in the area of regenerative medicine and examined some of the unconventional risks associated with these advances. To many, the insurance industry appears old fashioned and slow to embrace change. In reality, the global insurance industry is populated by many vibrant and forward-thinking practitioners who have repeatedly shown themselves capable of responding to the new and emerging risk exposures that scientific and technological endeavor inevitably generates. Highly evolved niches have been established in the insurance sector to meet the diverse risks presented by the computing, aerospace, and healthcare industries. Despite the near-revolutionary promise that regenerative medicine holds for 21st century healthcare, those who already provide insurance protection to the life science community will continue to develop the products and services required to meet the new perils that lie in wait. Insurance, like most other industries, is a commercial enterprise; in the midst of difficulty lies opportunity.

Index

A

Aastrom Biosciences, 104–105
Acute myocardial infarction (AMI)
    health care costs, 44, 323
    stem cell therapy, 55
Advanced BioHealing, 93
Advanced Cell Technology, 105
Advanced Therapy Medicinal Product (ATMP), 92
Advanced Tissue Sciences, 87, 93
Aggregate Therapeutics Inc., 45, 50–54
Alliance for Regenerative Medicine, 62
Allogeneic cell-based therapy
    advantages, 85
    characteristics, 114
    companies, 93–95
        business model, 86–93
        case history, 94–95
        considerations in developing, 86–93
            cell bank, 87–89
            cryopreservation, 90–91
            defining optimal sales channel, 92–93
            manufacturer of product, 89
            market focus, 92
            obtaining tissue, 87
            product assembly and manufacture, 89
            shelf life, 91–92
            shipping choices, 91
            site of manufacture, 90
    disadvantages, 85
    regulatory issues, 93
ALS. *See* Amyotrophic lateral sclerosis (ALS)
Alzheimer's disease, 16, 76, 104, 323, 369
AMI. *See* Acute myocardial infarction (AMI)
Amyotrophic lateral sclerosis (ALS), 105, 107
Apligraft, 62, 93
ARC. *See* Australian Research Council (ARC)
ASCC. *See* Australian Stem Cell Centre (ASCC)
ATMP. *See* Advanced Therapy Medicinal Product (ATMP)
Australia, 25, 36
    CRCs, 36
    stem cell research, 36–37
Australian Government Department of Health and Aging, 36
Australian Research Council (ARC), 36
Australian Stem Cell Centre (ASCC), 37–41
    current status, 41
    governance, 37
    intellectual property policy, 39, 40
    objective, 37–38
    organization of research, 39
    progress, 39–40
    recent developments, 40
    reporting responsibilities, 38
Autologous cell therapy
    allogeneic therapy *vs.*, 85, 114
    business model, 98–99
    challenges for development, 113
    characteristics, 1144
    clinical trials, 103, 104
    companies, 63, 104–108
        case study, 108–112
    cryopreservation, 103
    current treatments and future prospects, 103–104
    distribution and transportation, 103
    logistics, 103
    manufacturing process, 99, 101–103
    market, 103–108
    production process, 101–102
    quality control, 102
    regulatory issues, 100
    shelf life, 103
    storage, 103
    tissue source, 103
Azellon, 105

B

BioHeart, 105
Biological acellular matrices, 224–225
Biologics license application (BLA), 181
Biotechnology, 27–28
    funding, 68
BLA. *See* Biologics license application (BLA)
Bone marrow failure, 119
Brainstorm Cell Therapeutics, 105
Bush, George W., 16
Business model, 86–93, 161–164
    constraints to, 155–156
    cord blood banks, 122–126
        hybrid public-private, 125–126
        private, 123–125
        public, 122–123
    for investing in regenerative medicine, 6, 63–65
    pipeline trade-off, 162

C

California. *See also* California model
   state general fund, 33
California Institute for Regeneration Medicine (CIRM), 12, 216
   Center of Excellence, 32
   disease team program, 32
   establishment of, 33
   governing board, 28, 31, 32
   grant programs, 32
   Institute, 32
   major facilities, 19, 32
   monetary investments, 12
   Special Program, 32
California model, 17, 18–20
   components, 18–20
      collaborative funding agreements to enable globalization of effort, 20
      creating independent agency, 18
      funding derived from bonds, 18
      horizontally integrated pipeline from basic science through Phase II trials, 20
      large-scale, long-term portfolios, 20
      unlimited term, 20
   funding for large-scale biotech research, 26–27
   optimizing governmental cash flow of, 21–23
   providing models with enhanced opportunities, 23–24
   rationale, 21
   relationship of research complexity to capital, 24–26
California Stem Cell Research and Cures Act, 33
Canada
   aggregate therapeutic experiments, 49–54
      CECR and, 54
      financing for, 52–54
      SCN support for, 51
   agreements with scientific funding organizations, 25
   biotechnology sectors, 50
   disease incidence and health care cost in, 44
   Edmonton Protocol, 44
   emerging trends in stem cell research, 54–56
   stem cell companies, 55
   stem cell network, 45–49
      development of highly qualified workforce, 48
      networking and partnering, 48–49
      partnership and interactions, 49
      translational funding gap between therapeutic discovery and clinical use, 47
   TTO, 51
Canadian Cancer Society, 46
Cancer, 16, 135
   animal model, 241
   stem cells, 43, 47, 56, 106, 161
      iPS cells derived from, 153
   as WHO priority, 28
Capricor, 105
Cardio3 Biosciences, 105
CardioCure, 106
Carticel®, 62, 106, 216, 218, 220, 221, 353
CAT. *See* Committee on Advanced Therapies (CAT)
Cell(s), 5, 226–232
   culture optimization, 184–185
      automation and, 184
      direct labor substitution and, 184–185
      work streaming and, 185
   expansion, 231–232
   genetic stability, 187
   manipulation risks, 186
   procurement, 231
   product risk and benefit profile, 185–187
   for research and screening, 178–180
   for therapy, 180–187
   types, 228–231
   undesirable populations, 186–187
Cell bank, 87–89
Cell therapy(ies)
   adipose based, 181
   companies, 217–222
   defined, 214–215
   degree of cell manipulation in, 181
   ESC based, 181
   exogenous risks, 185–186
   intrinsic risks, 186–187
   manufacturing, 190–211
      facilities in, 197–198
      lessons learned, 191–193
      outsourcing considerations, 204–210
         autologous *vs.* allogeneic production, 207–208
         automation, 210
         bedside delivery, 209
         cost sensitivity, 204–207
         fresh *vs.* frozen products, 209–210
         salaries, 206
         transportation, 208
      patient components and product flow in, 198
      process descriptions and process flow in, 196–197
      product isolation and prevention of contamination in, 194–196
      program for, 194–198
      quality systems for, 199–204 (*See also* Quality system(s))
   models, 181
   products, 218–222
Cellerix, 93–94, 105–106
   acquisition strategy, 109

Index

marketing strategy, 108–109
product pipeline, 109–110
Celution technology, 63
Center for Biologics Evaluation and Research (CBER), 90
Centres of Excellence for Commercialization and Research (CERC), 54
CERC. *See* Centres of Excellence for Commercialization and Research (CERC)
Cerebral palsy, 122, 125
China, 25, 240, 268, 271
ChondroCelect®, 62, 108, 312, 313, 339–342
  efficacy, 340
  regulatory approval, 220
Chondrogen®, 61–62
CIRM. *See* California Institute for Regeneration Medicine (CIRM)
Clinical trial(s), 16, 20, 23, 65, 113, 121–122, 360
  ALS stem cell, 271
  authorization, 316, 317
  autologous cell therapies, 103, 104, 125
  cardiac repair, 7
  cartilage repair, 338
  challenges, 340
  diabetes, 16
  early stage, 25
  funding, 7, 24
  GMP antibodies, 186
  heart failure, 326
  late stage, 24, 90
  leukemia, 125, 326
  liability, 386
  limb ischemia, 326
  MSCs, 166, 172
  national consortium, 44
  neuroblastoma, 54
  as part of proprietary programs, 135, 136
  products for, 193
  registration, 172
  repeat, 192
  stroke, 54
  success rates from, 161
  support for, 216
  unauthorized, 309–310
  website, 226
  worldwide, 372
Collagen, 223
CombiCult™, 145
CombiScreen™, 145
Committee on Advanced Therapies (CAT), 100
Compat Select™, 185
Congenital anemia, 119
Congenital aplastic anemia, 119
Conjoined twins, 246–247
Cooperative Research Centers (CRCs), 36

Cord blood, 118–127
  for allogeneic hematopoietic stem cell transplantation, 118–121
    advantage over bone marrow donors, 120
    historical perspectives, 119–120
    indications for, 119
    retrospective studies, 121
  for autologous transplants, 121–122
  banks, 122–127
    business models for, 122–126
      hybrid public-private, 125–126
      private, 123–125
      public, 122–123
    ethical issues, 126–127
    lack of approval from specific professional sectors, 127
    licensing, 127
    manufacturing process, 196–197
CRCs. *See* Cooperative Research Centers (CRCs)
Crohn's disease, 93–94, 106, 110, 111, 112, 156
Cryopreservation, 90–91
Cx401, 65, 109, 110, 112
Cx501, 93, 106, 109, 110
Cx602, 109, 110, 111
Cx611, 94, 106, 109, 110, 111
Cx911, 94, 106, 110, 111
Cytori, 216

D

Dendritic cell manufacturing process, 197
Dermagraft®, 62, 90, 93, 220, 221, 326, 330, 362
Diabetes, 44
  in Asia, 151
  blindness and, 60
  pancreatic islet regeneration for, 259
  renal failure and, 60
  research, 106, 107
  stem cell therapy, 55, 107, 121, 164, 259
  stroke and, 60
  testing, 60
  Type I, 16, 156
  Type II, 16, 156
  as WHO priority, 28
Disease Team Program, 23, 25, 32
  grants, 28

E

Edmonton Protocol, 44
Embryo(s)
  for drug screening, 146
  iPS cells as, 245–246
  resolving definition of, 244
  stem cell from pre-implanted, 133, 227, 229, 243
Embryonic stem cell(s), 133, 229, 243
  in creation of transgenic animals, 145

degree of cell manipulation, 181
IND trials, 76
iPS vs., 230
procurement, 231
proliferative capacity, 144
Engine, 55
Entest Biomedical, 106
EPO. See Erythropoietin (EPO)
Epogen. See Erythropoietin (EPO)
Erythropoietin (EPO), 63
 CFU-E effect, 144
 chemotherapy-induced anemia and, 147
 function, 143
 global market, 142
 indications, 147
 receptor, 143
 recombinant, 143
ESCs. See Embryonic stem cell(s)
European Commission Directives and Legislation, 126–127
European Medicines Agency (EMA), 6
European Patent Convention (EPC), 240, 242
European Union
 patent system, 240–247
  duration of protection, 253
  future of, 253–254
  morality problem, 240–241
  parallel routes to protection, 247
  stem cell claims in granted patents, 250–253
  U. S. patent environment vs., 240
  WARF in, 243–244
 regenerative medicine in, 296–319
  ACTIVE treatments program, 339
  blood, 310
  cosmetics, 311
  licensing, 311–318
   centralized procedures, 312
   national procedures, 312
  medical devices, 306–308
   classes of, 307–308
   combination products, 308
   registering, 318
  medicinal products, 300–305
   advanced therapy, 300–302, 304–305, 314
   biological, 305
   defined, 297
   exemptions from marketing authorization, 314–316
   gene therapy, 301
   hierarchy, 305
   product classification, 302
   somatic cell therapy, 304
   substantial manipulation, 304–305
   tissue engineered, 301, 304
  regulatory guidance, 316–318
  tissues and cells, 308–310

European Union Law, 241–242
European Union Tissues and Cells Directives, 127

F

Fanconi's anemia, 119
Fibrin, 224
Filgrastin, 63
Food and Drug Administration (FDA), 6
 cord blood banking standards, 60
 new drug approvals, 60
Foundation for the Accreditation of Cellular Therapies (FACT), 126
Funding, 15–17
 biotechnology, 68
 California model, 12 (See also California model)
 clinical trials, 7, 24
 global, 28
 governmental, interface of private capital markets with, 26
 intellectual capital infrastructure for health care and, 14
 for large-scale biotech research, 26–27
 for millennium development goals, 29–30, 33
 scientific organizations, 25
 in U. S., 13–14, 326–328

G

Gamida Cell, 106
Gelatin, 223
Genetic disorders, 119
Genetic metabolic disorders, 119
Genzyme, 106, 216
Germany, 25, 343
Geron Corp., 273
 GMP antibodies, 186
 market approval applications, 221
GRNOPC1, 64, 273
Growth factors, 225–226

H

Harvard Oncomouse, 241
Harvard Stem Cell Institute, 135
Hemangioblast Program, 105
Hematopoietic stem cell(s), 117, 142, 145, 228, 231
Hemoglobinopathies, 119
HSCs. See Hematopoietic stem cell(s)
Human Tissue Act (2004), 127
Human Tissue Authority, 127
Hyaluronic acid, 224

Index

I

ICH. *See* International Convention on Harmonisation (ICH)
IGF. *See* Insulin-like growth factor (IGF)
Imaging methodologies, 6
Immune deficiency syndromes, 119
Immunology, 5–6
In vitro fertilization (IVF), 87, 242, 244, 274
India
    failure to comply with TRIPS, 283, 284
    patents in, 282
        applications, impact of policies on, 288–289
        legislation, 282–286
        recently granted, 287
        those pending examination, 290–291
Induced pluripotent stem cell(s), 4, 5, 154, 160, 229–230
    from cancer stem cells, 153
    clinical application, 133
        obstacles to, 133–134
    differentiation, 137, 230
    as embryos, 245–246
    ESCs *vs.*, 230
    industrialization, 45, 133
    intellectual property portfolio, 138–139
    patents, 138–139, 165, 262, 274–275, 276
    platform, 136
        game changing technology and, 139
        great execution and, 139
        outsize returns and, 139
        patents, 138–139
        risks associated with, 138
    production, 131–132, 134
    regenerative drug screening, 145
    risks associated with, 138
    for SMA, 136–137
    for toxicology screening, 134
    unresolved issue in creation, 231
    for in vitro drug screens, 134
Induced totipotent cell (iTS cell), 245
Industrial capital, 33
INFUSE bone graft, 63
Initial public offering (IPO)
    challenges to completing, 68–69
        company valuation, 69
        future drivers or milestones, 69
        investor appetite, 68–69
Initiative Proposition, 17, 33
Innovacell Biotechnologie, 106
Insception Biosciences, 55
Insulin-like growth factor (IGF), 225
Intellectual capital, 33
    asset, 33
    infrastructure, 14, 33
    medical research, 33

Intellectual property, 248–250. *See also* Patent(s)
    ASCC, 39, 40
    European patent environment, 240–247
        duration of protection, 253
        future of, 253–254
        morality problem, 240–241
        parallel routes to protection, 247
        stem cell claims in granted patents, 250–253
        U. S. patent environment *vs.*, 240
        WARF in, 243–244
    iPS cell portfolio, 138–139
    pooling, 45, 53
    protection, 386–387
Intellectual Property Office, 246
Intercytex, 94
International Convention on Harmonisation (ICH), 316
International Finance Faculty for Immunization (IFFIm), 31
    bond funding structure, 29
    booklet, 32
    creation, 29
    treasury manager, 30
International Juvenile Diabetes Research Foundations, 21
Investigational New Drug (IND), 33
Investment risks, 60–61
Invitrogen, 186
iPierian, 134, 139, 222, 248
    iPS cells, 137
    models, 135
    proprietary program, 136
IPO. *See* Initial public offering (IPO)
iPS cells. *See* Induced pluripotent stem cells
Italy, 344–345
iTS cell. *See* Induced totipotent cell (iTS cell)
IVF. *See* In vitro fertilization (IVF)

J

Japan, 20, 156, 316
    approvals for cell-based products, 93
    commercial cell therapy products, 220
    patent publications, 276
    patent status, 138–139, 248, 262, 267
    scientific funding organizations, 25
    tissue engineering, 278
Juvenile Diabetes Foundation, 46

K

Karocell Tissue Engineering, 106

L

Laboratory control system(s), 202–203
    equipment, 202

materials, 202
product tracking and labeling, 203
production, 202–203
Laminin, 224
Leukemia(s), 28, 108, 119, 227, 259
allografts *vs.* autografts for, 125
clinical trials, 125, 326
Lymphohistiocytic disorders, 119
Lymphomas, 119

M

Macular degeneration, 44, 105, 273
Major Facilities Grant Programs, 32
Matrices and scaffolds, 223–225
MedCell Bioscience, 106–107
Medical research
 aligning payments with benefit groups, 14
 funding, 15–17 (*See also* Funding)
 spreading costs over generations for, 15
Mesenchymal stem cell(s), 154, 167, 229
 animal studies, 168
 attributes, 164
 beneficial effects, 229
 in cardiac therapy, 94
 clinical trials, 166
 from clinical waste, 163
 commercialization, 61
 cytokine secretion, 229
 epithelial solid tumors and, 163
 HLA classification, 163
 immunogenicity, 208
 mechanism of activity, 229
 for meniscal tissue regeneration, 94
 procurement, 231
 starting material, 87
Motor Neurone Disease Association, 216
Mozobil®, 63
MSCs. *See* Mesenchymal stem cell(s)
Multiple myeloma, 119
Multiple sclerosis, 47
 health care cost, 44
 stem cell therapy, 55, 104, 107, 361
Myelodysplastic syndrome, 119
Myeloproliferative diseases, 119
Myoblast Program, 105
MyoCell, 105

N

National Academies of Science, Institute of Medicine, 126
National Health Service (NHS), 371–372
 Technology Adoption Centre, 371
National Institute for Health and Clinical Excellence (NICE), 338

National Institute of Neurological Disorders and Stroke, 137
National Institutes of Health (NIH)
 supplemental mandatory appropriation, 16, 33
National Juvenile Diabetes Research Foundation, 17
Neupogen®, 63
Neuroblastoma, 44, 47, 54, 124, 138
Neurotrophic factor cells, 105
New drug application (NDA), 181
NIH. *See* National Institutes of Health (NIH)
Northern Therapeutics, 55
Novo Nordisk, 271
Novocell, 260
NuPotential, 107
NurOwn™, 105

O

Ontaril, 111–112
 commercial strategy, 112
 production, 111
 regulatory strategy, 112
Ontario Institute of Cancer Research, 45–46
Opexa Therapeutics, 107
Organogenesis, 62, 94–95
Osiris Therapeutics, 94, 216
Osteocel®, 62, 221

P

Pancreatic islet regeneration, 259
Parkinson's disease, 104, 136, 178
 health care costs, 44
 pathology, 137
 therapy, 105, 107, 113, 259
Patent(s)
 applications, 267
 assignees, 269
 holders, 274
 iPS cells, 138–139, 165, 262, 274–275, 276
 protection, parallel routes to, 247
 strategic alliances and partnership, 270, 271–272
 strategies to protect, 267–268
Pattison Report, 248
PDGF. *See* Platelet-derived growth factor (PDGF)
Pharma
 accessing external and internal innovation as strategic response, 156–160
 business model
  constraints to, 155–156
  pipeline trade-off, 162
  for regenerative therapies, 161–164
 stem cell technologies first adopted by, 160–161
Physical infrastructure, 33
PIPE. *See* Public Investment in Private Equity (PIPE)

Index

Platelet-derived growth factor (PDGF), 225, 226, 308
Platelet-rich plasma (PRP), 226
Point-of-care, processing at, 63, 182, 185, 216, 231
Pre-implantation human embryo, 272
Preclinical studies, 25, 74, 78
Premarket approval application (PMA), 181
Priority Medicines Project, 28–29
Prochymal®, 61, 216
PRP. *See* Platelet-rich plasma (PRP)
Public Health and Safety Act, 215
Public Investment in Private Equity (PIPE), 71–72
Pulmonary hypertension, 44, 55

Q

Quality system(s), 199–204
   considerations, 193–194
   elements, 199–204
      CAPA programs as, 201
      change controls as, 199–200
      document controls as, 199
      internal audits as, 200
      job-specific skills as, 201
      laboratory control systems as, 202–203
         (*See also* Laboratory control system(s))
      non-conforming materials and product systems as, 201
      outcome analyses as, 203–204
      personnel qualifications as, 200
      quality agreements as, 200
      supplier qualifications as, 200

R

Raw material risks, 186
Regenerative drugs. *See also* Regenerative medicine
   alignment with pharma model, 151
   convention cell therapy *vs.*, 147–148
   discovery and development, 148–149
   indications, 147
   manufacture, storage, and distribution, 149–150
   market acceptance, 148
   mechanism of action, 146
   natural and synthetic, 143
   path to market, 148
   product regulation, marketing, and reimbursement, 150–151
   screening, 143–144
      animal models in, 146
      stem cells for, 144–145
Regenerative medicine
   business, 215–218
      model, 6, 63–65, 161–164
      non-cell-based segment, 218, 222
      stakeholders, 218–223
   challenges, 5–6, 178

clinical applications, 6–7
companies, 217–218
   budgeting for disaster, 389
   business plan, 180–181
   collaboration with insurers, 389
   criteria for acquisition of, 62
   timelines, 389
core components, 222–232
   cells, 226–232
   growth factors, 225–226
   matrices and scaffolds, 223–225
costs, 7, 62–63
defined, 3, 214
funding, 7
global market potential, 61
insurance for, 379–389
   challenges, 382–383
   critical elements of portfolios, 383–387
      changes in environmental conditions, 384
      clinical trials, 386
      directors and officers liability coverage, 385
      intellectual property protection, 386–387
      machinery breakdown and transit risks, 386
      research and development operations, 384
      security and expertise, 384
      self-insurance option, 387
      supplier dependencies, 385–386
      terrorism and attacks by activists, 386
      uncoverable expenses, 384–385
   roles of risk management and business continuity management in, 387–389
intellectual property, 248–250 (*See also* Intellectual property)
investing in
   business model for, 6, 63–65
   considerations for, 59–62
      acquisition criteria, 62
      economic and health costs, 59–60
      exit options, 61–62
      risk *vs.* reward, 60–61
market potential, 63–65
opportunities, 4–5, 178
organizations, 381
promise, 259–260
publicity, 370
regulatory hurdles, 6
reimbursement status
   in Austria, 345
   in Belgium, 345
   case study, 339–342
   in Denmark, 346
   in Europe, 337–347

in Finland, 346
in France, 344
in Germany, 343
in Netherlands, 344
in Norway, 346
in Portugal, 346
in Spain and Italy, 344–345
in Sweden, 346–347
in United Kingdom, 338, 343, 374–375
in the United States, 351–366
translational gap between therapeutic discovery and clinical use, 47
Regenerative Medicine Oversight Committee, 62
Regenerative technology. *See also* Biotechnology
approaches for claiming, 260–262
ReNeuron, 94
Replacement organs, 275
academic patents, 278
industry patents, 277
Retinal Pigmented Epithelial Cell Program, 105
Rheumatoid arthritis, 107, 158, 361, 365

S

SCN. *See* Stem cell network (SCN)
Severe aplastic anemia, 119
Shelf life
allogeneic cell-based therapy, 91–92
for clinic, 92
for manufacturing, 91
for quality, 92
for shipping, 92
autologous cell therapy, 103
Silk, 223–224
SMA. *See* Spinal muscular atrophy (SMA)
Solving Kids' Cancer, 47
Spain, 25, 344–345
Spinal muscular atrophy (SMA), 136–137
genetic profile, 136–137
incidence, 136
therapy, 137
SSEA-4 Dynabeads™, 186
Stem cell(s)
classification, 162
embryonic, 229
hematopoietic, 142, 145, 228, 231
induced pluripotent, 4, 5, 154, 160, 229–230
clinical application, 133
obstacles to, 133–134
differentiation, 137, 230
as embryos, 245–246
ESC *vs.*, 230
industrialization, 45, 133
intellectual property portfolio, 138–139
patents, 138–139, 165, 262, 274–275, 276
platform, 136

game changing technology and, 139
great execution and, 139
outsize returns and, 139
patents, 138–139
risks associated with, 138
production, 131–132, 134
regenerative drug screening, 145
risks associated with, 138
for SMA, 136–137
for toxicology screening, 134
unresolved issue in creation, 231
for in vitro drug screens, 134
mesenchymal, 229 (*See also* Mesenchymal stem cell(s))
neural, 230–231
research
challenges and opportunities, 112
emerging trends in, 54–56
chemical biology, 54–55
commercial growth, 55
internationalism, 56
philanthropy and championing, 55–56
Stem cell companies
building new franchises, 164–168
delivery, 168
logistics, 168
manufacturing, 167–168
patents, 165–166
pricing and reimbursement, 168
product concept, indications, and markets, 166
regulatory requirements, 166–167
Canadian, 55
dilution, 73
economic times and, 73
fundamentals, 72–73
investors, 73
public markets and, 68–81
case study, 79–81
market reaction to, 72–73
options for, 68–72
bought deal, 72
convertible debt deal, 71
empty shell merger, 70
fully marketed follow-on, 71
IPO, 68–69
PIPE, 71–72
registered direct transaction, 72
reverse merger, 69–70
synergistic merger, 70
virgin shell merger, 70
valuation, 74–79
cash and, 74
data and, 74–75
execution of goals and, 74
hope and, 75–76

Index

key drivers of, 74–76
metrics, 76–79
Stem-cell-derived therapy, 64. *See also* Cell therapy(ies)
Stem Cell Foundation, 216
Stem Cell Initiative, 248
Stem cell network (SCN), 45–49
   development of highly qualified workforce, 48
   networking and partnering, 48–49
   partnership and interactions, 49
   translational funding gap between therapeutic discovery and clinical use, 47
Stem Cell Therapeutics, 55
Stem cell transplantation, 117, 215
Stempeutics Research, 107
Stroke, 54
   diabetes and, 60
   health care costs, 44, 325
   stem cell therapy, 47, 55, 94, 104, 122, 259
   as WHO priority, 28
Supplementary Mandatory Appropriations Bill, 16, 33

T

T2 Cure GmbH, 107
TCA Cellular Therapy, 107
Tengion, 107–108
TGF-B. *See* Transforming growth factor-beta (TGF-B)
Thrombopoietin (TPO), 143
TiGenix, 108. *See also* ChondroCelect®
Tissue engineering, 214, 301, 304
   academic patents, 278
   global market potential, 61
   industry patents, 277
TPO. *See* Thrombopoietin (TPO)
Trade-related aspects of intellectual property (TRIPS), 240, 281, 283, 284
Transforming growth factor-beta (TGF-B), 225
Transmissible spongiform encephalopathy (TSE), 89
TRIPS. *See* Trade-related aspects of intellectual property (TRIPS)
Tristem, 108
TSE. *See* Transmissible spongiform encephalopathy (TSE)
TxCell, 108

U

U. S. Food and Drug Administration. *See* Food and Drug Administration (FDA)
U. S. Trade and Patent Office (USPTO), 258
   patent applications *vs.* granted patents, 267
   policy regarding regenerative medicine, 262, 267
   WARF patents and, 263
United Kingdom, 4, 156
   fertilization and embryology laws, 244
   health care costs, 60
   hematopoietic stem cell transplantation, 117
   Intellectual Property Office, 246
   NICE, 338
   regenerative medicine in, 61
      delivering solutions to patients, 372–375
      infrastructure, 372–375
      key questions requiring answers, 370–371
      NHS as innovator, 371–372
      scientific funding organizations, 25
   regulatory environment, 371
   reimbursement in, 338, 343, 374–375
   scientific funding organizations, 25
   Stem Cell Initiative, 248
United States
   health care costs, 59–60
   patent system, 260
      issues affecting, 262–267
         competition, 263, 266–267
         early stage technologies and speed, 262
         scrutiny, 262–263
      office (*See* U. S. Trade and Patent Office (USPTO))
   regenerative medicine in
      federal funding, 326–328
      overall political environment for, 331–332
      regulatory issues, 328–330
      reimbursement (*See* United States, reimbursement in)
   reimbursement in, 351–366
      bundled *vs.* separate payment considerations, 356
      clinical evidence requirements for, 364
      coding and payment considerations, 353–358
      core elements of, 354
      cost and economic considerations, 365–366
      HTA assessment and, 358–365
      regulatory pathway and, 352–353
      site of care and, 355, 356
USPTO. *See* U. S. Trade and Patent Office (USPTO)

V

Valuation of stem cell companies, 74–79
   cash and, 74
   data and, 74–75
   execution of goals and, 74
   hope and, 75–76

IPO and, 69
key drivers of, 74–76
metrics, 76–79
    cash flow analysis, 78
    comps analysis, 78
    multiple analysis, 77–78
Valvelta, 62
Vascular endothelial growth factor (VEGF), 225
Vascularization, 5
VEGF. *See* Vascular endothelial growth factor (VEGF)
Venture funds, 61
Verio Therapeutics, 55

W

WHO. *See* World Health Organization (WHO)
Wisconsin Alumni Research Foundation (WARF)
    background of case against, 242–243
    in European patent environment, 243–244
    issues at heart of decision concerning, 246
    patents, 270, 272
World Bank, 30
World Health Organization (WHO), 28–29, 59
    coding system for disease classification, 354
    millennium development goals, 33
World Health Statistics 2008 report, 60